Mathematical Methods in Science and Engineering

Mathematical Methods in Science and Engineering

Edited by Alan Fraser

CLANRYE INTERNATIONAL
www.clanryeinternational.com

Clanrye International,
750 Third Avenue, 9th Floor,
New York, NY 10017, USA

ISBN: 978-1-64726-132-0

Cataloging-in-Publication Data

Mathematical methods in science and engineering / edited by Alan Fraser.
p. cm.
Includes bibliographical references and index.
ISBN 978-1-64726-132-0
1. Science--Mathematical models. 2. Engineering--Mathematical models. 3. Science. 4. Engineering.
5. Mathematics. I. Fraser, Alan.
Q158.5 .M38 2022
016.5--dc23

For information on all Clanrye International publications
visit our website at www.clanryeinternational.com

Clanrye International

Contents

Preface

This book aims to highlight the current researches and provides a platform to further the scope of innovations in this area. This book is a product of the combined efforts of many researchers and scientists, after going through thorough studies and analysis from different parts of the world. The objective of this book is to provide the readers with the latest information of the field.

Mathematics plays an important role in developing hypotheses, laws and theories in science and engineering. It is used in quantitative scientific modeling which generates new hypotheses and predictions. It is frequently used in collecting and observing measurements. A branch of mathematics known as statistics plays an important role in summarizing and analyzing data. This data allows scientists to evaluate the reliability and variability of the results of their experiments. Both science and engineering apply computational science and mathematics to simulate real-world situations. Mathematics in engineering includes the applications of differential equations, real and complex analysis, approximation theory, Fourier analysis, potential theory as well as probability and linear algebra. The topics included in this book on mathematical concepts in science and engineering are of utmost significance and bound to provide incredible insights to readers. It presents researches and studies performed by experts across the globe. Those in search of information to further their knowledge will be greatly assisted by this book.

I would like to express my sincere thanks to the authors for their dedicated efforts in the completion of this book. I acknowledge the efforts of the publisher for providing constant support. Lastly, I would like to thank my family for their support in all academic endeavors.

Editor

1

A Numerical Computation Approach for the Optimal Control of ASP Flooding Based on Adaptive Strategies

Shurong Li[1] **and Yulei Ge**[2]

[1]*Automation School, Beijing University of Posts and Telecommunications, Beijing 100876, China*
[2]*College of Information and Control Engineering, China University of Petroleum (East China), Qingdao 266580, China*

Correspondence should be addressed to Shurong Li; lishurong@bupt.edu.cn

Academic Editor: Łukasz Jankowski

A numerical computation approach based on constraint aggregation and pseudospectral method is proposed to solve the optimal control of alkali/surfactant/polymer (ASP) flooding. At first, all path constraints are aggregated into one terminal condition by applying a Kreisselmeier-Steinhauser (KS) function. After being transformed into a multistage problem by control vector parameter, a normalized time variable is introduced to convert the original problem into a fixed final time optimal control problem. Then the problem is discretized to nonlinear programming by using Legendre-Gauss pseudospectral method, whose numerical solutions can be obtained by sequential quadratic programming (SQP) method through solving the KKT optimality conditions. Additionally, two adaptive strategies are applied to improve the procedure: (1) the adaptive constraint aggregation is used to regulate the parameter ρ in KS function and (2) the adaptive Legendre-Gauss (LG) method is used to adjust the number of subinterval divisions and LG points. Finally, the optimal control of ASP flooding is solved by the proposed method. Simulation results show the feasibility and effectiveness of the proposed method.

1. Introduction

With the exploitation of oil, most oil fields of China have stepped into the high water cut period [1]. The production of oil cannot satisfy our demands for daily life and economic development. A series of tertiary oil recovery technologies such as chemical flooding [2], microorganism flooding [3], and carbon dioxide flooding [4] are put to use. The newly emerging alkali/surfactant/polymer (ASP) flooding which is an important tertiary oil recovery technology can enhance oil production obviously. The basic idea is to utilize three displacing agents (alkali, surfactant, and polymer), whose synergistic effects are important to enhanced oil recovery, to change the physicochemical property [5]. Since the price of displacing agents is high, how to determine the injection strategy to maximize the profit as much as possible is always a challenging problem.

The essence of determining the injection strategy for ASP flooding is an optimal control problem. In current industrial application, the index comparison method is usually adopted to select the injection strategy, in which the best one is chosen among many given feasible strategies by simulating on the numerical simulation software according to a defined index [6, 7]. This method is simple and easy to be operated, but it is too dependent on the experience of manipulators. The optimization result is not the optimum. The optimal control technology, which is first used in oil exploitation in [8], works on searching for the optimal control strategy from all feasible solutions with considering the process dynamic characteristics, and it can realize the simultaneous optimization of all variables. In addition, the whole theory is proved by mathematical method; it is scientific and reasonable. Therefore, it is suitable for the control of ASP flooding.

Some scholars have studied the optimal control for oil exploitation. Jansen et al. [9] studied the problems of optimal control and nonlinear model predictive control for water flooding, in which the control variables are bottom hole pressure and flooding rate, and the index is the maximum of net present value (NPV). Lei et al. [10] presented mixed-integer iterative dynamic programming (IDP) to optimize the polymer flooding and got good result. Ramirez et al. [11, 12] optimized the injection strategy for enhanced oil recovery of surfactant flooding with optimal control theory, in which the necessary condition of optimal control was deduced on the basis of maximum principle before being solved by gradient method. Furthermore, this method was applied to carbon dioxide flooding, nitrogen flooding, and binary system flooding [13]. Zerpa et al. [14] used field scale numerical simulation and multiple surrogates to optimize alkaline–surfactant–polymer flooding process based on UTCHEM. Ge et al. [15] developed an approximate dynamic programming method to solve the optimization of ASP flooding, in which an Actor-Critic algorithm is introduced to search the optimal injection strategy. Most of the researches are about water flooding and single chemical flooding. The optimal control for ASP flooding needs to be further studied urgently.

With regard to the research on optimal control methods, although extensive research on optimal control has been conducted, large-scale nonlinear dynamic models processing and efficient solution methods for optimal control problems are still two main barriers for the widespread industrial application of optimal control [16, 17]. Numerical methods for optimal control can be divided into two classes, which are direct methods and indirect ones. Direct methods, which include the control vector parameterization (CVP) [18], the direct multiple shooting [19], and the full discretization [20], are closely related to the discretization of the original control problem and the application of nonlinear programming technique. These methods are very popular and well suited for many practical problems [21]. What differs in these methods is how to select the kind of variables to be discretized and how to approximate the system equations. However, direct methods may bring about unsatisfactory results especially when the control variables are discontinuous, such as with switching points and singular arcs [22]. For these reasons, the reasonable discretization methods for variables and approximation methods for system equations are quite important.

As to the discretization, if the discretization grid is too coarse, this will lead to some local controls being not in the right place and the accuracy cannot meet the prespecified requirements. What is more, if the discretization grid is too fine, the computational cost will be very high and the robustness will be poor. Generally speaking, the discretization grid is determined by visually examining the obtained result which is balanced between efficiency and precision. For example, an adaptive CVP method was proposed in [23], in which the problem was discretized adaptively over time spans. The discretization was sequentially refined based on a wavelet-based analysis of the optimal solution which was obtained in the previous optimization step. Thus, the efficiency is enhanced with less loss of accuracy.

As to the approximation methods for system equations, pseudospectral method, which is a major kind of direct method, has been widely used in recent years [24]. In this method, states and controls are approximated by a series of orthogonal basis functions, such as Lagrange, Laguerre, and Legendre polynomials. Some famous pseudospectral methods have been developed in the past decades, for example, the Gauss pseudospectral method (GPM) [25], the Lobatto pseudospectral method (LPM) [26], and the Radau pseudospectral method (RPM) [27]. The most obvious difference between them is the selection of interpolation point. For smooth optimal control problems, these methods can obtain good accuracy and performance compared with traditional direct methods. But when the control is discontinuous, for example, the three-slug injection strategy of ASP flooding, there may be some problems especially in the discontinuous points. So, many adaptive methods are proposed to cope with the interpolation points and subintervals. Darby et al. [28] developed an adaptive method, in which the difference between a Legendre-Gauss-Radau approximation to the integral of the dynamics and an interpolated value of state was used to estimate the approximation error. According to this error, the interpolation points were selected under a given accuracy requirement.

Constraint handling is an important step for a control problem before being optimized. In general, the constraints are disposed by penalty function method, in which a penalty factor is introduced to convert the original problem into an unconstrained optimization problem [29]. But there is no definite theory to determine the penalty factor, which is usually obtained by experience. If the value of the factor is not chosen scientifically, this will lead to adverse impact on the optimization. In [30], the constraint aggregation method, which was expressed as the Kreisselmeier-Steinhauser (KS) function, was first proposed to transform the path constraints into a terminal condition. To regulate the parameter ρ, an adaptive constraint aggregation method was presented by Zhang et al. [31], in which the value of ρ was changed according to the partial differential automatically.

Many researches about the optimization algorithms have been studied. Fonseca proposed a stochastic simplex approximate gradient (StoSAG) for optimization under uncertainty, in which the gradient is estimated approximately by an ensemble of randomly chosen control vectors, known as Ensemble Optimization (EnOpt) in the oil and gas reservoir simulation community [32]. Chen and Reynolds studied the optimal control of inflow control valves and well operating conditions for the water-alternating-gas injection process [33]. Shirangi et al. developed a new methodology for the joint optimization of economic project life and time-varying well controls [34]. Sequential quadratic programming (SQP) is an effective method to solve constrained optimization problems, which has good global convergence and more than one order of local convergence and does not need to construct the penalty function factor when dealing with constraints [35]. It is suitable for the optimal control of ASP flooding. In [36], an approximate feasible direction method was put forward in which the constraint aggregation was adopted to transform all path constraints into a generalized constraint. Then SQP was applied to solve this optimal control problem. Bernardo et al. [37] presented a slug optimization method for water flooding in which the Kriging interpolation was utilized to build the model between variables and index before being optimized by SQP.

To solve the optimal control for alkali/surfactant/polymer (ASP) flooding which has discontinuous control variables, a methodology based on constraint aggregation and pseudospectral method with adaptive strategies is proposed in this paper. The rest of this article is outlined as follows. In Section 2, the specific model of ASP flooding is given and the optimal control for ASP flooding is converted into a more general problem in mathematical form. Section 3 introduces the details of the proposed methodology in this paper, such as processing the path constraints with constraints aggregation method, converting original problem into a multistage problem with CVP, introducing the new time variable, and discretizing the control problem with LG pseudospectral method. The KKT optimality condition of this optimal control problem is provided in Section 3.4. In Section 4, the adaptive strategies for the constraints aggregation and LG pseudospectral method are presented to regulate the control method. Then SQP is introduced to solve the problem. Section 5 applies the method proposed in this paper to solve the optimal control problem for ASP flooding and Section 6 contains some discussion and concluding remarks. Finally, some essential information for ASP flooding is given in the Appendix.

2. Problem Formulations

As to the optimal control problem for ASP flooding, the control variables are the injection concentrations of three displacing agents (alkali, surfactant, and polymer) at all injection wells; the states cover pressure, grid concentration, and water saturation; and the performance index is the maximal NPV. This is a complex distributed parameter control problem aiming at getting the optimal injection strategy to fulfill the maximal profit. Since the three-dimensional model for ASP flooding, which includes a series of divergences and cross terms, is too complex, we only consider the one-dimensional model in this paper. In this section, we will obtain the optimal injection strategy with the method proposed in this paper.

2.1. Optimal Control Model Description for One-Dimensional ASP Flooding. In view of [5, 6, 38–40], we can make the following descriptions.

Considering a long tube core with diameter d and length L, inject the displacing agents liquid with flow Q from one side at a constant speed; the core porosity is ϕ and the residual oil saturation is S_{or}. On the basis of the model assumptions in [39], we add the following assumptions:

(a) All adsorption processes satisfy Langmuir isothermal adsorption equation.

(b) The displacing agents exist in water phase, the adsorption satisfies the generalized FICK law, and the balance is established momentarily.

On the basis of the oil/water seepage continuity equations and adsorption diffusion equations of displacing agents, combining mass balance conditions, we can obtain the following model with interaction of alkali, surfactant, and polymer fully considered.

The seepage continuity equation is

$$\frac{\partial S_w}{\partial t} = -\frac{Q}{A\phi}\frac{\partial f_w}{\partial z}. \tag{1}$$

The adsorption diffusion equation of surfactant is

$$\phi S_w \frac{\partial C_s}{\partial t} = -v_w \frac{\partial C_s}{\partial z} + \frac{\partial}{\partial z}\left(D_s \phi \frac{\partial C_s}{\partial z}\right) - \rho_r \frac{\partial \Gamma_s}{\partial t}. \tag{2}$$

The adsorption diffusion equation of polymer is

$$\phi S_w \frac{\partial C_p}{\partial t} = -v_w \frac{\partial C_p}{\partial z} + \frac{\partial}{\partial z}\left(D_p \phi \frac{\partial C_p}{\partial z}\right) - \rho_r \frac{\partial \Gamma_p}{\partial t}. \tag{3}$$

The adsorption diffusion equation of alkali is

$$\phi S_w \frac{\partial C_a}{\partial t} = -v_w \frac{\partial C_a}{\partial z} + \frac{\partial}{\partial z}\left(D_a \phi \frac{\partial C_a}{\partial z}\right) - \rho_r \frac{\partial \Gamma_a}{\partial t} - R_a, \tag{4}$$

where a denotes the alkali, s denotes the surfactant, p denotes the polymer, S_w is the water saturation, A is the core cross section area, f_w is the moisture content, v_w denotes the seepage speed of water phase, and C_a, C_s, C_p and D_a, D_s, D_p denote the concentration and diffusion coefficient of alkali, surfactant, and polymer, respectively. ρ_r denotes the core density, $\Gamma_a, \Gamma_s, \Gamma_p$ are the adsorbing capacity of core for different displacing agents, and R_a is the alkali consumption.

The initial conditions are

$$S_w(z,t)\big|_{t=0} = 1 - S_{or}, \tag{5a}$$

$$\mathbf{C}_\Theta\big|_{t=0} = 0, \quad \Theta = \{a, s, p\}. \tag{5b}$$

The boundary conditions are

$$f_w\big|_{z=0} = 1.0, \tag{6a}$$

$$S_w\big|_{z=0} = 1 - S_{or}, \tag{6b}$$

$$\begin{aligned} \left.\frac{\partial \mathbf{C}_\Theta}{\partial z}\right|_{z=0} &= \frac{v_w}{D_\Theta \phi}\left(\mathbf{C}_\Theta - \mathbf{u}_\Theta\right), \\ \left.\frac{\partial \mathbf{C}_\Theta}{\partial z}\right|_{z=L} &= 0, \end{aligned} \tag{6c}$$

where $\mathbf{u}_\Theta$ denotes the injection concentration of displacing agents, which is the control variable.

In application, the slug injection strategy is usually adopted. Suppose that there are P slugs,

$$\mathbf{u}_\Theta(t) = \begin{cases} \mathbf{u}_{\Theta i}, & t_{i-1} \le t \le t_i,\ i = 1, 2, \ldots, P, \\ 0, & t_P \le t \le t_f, \end{cases} \tag{7}$$

where t_i denotes the time node, the length of every slug is $T_i = t_i - t_{i-1}$, and $\mathbf{u}_{\Theta i}$ denotes the injection concentration of displacing agent Θ in slug i.

Furthermore, the dosage limit of displacing agents is

$$\sum_{i=1}^{P}\left(\mathbf{u}_{\Theta i} \cdot T_i\right) \le \mathbf{M}_{\Theta p}, \tag{8}$$

where $\mathbf{M}_{\Theta p}$ denotes the maximum usage of displacing agents Θ.

The injection concentration and slug size limitations are

$$0 \le \mathbf{u}_{\Theta i} \le \mathbf{u}_{\Theta \max},$$
$$\sum_{i=1}^{P} T_i = t_P. \tag{9}$$

The other physicochemical algebraic equations can be found in the Appendix.

The maximum net present value (NPV) is chosen as the index; the specific description is

$$\max J_{\text{NPV}} = \int_0^{t_f} (1+\chi)^{-t} \left[\xi_o (1-f_w) q_{\text{out}} - \boldsymbol{\xi}_\Theta q_{\text{in}} \mathbf{u}_\Theta\right] dt, \tag{10}$$

where χ denotes the discount rate, q_{in} and q_{out} denote the volume flow rate of injection and production, and $\boldsymbol{\xi}_\Theta$ denotes the price of three displacing agents.

2.2. Model Transformation. To make the problem more concise, we use $\mathbf{x} \in R^{n_w}$ $(C_\Theta, S_w, \Gamma_\Theta)$ to denote all states and $\mathbf{u} \in R^{n_u}$ (the injection concentration of all displacing agents $\mathbf{u}_\Theta$) to denote all control variables. Since the ASP flooding model contains many partial differentials with respect to time variable and spatial variable, it is difficult to solve with conventional methods. For simplicity, the finite difference method is used to discretize the spatial grid [6]. Thus there is only the differential with respect to the time variable. Then the optimization for ASP flooding can be reformulated as in the following continuous Bolza problem [41]:

$$\max_{\mathbf{u}(t),t_f} J\left(\mathbf{u}\left(t_f\right), t_f\right) = \theta\left(\mathbf{x}\left(t_f\right), t_f\right) + \int_{t_0}^{t_f} \phi\left(\mathbf{x}(t), \mathbf{u}(t), t\right) dt, \tag{11a}$$

$$\text{s.t.} \quad \dot{\mathbf{x}}(t) = \mathbf{f}\left(\mathbf{x}(t), \mathbf{u}(t), t\right),$$
$$\mathbf{x}\left(t_0\right) = \mathbf{x}_0, \tag{11b}$$

$$\mathbf{h}\left(\mathbf{x}(t), \mathbf{u}(t), t\right) \le 0, \tag{11c}$$

$$\mathbf{N}\left(\mathbf{x}\left(t_f\right), t_f\right) \le 0, \tag{11d}$$

$$\mathbf{u}_{\text{LB}} \le \mathbf{u}(t) \le \mathbf{u}_{\text{UB}}, \tag{11e}$$

where $\mathbf{x}(t) \in R^{n_x}$ denotes the state variables including water saturation, pressure, and grid concentration. The initial state is $\mathbf{x}_0$. $\mathbf{u}(t) \in R^{n_u}$ is the control variable including the injection concentration of displacing agents (alkali, surfactant, and polymer). The system process is described by function $\mathbf{f}(\bullet) \in R^{n_f}$. The path constraints and terminal constraints are expressed as $\mathbf{h}(\bullet) \in R^{n_h}$ and $\mathbf{N}(\bullet) \in R^{n_N}$, respectively. The whole process is optimized on the time interval $t \in [t_0, t_f]$. The interval of control is $[\mathbf{u}_{\text{LB}}, \mathbf{u}_{\text{UB}}]$ associated with the lower bound and upper bound, which can be reformulated to two path constraints, $-\mathbf{u}(t) + \mathbf{u}_{\text{LB}} \le 0$ and $\mathbf{u}(t) - \mathbf{u}_{\text{UB}} \le 0$. The equality constraints can be converted into the inequality constraints by introducing the relaxing factors. For example, an equation $g(x) = 0$ can be expressed as $g(x) - \varsigma \le 0$, in which ς denotes the relaxing factor.

Then we take a series of measures to process this problem. The details will be presented in the following contents.

3. The Numerical Computation Approach Based on Constraint Aggregation and Pseudospectral Discretization

In this section, all path constraints are transformed into one terminal constraint by the constraints aggregation method. Furthermore, CVP is introduced to convert the original problem into a multistage problem (MSP) to cope with the situation of discontinuous control variables after the new time variable is brought in. Finally, to solve the optimal control problem, LG pseudospectral method is adopted to transform the MSP into a general discrete nonlinear programming problem (NLP) [42].

3.1. Constraint Aggregation. The main idea of constraints aggregation is the KS function, which was developed by Kreisselmeier and Steinhauser in 1976 [30]. The mathematical form is

$$\text{KS}\left(h(x), \rho\right) = \frac{1}{\rho} \ln \left[\sum_{j=1}^{N_c} e^{\rho h_j(x)}\right], \tag{12}$$

where N_c denotes the number of path constraints $h_j(x)$. ρ is an approximation parameter.

KS function can generate one envelope surface which is C_1-consecutive. This can estimate the maximal function of $\{h(x)\}$ conservatively. The properties are as follows [43].

(a) For an arbitrary $\rho > 0$, there exist the following formulas:

$$\max\left[h(x)\right] \le \text{KS}\left(h(x), \rho\right) \le \max\left[h(x)\right] + \frac{1}{\rho} \ln N_c,$$
$$\lim_{\rho\to\infty} \text{KS}\left(h(x), \rho\right) = \text{KS}\left(h(x), \rho\right). \tag{13}$$

(b) If $\rho_2 > \rho_1 > 0$, then $\text{KS}(h(x), \rho_1) \ge \text{KS}(h(x), \rho_2)$.

(c) $\text{KS}(h(x), \rho)$ is a convex function, if and only if it is convex for all constraints $h(x)$.

From the above properties, KS function is the underestimation for feasible region. The bigger the value of ρ is, the more accurate the estimation for the maximal constraint is. This can be obviously shown in Figure 1. What is more, if the original problem is convex, the problem after being processed by KS function is convex too.

The path constraints in (11a), (11b), (11c), (11d), and (11e) can be transformed into a terminal constraint using KS function; the detailed process is as follows [44].

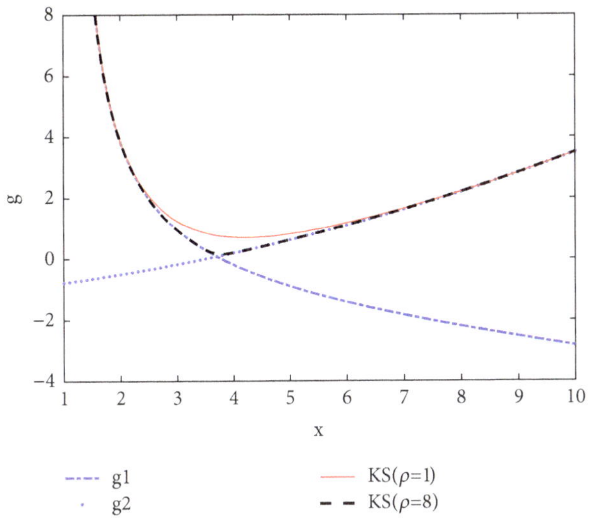

FIGURE 1: Estimate ability of KS function for $\max[g_1(x) = 5/\log(x) - 0.2x - 3, g_2(x) = x^2/40 + x/5 - 1]$.

Considering the point constraints on the path constraints in (12), $t_i \in [t_0, t_f]$, then

$$\mathrm{KS}(h(x), \rho) = \frac{1}{\rho} \ln \left[\sum_{j=1}^{N_c} \sum_{i=1} e^{\rho h_j(x(t_i))} \right]. \quad (14)$$

Transform it into the continuous form by introducing the impulse function,

$$\mathrm{KS}(h(x), \rho) = \frac{1}{\rho} \ln \left[\int_{t_0}^{t_f} \sum_{j=1}^{N_c} \sum_{i=1} e^{\rho h_j(x(t_i))} \delta(t - t_i)\, dt \right], \quad (15)$$

where $\delta(t - t_i)$ is the impulse function.

Define a new state p,

$$\dot{p} = f_p = \sum_{j=1}^{N_c} \sum_{i=1} e^{\rho h_j(x(t_i))} \delta(t - t_i), \quad p(t_0) = 0. \quad (16)$$

Substitute (16) into (15); then

$$\mathrm{KS}(h(x), \rho) = \frac{1}{\rho} \ln \left[p(t_f) - p(t_0) \right]. \quad (17)$$

Since $\mathbf{h}(\mathbf{x}(t), \mathbf{u}(t), t) \leq 0$, the constraints processed meet the following condition:

$$\mathrm{KS}(h(x), \rho) \leq 0. \quad (18)$$

In consequence, the path constraints can be transformed into the following terminal time constraint:

$$p(\mathbf{x}(t_f), \mathbf{u}(t_f), t_f) - 1 \leq 0. \quad (19)$$

Moreover, a new state variable q is introduced to the index in (11a), which satisfies

$$\dot{q} = f_q = \phi(\mathbf{x}(t), \mathbf{u}(t), t), \quad q(t_0) = 0. \quad (20)$$

Then

$$q(\mathbf{x}(t_f), \mathbf{u}(t_f), t_f) = \int_{t_0}^{t_f} \phi(\mathbf{x}(t), \mathbf{u}(t), t)\, dt. \quad (21)$$

Let $\boldsymbol{\omega} = [\mathbf{x}, p, q]^T$, $\mathbf{F} = [\mathbf{f}, f_p, f_q]^T$; the original problem can be transformed into the following form:

$$\max_{\mathbf{u}(t), t_f} J(\mathbf{u}(t), t_f) = \theta(\mathbf{x}(t_f), t_f) + q(\mathbf{x}(t_f), \mathbf{u}(t_f), t_f), \quad (22a)$$

$$\text{s.t.} \quad \dot{\boldsymbol{\omega}}(t) = \mathbf{F}(\mathbf{x}(t), \mathbf{u}(t), t), \quad (22b)$$

$$p(\mathbf{x}(t_f), \mathbf{u}(t_f), t_f) - 1 \leq 0, \quad (22c)$$

$$\mathbf{N}(\mathbf{x}(t_f), t_f) \leq 0, \quad (22d)$$

$$\mathbf{x}(t_0) = \mathbf{x}_0, \quad p(t_0) = q(t_0) = 0. \quad (22e)$$

3.2. Multistage Problem Formulation. To cope with the discontinuity points of control variables preferably, CVP is adopted to convert the original problem into a MSP [23]. The main idea of this process is to approximate the control with piecewise function according to the slug injection strategy. Discretize the whole time domain $[t_0, t_f]$ into K uneven time periods. The specific time control nodes are

$$t_0 < t_1 < t_2 < \cdots < t_K = t_f. \quad (23)$$

Approximate the continuous control variable with piecewise function in every period; that is,

$$u(t) = \widehat{u}_k, \quad t \in [t_{k-1}, t_k], \; k = 1, 2, \ldots, K. \quad (24)$$

Furthermore, we deal with the time variable by defining a new time variable τ:

$$\tau = \frac{t - t_{k-1}}{t_k - t_{k-1}}, \quad k = 1, 2, \ldots, K. \quad (25)$$

Then every original period $t \in [t_{k-1}, t_k]$ is transformed into $\tau \in [0, 1]$.

Take derivation of (25) with respect to τ; then,

$$d\tau = \frac{1}{t_k - t_{k-1}} dt, \quad 1 \leq k \leq K. \quad (26)$$

Define $\mathbf{x}^{(k)}(\tau), \mathbf{u}^{(k)}(\tau)$ as the state and control on kth period, respectively. Then, (22a), (22b), (22c), (22d), and (22e) can be reformulated as

$$\max_{\mathbf{u}^{(k)}(\tau),t_k} \quad J\left(\mathbf{u}^{(k)}(\tau), t_k\right)$$

$$= \theta\left(\mathbf{x}^{(K)}(1), 1\right) + q\left(\mathbf{x}^{(K)}(1), \mathbf{u}^{(K)}(1), 1\right), \tag{27a}$$

$$\text{s.t.} \quad \dot{\boldsymbol{\omega}}^{(k)}(\tau) = \left(t_k - t_{k-1}\right)\mathbf{F}\left(\mathbf{x}^{(k)}(\tau), \mathbf{u}^{(k)}(\tau), \tau\right), \tag{27b}$$

$$p\left(\mathbf{x}^{(K)}(1), \mathbf{u}^{(K)}(1), 1\right) - 1 \le 0, \tag{27c}$$

$$\mathbf{N}\left(\mathbf{x}^{(K)}(1), 1\right) \le 0, \tag{27d}$$

$$\mathbf{x}^{(1)}(0) = \mathbf{x}_0, \quad p^{(1)}(0) = q^{(1)}(0) = 0, \tag{27e}$$

$$\mathbf{x}^{(k)}(1) = \mathbf{x}^{(k+1)}(0), \quad 1 \le k \le K-1, \tag{27f}$$

where $\boldsymbol{\omega} = [\mathbf{x}, p, q]^T$ and $\mathbf{F} = [\mathbf{f}, f_p, f_q]^T$. Equation (27f) is the linkage constraint which is used to keep the continuity of state variables at the interface of all subintervals. The MSP in (27a), (27b), (27c), (27d), (27e), and (27f) consists of K subproblems corresponding to K subintervals. The final control is obtained by solving K optimal control problems.

3.3. Legendre-Gauss Pseudospectral Discretization. For being solved by well-developed optimization algorithms, Legendre-Gauss (LG) pseudospectral discretization is applied to transform the original optimal control problem into a NLP on every subinterval [26, 28, 42].

Construct N_k+1-order Lagrange interpolation polynomials with N_k LG points to approximate the states and controls in the subinterval $k \in [1, 2, \ldots, K]$; then,

$$\boldsymbol{\omega}^{(k)}(\tau) \approx \mathbf{W}^{(k)}(\tau) = \sum_{j=0}^{N_k} \mathbf{W}_j^{(k)} L_j^{(k)}(\tau). \tag{28a}$$

$$L_j^{(k)}(\tau) = \prod_{l=0, l\ne j}^{N_k} \frac{\tau - \tau_l^{(k)}}{\tau_j^{(k)} - \tau_l^{(k)}}, \quad j = 0, 1, 2, \ldots, N_k. \tag{28b}$$

$$\mathbf{u}^{(k)}(\tau) \approx \mathbf{U}^{(k)}(\tau) = \sum_{i=0}^{N_k} \mathbf{U}_i^{(k)} \widehat{L}_i^{(k)}(\tau), \tag{29a}$$

$$\widehat{L}_i^{(k)}(\tau) = \prod_{l=0, l\ne i}^{N_k} \frac{\tau - \tau_l^{(k)}}{\tau_i^{(k)} - \tau_l^{(k)}}, \quad i = 1, 2, \ldots, N_k, \tag{29b}$$

where $\tau_0^{(k)}, \tau_1^{(k)}, \ldots, \tau_{N_{k-1}}^{(k)}$ denote the LG interpolation points that are defined on every subinterval $\tau^{(k)} \in [t_{k-1}, t_k)$. $\mathbf{W}^{(k)}(\tau)$, $\mathbf{U}^{(k)}(\tau)$ are the approximations of state and control on corresponding subinterval. Note that the interpolation points of state are one more than that of control.

Take differential of state with respect to τ; then,

$$\dot{\mathbf{W}}^{(k)}(\tau) = \sum_{j=0}^{N_k} \mathbf{W}_j^{(k)} \dot{L}_j^{(k)}(\tau). \tag{30}$$

The differential of Lagrange interpolation polynomials at every LG point can be expressed by a differential calculation matrix $D \in R^{N_k \times (N_k+1)}$ which can be described as

$$D_{sj}^{(k)} = \dot{L}_j^{(k)}(\tau) = \sum_{c=0}^{N_k} \frac{\prod_{l=0, l\ne j,c}^{N_k} \tau_s - \tau_l^{(k)}}{\prod_{l=0, l\ne j}^{N_k} \tau_j^{(k)} - \tau_l^{(k)}}, \quad s = 0, 1, 2, \ldots, N_{k-1}. \tag{31}$$

Then the dynamic constraints in (27b) can be transformed into the following algebraic constraints:

$$\sum_{j=0}^{N_k} D_{sj}^{(k)} \mathbf{W}_j^{(k)} - \left(t_k - t_{k-1}\right)\mathbf{F}\left(\mathbf{X}_s^{(k)}(\tau), \mathbf{U}_s^{(k)}(\tau), \tau\right) = 0. \tag{32}$$

Then the optimal control problem can be reformulated into a new NLP:

$$\max_{\mathbf{U}^{(k)}(\tau),t_k} \quad J\left(\mathbf{U}^{(k)}(\tau), t_k\right) = \theta\left(\mathbf{X}_{N_k}^{(K)}(1)\right) + q\left(\mathbf{X}_{N_k}^{(K)}(1), \mathbf{U}_{N_k}^{(K)}(1)\right), \tag{33a}$$

$$\text{s.t.} \quad \sum_{j=0}^{N_k} D_{sj}^{(k)} \mathbf{W}_j^{(k)} - \left(t_k - t_{k-1}\right)\mathbf{F}\left(\mathbf{X}_s^{(k)}(\tau), \mathbf{U}_s^{(k)}(\tau), \tau\right) = 0, \quad s = 0, 1, 2, \ldots, N_{k-1}, \tag{33b}$$

$$p\left(\mathbf{X}_{N_k}^{(K)}(1), \mathbf{U}_{N_k}^{(K)}(1)\right) - 1 \le 0, \tag{33c}$$

$$\mathbf{N}\left(\mathbf{X}_{N_k}^{(K)}(1)\right) \le 0, \tag{33d}$$

$$\mathbf{X}_0^{(1)}(0) = \mathbf{x}_0, \quad p_0^{(1)}(0) = q_0^{(1)}(0) = 0, \tag{33e}$$

$$\mathbf{X}_{N_k}^{(k)}(1) = \mathbf{X}_0^{(k+1)}(0). \tag{33f}$$

Since the same variable is calculated two times at $\mathbf{X}_{N_k}^{(k)}(1)$ and $\mathbf{X}_0^{(k+1)}(0)$, respectively, (33b) can be reformulated as follows:

$$\sum_{j=0}^{N_{k-1}} D_{sj}^{(k)} \mathbf{W}_j^{(k)} + D_{sN_k}^{(k)} \mathbf{W}_0^{(k+1)} - \left(t_k - t_{k-1}\right)\mathbf{F}\left(\mathbf{X}_s^{(k)}(\tau), \mathbf{U}_s^{(k)}(\tau), \tau\right) = 0, \quad 1 \le k \le K-1, \tag{34a}$$

$$\sum_{j=0}^{N_{k-1}} D_{sj}^{(K)} \mathbf{W}_j^{(K)} + D_{sN_k}^{(K)} \mathbf{W}_{N_k}^{(K)} - (t_K - t_{K-1}) \mathbf{F}\left(\mathbf{X}_s^{(K)}(\tau), \mathbf{U}_s^{(K)}(\tau), \tau\right) = 0. \tag{34b}$$

After being discretized by LG pseudospectral method, the final NLP is obtained which consists of equations (33a), (33c)~(33f), (34a), and (34b). The solution of this NLP is the approximation solution of original optimal control problem in (11a), (11b), (11c), (11d), and (11e). This is easy to be solved by general optimization algorithms.

3.4. The KKT Optimality Conditions for Nonlinear Programming. Introducing the Lagrange multiplier vectors, the following augmented performance index can be drawn:

$$\begin{aligned} J^*\left(\mathbf{U}^{(k)}(\tau), t_k\right) &= \theta\left(\mathbf{X}_{N_k}^{(K)}(1)\right) \\ &+ q\left(\mathbf{X}_{N_k}^{(K)}(1), \mathbf{U}_{N_k}^{(K)}(1)\right) - \boldsymbol{v}^T \bullet \mathbf{N}\left(\mathbf{X}_{N_k}^{(K)}(1)\right) - \alpha \\ &\bullet \left[p\left(\mathbf{X}_{N_k}^{(K)}(1), \mathbf{U}_{N_k}^{(K)}(1)\right) - 1\right] - \sum_{k=1}^{K-1} \sum_{s=0}^{N_{k-1}} \nu_s^{(k)} \\ &\bullet \left[\sum_{j=0}^{N_{k-1}} D_{sj}^{(k)} \mathbf{W}_j^{(k)} + D_{sN_k}^{(k)} \mathbf{W}_0^{(k+1)} - (t_k - t_{k-1}) \mathbf{F}\left(\mathbf{X}_s^{(k)}(\tau), \mathbf{U}_s^{(k)}(\tau), \tau\right)\right] - \sum_{s=0}^{N_{k-1}} \nu_s^{(K)} \\ &\bullet \left[\sum_{j=0}^{N_{k-1}} D_{sj}^{(K)} \mathbf{W}_j^{(K)} + D_{sN_k}^{(K)} \mathbf{W}_{N_k}^{(K)} - (t_K - t_{K-1}) \mathbf{F}\left(\mathbf{X}_s^{(K)}(\tau), \mathbf{U}_s^{(K)}(\tau), \tau\right)\right], \end{aligned} \tag{35}$$

where $\boldsymbol{v} \in R^N$, $\alpha \in R^n$, and $\nu \in R^{n_w}$ denote the Lagrange multipliers associated with discretized original terminal constraints of (33d), discretized aggregated terminal constraints of (33c), and discretized process constraints of (34a) and (34b) in corresponding subintervals, respectively.

Take partial derivative of (35) with respect to $\mathbf{X}^{(k)}(\tau), \mathbf{X}_{N_k}^{(K)}(1), \mathbf{U}^{(k)}(\tau), \mathbf{U}_{N_k}^{(K)}(1), \boldsymbol{v}, t_f, \nu^{(k)}, \nu^{(K)}, \alpha$, and equating them to zero, then the KKT optimal condition is concluded as (36a), (36b), (36c), (36d), (36e), (36f), and (36g).

$$\frac{\partial q}{\partial \mathbf{X}_{N_k}^{(K)}(1)} - \boldsymbol{v}^T \bullet \frac{\partial \mathbf{N}}{\partial \mathbf{X}_{N_k}^{(K)}(1)} - \alpha \bullet \frac{\partial p}{\partial \mathbf{X}_{N_k}^{(K)}(1)} + \nu_{N_{k-1}}^{(K)} D_{N_{k-1}N_k}^{(K)} + \frac{\partial \theta}{\partial \mathbf{X}_{N_k}^{(K)}(1)} = 0, \tag{36a}$$

$$\frac{\partial q}{\partial \mathbf{U}_{N_k}^{(K)}(1)} - \boldsymbol{v}^T \bullet \frac{\partial \mathbf{N}}{\partial \mathbf{U}_{N_k}^{(K)}(1)} - \alpha \bullet \frac{\partial p}{\partial \mathbf{U}_{N_k}^{(K)}(1)} = 0, \tag{36b}$$

$$\sum_{j=0}^{N_{k-1}} D_{sj}^{(k)} = (t_k - t_{k-1}) \frac{\partial \mathbf{F}_k}{\partial \mathbf{X}_s^{(k)}(\tau)}, \quad 1 \le k \le K-1, 0 \le s \le N_{k-1}, \tag{36c}$$

$$\sum_{j=0}^{N_{k-1}} D_{sj}^{(K)} = (t_K - t_{K-1}) \frac{\partial \mathbf{F}_K}{\partial \mathbf{X}_s^{(K)}(\tau)}, \quad 0 \le s \le N_{k-1}, \tag{36d}$$

$$\frac{\partial \mathbf{F}_k}{\partial \mathbf{U}_s^{(k)}(\tau)} = 0, \quad 1 \le k \le K, 0 \le s \le N_k, \tag{36e}$$

$$\boldsymbol{v}^T \bullet \mathbf{N}\left(\mathbf{X}_{N_k}^{(K)}(1)\right) = 0, \tag{36f}$$

$$\alpha \bullet \left[p\left(\mathbf{X}_{N_k}^{(K)}(1), \mathbf{U}_{N_k}^{(K)}(1)\right) - 1\right] = 0, \tag{36g}$$

where $\mathbf{F}_k \equiv \mathbf{F}(\mathbf{X}_s^{(k)}(\tau), \mathbf{U}_s^{(k)}(\tau), \tau)$, $1 \le k \le K$.

4. The Adaptive Strategies for Solving the Control Problem

Although the original optimal problem has been transformed into a general NLP, there are still many problems that need to be coped with, such as how to adjust the parameter ρ in KS function, how to determine the subintervals, and how to regulate the degree of approximating in LG pseudospectral discretization. In view of these situations, two adaptive strategies are applied in constraint aggregation and LG pseudospectral discretization, respectively.

4.1. Adaptive Constraint Aggregation Strategy. When ρ tends to infinity, the value of KS function is the value of maximal constraint in theory. But if ρ is too large, there will be angular points at intersection points of multiple constraints of original problems. The continuity of first-order derivative will be decreased. After being aggregated, the second-order derivative will become morbid at angular points which will lead to slow convergence or even no convergence. This can be well explained in Figure 2. To keep good smoothness of KS function, the value of ρ should be decreased adequately. But this may lead to the feasible region of KS function being less than the real feasible region and the actual optimal control being unobtainable when the actual optimal state is on the boundary. The adaptive strategy in [44] can cope with this well. In this method, the calculation can be carried out with a big value of ρ in the feasible region or near a single constraint, and ρ can be decreased adaptively according to the first-order derivative at intersection points of multiple constraints.

Take partial differential of (12) with respect to ρ; then,

$$\frac{\partial \mathrm{KS}}{\partial \rho} = \frac{1}{\rho}\left(h_{\max}(x) - \mathrm{KS} - M(h(x))\right), \tag{37a}$$

$$M(h(x)) = \frac{\sum_{j=1}^{N_c} e^{\rho h_j(x)}\left(h_{\max}(x) - h_j(x)\right)}{\sum_{j=1}^{N_c} e^{\rho h_j(x)}}. \tag{37b}$$

Here, $h_{\max}(x)$ is the maximal constraint among all path constraints.

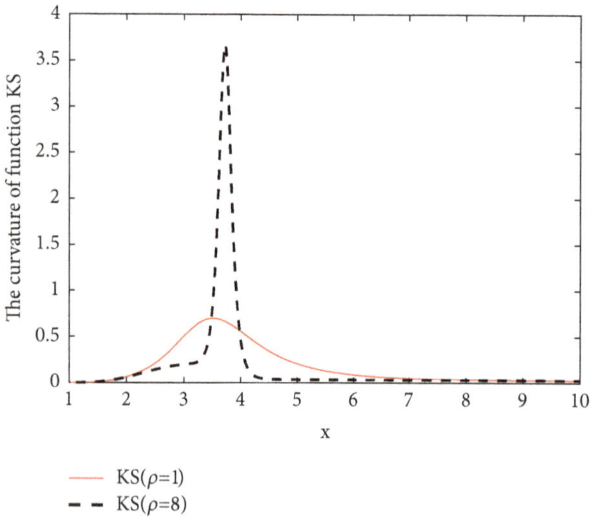

FIGURE 2: The curvature of KS function corresponding to $\max[g_1(x) = 5/\log(x) - 0.2x - 3, g_2(x) = x^2/40 + x/5 - 1]$.

From property (a) of KS function in Section 3.1, if all constraints satisfy $h_j(x) = h_{\max}(x)$, then

$$\mathrm{KS} = h_{\max}(x) + \frac{1}{\rho}\ln m. \quad (38)$$

Equation (37a) can be written as

$$\mathrm{KS}_1' = \frac{1}{\rho}\left(\frac{1}{\rho}\ln m - M(h(x))\right). \quad (39)$$

If the current state is near the intersection point of k constraints, (37a) can be similarly written as

$$\mathrm{KS}_u' = -\frac{1}{\rho}M(h(x)) - \frac{1}{\rho^2}\ln k. \quad (40)$$

Likewise, if the current state approximates only one constraint, (37a) can be transformed into another form:

$$\mathrm{KS}_d' = -\frac{1}{\rho}M(h(x)). \quad (41)$$

Define $\kappa(x,\rho) = (\mathrm{KS}_u' - \mathrm{KS}_d')/(\mathrm{KS}_1' - \mathrm{KS}_d')$, $\kappa(x,\rho) \in [0,1]$. It reflects the relation between the derivative of KS function and the number of constraints. When in use, we can give one ρ^* as the fiducial value and calculate κ as the distance parameter between the current state and the intersection point of multiple constraints at each iteration. Then the following adaptive constraints aggregation strategy can be defined:

$$\rho = \underline{\rho} + \Delta\rho\left(1 - \kappa\left(x, \rho^*\right)\right), \quad (42)$$

where $\underline{\rho}$ is the lower bound of ρ and $\Delta\rho > 0$ is the regulating range. The choice of ρ^* has to ensure that enough regulating range can be obtained near the intersection point of multiple constraints.

4.2. Adaptive LG Discretization Strategy. In LG pseudospectral method, enough interpolation points have to be chosen to meet given accuracy, but this will lead to the burdensome calculation. To regulate the subintervals reasonably, a mesh refinement algorithm that refers to [28, 42] is adopted in this section.

For subinterval k, select N_k LG points $(\widetilde{\tau}_1^{(k)}, \widetilde{\tau}_2^{(k)}, \ldots, \widetilde{\tau}_{B_k}^{(k)})$. Let $B_k = N_k + 1$, $\widetilde{\tau}_1^{(k)} = \tau_1^{(k)} = t_{k-1}$, and $\widetilde{\tau}_{B_k}^{(k)} = t_k$. According to (27b) and (28a), if the differential of state $\widehat{\mathbf{W}}^{(k)}$ accords with the dynamics at every LG point $\widetilde{\tau}_i^{(k)}$, we have

$$\widehat{\mathbf{W}}^{(k)}\left(\tau_i^{(k)}\right) = \mathbf{W}^{(k)}\left(\tau_{k-1}\right) + \left(t_k - t_{k-1}\right) \cdot \sum_{l=1}^{B_k} \widetilde{I}_{jl}^{(k)} \mathbf{F}\left(\mathbf{X}^{(k)}\left(\widetilde{\tau}_l^{(k)}\right), \mathbf{U}^{(k)}\left(\widetilde{\tau}_l^{(k)}\right), \widetilde{\tau}_l^{(k)}\right), \quad (43)$$

where $\widetilde{I}_{jl}^{(k)}$, $j,l = 1,2,\ldots,B_k$, denotes the $B_k \times B_k$ LG integration matrix which is similar to (30).

Define the absolute error and relative error as follows:

$$E_i^{(k)}\left(\widetilde{\tau}_l^{(k)}\right) = \left|\widehat{\mathbf{W}}_i^{(k)}\left(\widetilde{\tau}_l^{(k)}\right) - \mathbf{W}_i^{(k)}\left(\widetilde{\tau}_l^{(k)}\right)\right|. \quad (44)$$

$$e_i^{(k)}\left(\widetilde{\tau}_l^{(k)}\right) = \frac{E_i^{(k)}\left(\widetilde{\tau}_l^{(k)}\right)}{1 + \max_{j\in[1,2,\ldots,B_k]}\left|\mathbf{W}_i^{(k)}\left(\widetilde{\tau}_j^{(k)}\right)\right|}, \quad (45)$$

where $l = 1,2,\ldots,B_k$; $i = 1,2,\ldots,n_{x+2}$.

Then the maximal relative error can be written as

$$e_{\max}^{(k)} = \max_{i\in[1,2,\ldots,n_{x+2}], l\in[1,2,\ldots,B_k]} e_i^{(k)}\left(\widetilde{\tau}_l^{(k)}\right). \quad (46)$$

Assume that the given accuracy tolerance is ε. For subinterval k, if $e_{\max}^{(k)} \le \varepsilon$, we can conclude that the current LG points are reasonable. If $e_{\max}^{(k)} > \varepsilon$, we can adopt the following equation to update the LG points [28]:

$$N_k' = N_k + \mathrm{ceil}\left(\log_{N_k}\left(\frac{e_{\max}^{(k)}}{\varepsilon}\right)\right). \quad (47)$$

Here ceil is the operator that rounds to the next highest integer.

Assume that $N_{\min}$, $N_{\max}$ are the lower bound and upper bound of the number of LG points in subinterval k. If the regulated number N_k' satisfies $N_k' \le N_{\max}$, then N_k' is adopted to update the number of LG points. On the other hand, if $N_k' > N_{\max}$ which means the LG points are too many, then the current subintervals should be further refined. We use K' to update the number of subintervals [42]:

$$K' = \max\left(\mathrm{ceil}\left(\frac{N_k'}{N_{\min}}\right), 2\right). \quad (48)$$

From the above, the adaptive LG discretization strategy has been achieved. It balances the accuracy and computation burden by updating the number of LG interpolation points and refining the subintervals.

4.3. Control Method. In this section, the SQP [35, 36] is adopted to solve the NLP that is described as (33a), (33b), (33c), (33d), (33e), (33f), (34a), and (34b). The specific procedures are as follows:

(1) Initialize the parameters $\rho^*, \underline{\rho}, \Delta\rho, N_k, N_{\max}, N_{\min}$, the number of subintervals K, the initial state $\mathbf{W}_0^{(1)}(0) = \mathbf{0}$, initial control $\mathbf{U}_0^{(1)}(0) = \mathbf{0}$, the accuracy tolerance $\varepsilon_J, \varepsilon_s$, and the current iteration $r = 1$.

(2) Transcribe to NLP of (33a), (33b), (33c), (33d), (33e), (33f), (34a), and (34b) with current subintervals.

(3) Calculate the value of ρ in step r according to (42) on the basis of the partial differentials of KS function.

(4) Calculate the states and performance index J_k of each subinterval at each LG point in step r according to (35) and (43).

(5) Estimate the error $e_{\max}^{(k)}$ for each subinterval.

(6) Assess the number of LG points: if $e_{\max}^{(k)} \le \varepsilon_s$, go to step (7); if not, update the number of LG points as (47).

(7) Assess the refinement of subintervals: if $N_k' \le N_{\max}$, go to step (8); if not, update the subintervals as (48).

(8) Determine the search direction and search step length, solve the problem in (35) and (36a), (36b), (36c), (36d), (36e), (36f), and (36g), and calculate the optimal control $\mathbf{U}_{k+1}$ and index J_{k+1}.

(9) If $|J_{k+1} - J_k| < \varepsilon_J$, finish the computing process and output the optimal control $\mathbf{U}_{k+1}$ and index J_{k+1}. If not, go to step (2) and let $k = 1$; carry on the procedure until $|J_{k+1} - J_k|$ satisfies the given accuracy.

5. Numerical Solution of Optimal Control for ASP Flooding

For a real ASP flooding optimization problem in Section 2.1, carry out a series of operations—constraints aggregation, multistage problem formulation, Legendre-Gauss pseudospectral discretization, and adaptive strategies—and get the KKT optimality conditions. Then SQP is applied to solve the KKT conditions as in Section 4.3 to get the optimal injection strategy. After dimensionless processing, the optimal injection strategy for ASP flooding is solved by the proposed method in this paper. To explain the result better, the experiment optimization method and CVP method are adopted to solve the same problem and to be compared with the proposed method. All the simulations are carried out on software Matlab R2011b.

In this section, three-slug injection is used in simulation. There are twelve optimization variables which are $u_{\Theta 1}, u_{\Theta 2}, u_{\Theta 3}, T_\Theta$. Set $t_f = 2$, $t_P = 0.5$, $u_{a\max} = 2$, $u_{s\max} = 0.5$, $u_{p\max} = 1.5$, $M_p = 0.6$, $M_a = 0.6$, and $M_s = 0.2$. Give lots of different control inputs, and compute the performance index. Through the comparison of experiment results many times, we select the best injection concentration and slug size with the maximum profit. We call these results obtained with this method as experiment optimized results (EOS). The injection strategy is $u_a^0 = (1.2, 1, 0.8)$, $u_s^0 = (0.5, 0.4, 0.2)$, and $u_p^0 = (1, 0.6, 0.4)$, the slug size is $T^0 = (0.2, 0.2, 0.1)$, and the performance index is $J^0 = 0.2538$.

Solve this optimization problem with CVP method and the method proposed in this paper, respectively. The parameters are set as follows:

$$\begin{aligned} \underline{\rho} &= 200, \\ \Delta\rho &= 100, \\ \rho^* &= 50, \\ N_{\min} &= 4, \\ N_{\max} &= 25, \\ \varepsilon_s &= 10^{-3}, \\ \varepsilon_J &= 10^{-4}, \\ K &= 3. \end{aligned} \tag{49}$$

Some essential data for solving the optimization model is shown in Table 1.

Table 1: Partial data for solving ASP flooding model.

Sign	Value	Sign	Value	Sign	Value
M	8.3	N	8.33	L	1
S_{or0}	0.35	S_{or}	0.02	ϕ	0.4728
ap_1	7	ap_2	4	ap_3	40
μ_{w0}	0.5	$\mu_{w\max}$	18	μ_o	20
v_w	0.0045	ρ_r	2.0	ρ_s	1.0

Note. Table 1 is reproduced from Ge et al. (2016), (under the Creative Commons Attribution License/public domain). The specific variable declaration and unit can be found in reference [6].

After being optimized, the optimal injection strategy with CVP is $u_a^* = (1.862, 0.982, 0.826)$, $u_s^* = (0.461, 0.350, 0.403)$, $u_p^* = (1.415, 1.171, 1.024)$, and $T^0 = (0.22, 0.21, 0.07)$, and the dimensionless profit is $J^* = 0.26385$. The optimal injection strategy with the proposed method is $u_a^* = (1.869, 0.981, 0.829)$, $u_s^* = (0.463, 0.35, 0.402)$, $u_p^* = (1.418, 1.168, 1.024)$, and $T^0 = (0.23, 0.19, 0.08)$, and the dimensionless profit is $J^* = 0.26467$. The results are shown in Figures 3–6.

In these figures, the blue chain dotted line, the red imaginary line, and the black dotted line denote the results for EOS, CVP, and proposed method, respectively. Particularly, to distinguish the results in Figure 6, the black line with circle is used to denote the result of proposed method. According to these figures, the result of the proposed method in this paper is very close to that of CVP method, which are both better than EOS method. Regardless of the optimized injection concentration for three displacing agents or the optimized performance index, the values are nearly the same. The performance index optimized by the proposed method is increased by 0.01087 compared with that of EOS. This can fully prove the effectiveness of the proposed method. In addition, the switch time points and the running time are

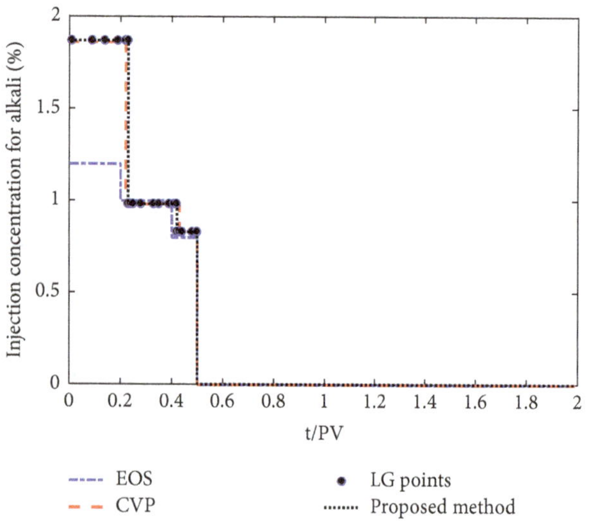

FIGURE 3: Comparison results of injection concentration for alkali.

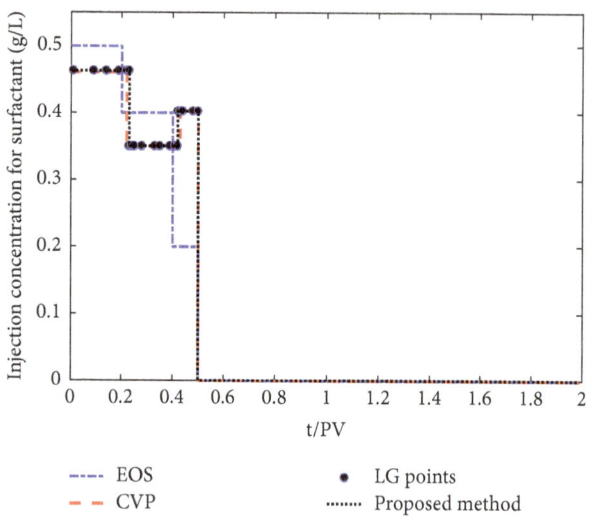

FIGURE 5: Comparison results of injection concentration for surfactant.

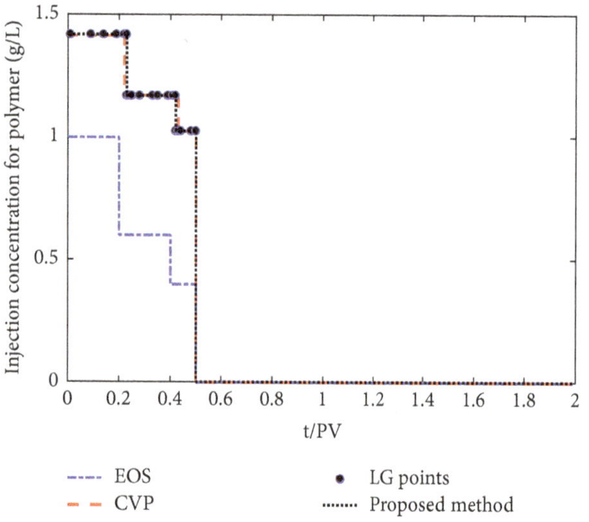

FIGURE 4: Comparison results of injection concentration for polymer.

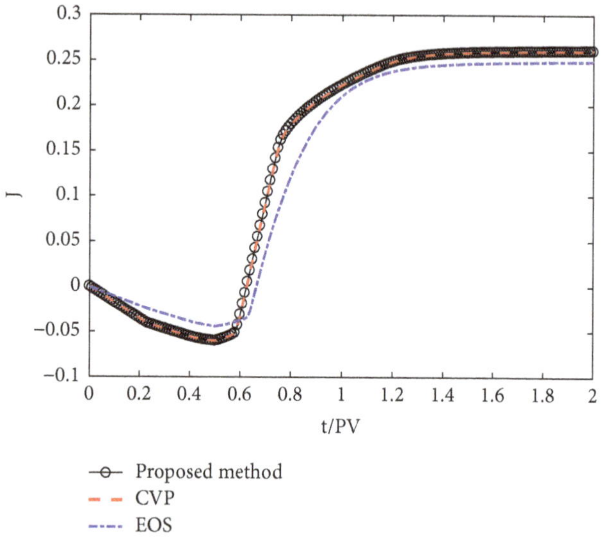

FIGURE 6: Comparison results of performance index.

displayed in Table 2. From Figures 3–5, we can find that the LG points are very intensive at the discontinuous points. In view of the switch time points in Table 2, the disposal of the proposed method is much better than CVP method and EOS. The index is further enhanced by 0.00082. This is because of the introduction of adaptive strategies and multistage problem formulation, which can adjust the discontinuous point more precisely. What is more, the running time of the proposed method is less than the other two methods. What is noteworthy is that because of the introduction of approximation, the results may have tiny error compared with the theoretical value.

To discuss the performance of the proposed method further, ASP flooding problem is optimized by different accuracy tolerances. The corresponding results are shown in Table 3.

From Table 3, the LG points, iterations, and running times increase evidently with the accuracy tolerance refining. And the maximum of error in every step increases at the same time. This is because with the increasing of accuracy more LG points are needed to be introduced to realize the approximation of states and controls and the disposal of discontinuous points, and more iterations are needed to search for optima. But the performance index is almost invariant. So we can conclude that the result is the optimal one. Though a better accuracy is desired, this will lead to an increase in the computation burden. So, there is a need a balance accuracy and running time in application.

6. Conclusions

In this paper, a method based on constraint aggregation and LG pseudospectral method with adaptive strategies

TABLE 2: Comparison of three methods (proposed method, CVP, and EOS).

	Switch time points			Performance index	CPU times(s)
	T_1	T_2	T_3		
Proposed method	0.23	0.42	0.5	0.26467	89.57
CVP method	0.22	0.43	0.5	0.26385	126.36
EOS method	0.2	0.4	0.5	0.2538	

Note. For the need of production, the switch time points for alkali, surfactant and polymer have to be kept the same.

TABLE 3: Comparison of the proposed method on accuracy and speed under different accuracy tolerances ε.

ε	Number of LG points	Number of iterations	$e^{(k)}_{\max}$	Performance index	CPU times (s)
10^{-3}	20	15	4.78×10^{-4}	0.26467	89.57
10^{-4}	23	16	5.19×10^{-5}	0.26467	99.32
10^{-5}	29	19	2.97×10^{-6}	0.26468	127.11
10^{-6}	38	23	7.08×10^{-7}	0.26468	184.59

Note. Keep $N_{\min} = 4$ for all calculations.

has been proposed to solve the optimal control for ASP flooding whose control variables are discontinuous. Firstly, we extended this problem to a class of optimal control problems. Then we introduced the constraint aggregation to convert all path constraints into one terminal condition and discretized the system equations with LG pseudospectral method. For simplicity, we also developed a new time variable and transformed the original problem into a multistage problem with CVP. Furthermore, we used two adaptive strategies to decrease the approximation error with regulating ρ of KS function, the LG points, and the refinement of subintervals adaptively. Finally, the optimal control problem for ASP flooding was solved by the proposed method. To demonstrate the performance better, we also solved this problem with CVP and experiment optimized result method. Numerical simulations showed that the proposed method can complete this well with less time and no loss of accuracy. The multiobjective optimization problem (the maximal oil production, minimal cost, etc.) and uncertainty problem (the uncertainty of geological parameters, oil price, external disturbance, etc.) for ASP flooding are very important to enhance the oil recovery and make the optimal injection strategy. These are the next research direction in the future.

Appendix

Physicochemical Algebraic Equations for ASP Flooding

The viscosity of displacing fluid is affected by polymer primarily. After being dissolved in water, the viscosity will be changed as follows [6]:

$$\mu_w = \mu_{w0}\left[1 + \left(\mathrm{ap}_1 C_p + \mathrm{ap}_2 C_p^2 + \mathrm{ap}_3 C_p^3\right)\right] C_{\mathrm{sep}}^{\mathrm{sp}}, \qquad (\text{A.1})$$

where C_{sep} is the salinity. $\mathrm{ap}_1, \mathrm{ap}_2, \mathrm{ap}_3, \mathrm{sp}$ are the constant parameters. μ_{w0}, μ_w are the viscosity of water and displacing fluid, respectively.

The alkali consumption is [38]

$$R_a = -\phi S_w \frac{\partial}{\partial t}\left(r_1 + r_2 + \cdots\right), \qquad (\text{A.2})$$

where r_i denotes the alkali consumption per unit volume of reaction i.

The adsorption retention loss of surfactant is [39]

$$\Gamma_s = \Gamma_s^0 \frac{a_s C_s}{1 + a_s C_s}\left(1 - b_s \frac{\mathrm{pH\text{-}7}}{\mathrm{pH}_{\max}\text{-}7}\right), \qquad (\text{A.3})$$

where Γ_s^0, a_s are related to ionic strength, pH denotes the pH value of the liquor, and b_s is the coefficient.

The adsorption retention loss of polymer is [39]

$$\Gamma_p = \Gamma_p^{\max} \frac{a_1 C_p}{1 + b_1 C_p}, \qquad (\text{A.4})$$

where $\Gamma_p^{\max}$ is the maximum adsorbing capacity of core at different salinity. a_1, b_1 denote the adsorption equilibrium constant.

The oil and water relative permeability k_{ro}, k_{rw} can be described as

$$K_{rw,ro} = \mathrm{A} \cdot \left(1 - S_w\right)^{\mathrm{B}} \cdot \left(S_w - \mathrm{C}\right)^{\mathrm{D}}, \qquad (\text{A.5})$$

where A, B, C, D are the identification coefficients, and the specific method is shown in [45, 46].

Nomenclature

d: Diameter for the core, m
u_Θ: Concentration for displacing agents Θ, $\mathrm{g{\cdot}L^{-1}}$
D_Θ: Diffusion coefficient for displacing agents Θ, $\mathrm{m^2{\cdot}s^{-1}}$
L: Length of the core, m
M_Θ: Dosage constraints for displacing agents Θ, kg
p: Pressure, MPa
Q: Volume flow rate, $\mathrm{m^3}$
S_w: Water saturation
S_{or}: Residual oil saturation

v_w: Flow velocity, m·s^{-1}
ϕ: Porosity of the core
ρ: Density, kg·m^{-3}
μ: Viscosity, μm^2.

Subscripts

a: Alkali
o: Oil
p: Polymer
s: Surfactant
w: Water
Θ: Mathematical set for alkali, surfactant, and polymer.

Acknowledgments

This work is supported by the National Natural Science Foundation of China under Grants nos. 60974039 and 61573378, the Natural Science Foundation of Shandong Province under Grant no. ZR2011FM002, and the Fundamental Research Funds for the Central Universities under Grant no. 15CX06064A.

References

[1] A. Kamari, M. Nikookar, L. Sahranavard, and A. H. Mohammadi, "Efficient screening of enhanced oil recovery methods and predictive economic analysis," *Neural Computing and Applications*, vol. 25, no. 3-4, pp. 815–824, 2014.

[2] Q. Liu, M. Dong, W. Zhou, M. Ayub, Y. P. Zhang, and S. Huang, "Improved oil recovery by adsorption-desorption in chemical flooding," *Journal of Petroleum Science and Engineering*, vol. 43, no. 1, pp. 75–86, 2004.

[3] Q. X. Feng, H. J. Yang, T. N. Nazina, J. Q. Wang, Y. H. She, and F. T. Ni, "Pilot test of indigenous microorganism flooding in Kongdian Oilfield," *Petroleum Exploration & Development*, vol. 32, no. 5, pp. 125–128, 2005.

[4] J. Qin, H. Han, and X. Liu, "Application and enlightenment of carbon dioxide flooding in the United States of America," *Petroleum Exploration and Development*, vol. 42, no. 2, pp. 232–240, 2015.

[5] H. B. Li, *The New Progress and Field Test Research for ASP Flooding*, Science Press, Beijing, China, 2007.

[6] S. R. Li and X. D. Zhang, *Optimal Control of Polymer Flooding for Enhanced Oil Recovery*, China university of petroleum press, Dongying, China, 2013.

[7] Y. Ge, S. Li, S. Lu, P. Chang, and Y. Lei, "Spatial-temporal ARX modeling and optimization for polymer flooding," *Mathematical Problems in Engineering*, vol. 2014, Article ID 713091, 10 pages, 2014.

[8] A. S. Lee and J. S. Aronofsky, "A linear programming model for scheduling crude oil production," *Journal of Petroleum Technology*, vol. 10, no. 7, pp. 51–54, 1958.

[9] J. D. Jansen, O. H. Bosgra, and P. M. J. Van den Hof, "Model-based control of multiphase flow in subsurface oil reservoirs," *Journal of Process Control*, vol. 18, no. 9, pp. 846–855, 2008.

[10] Y. Lei, S. R. Li, X. D. Zhang, Q. Zhang, and L. Guo, "Optimal control of polymer flooding based on mixed-integer iterative dynamic programming," *International Journal of Control*, vol. 84, no. 11, pp. 1903–1914, 2011.

[11] W. F. Ramirez, Z. Fathi, and J. L. Cagnol, "Optimal injection policies for enhanced oil recovery: part 1 - theory and computational strategies," *Society of Petroleum Engineers Journal*, vol. 24, no. 3, pp. 328–332, 1984.

[12] Z. Fathi and W. F. Ramirez, "Use of optimal control theory for computing optimal injection policies for enhanced oil recovery," *Automatica*, vol. 22, no. 1, pp. 33–42, 1986.

[13] W. F. Ramirez, *Application of optimal control to enhanced oil recovery*, Elsevier, Amsterdam, The Netherlands, 1987.

[14] L. E. Zerpa, N. V. Queipo, S. Pintos, and J.-L. Salager, "An optimization methodology of alkaline-surfactant-polymer flooding processes using field scale numerical simulation and multiple surrogates," *Journal of Petroleum Science and Engineering*, vol. 47, no. 3, pp. 197–208, 2005.

[15] Y. L. Ge, S. R. Li, and P. Chang, "An approximate dynamic programming method for the optimal control of Alkai-Surfactant-Polymer flooding," *Journal of Process Control*, vol. 64, pp. 15–26, 2018.

[16] T. Binder, L. Blank, H. Bock G et al., *Introduction to Model Based Optimization of Chemical Processes on Moving Horizons*, Springer, Berlin, Germany, 2001.

[17] S. Engell, "Feedback control for optimal process operation," *Journal of Process Control*, vol. 17, no. 3, pp. 203–219, 2007.

[18] V. S. Vassiliadis, R. W. H. Sargent, and C. C. Pantelides, "Solution of a class of multistage dynamic optimization problems. 1. problems without path constraints," *Industrial & Engineering Chemistry Research*, vol. 33, no. 9, pp. 2111–2122, 1994.

[19] U. Ali and Y. Wardi, "Multiple shooting technique for optimal control problems with application to power aware networks," *IFAC*, vol. 48, no. 27, pp. 286–290, 2015.

[20] L. T. Biegler, "Solution of dynamic optimization problems by successive quadratic programming and orthogonal collocation," *Computers & Chemical Engineering*, vol. 8, no. 3-4, pp. 243–247, 1984.

[21] S. A. Taghavi, R. E. Howitt, and M. A. MariÑ, "Optimal Control of Ground-Water Quality Management: Nonlinear Programming Approach," *Journal of Water Resources Planning & Management*, vol. 120, no. 6, pp. 962–982, 2014.

[22] B. Srinivasan, S. Palanki, and D. Bonvin, "Dynamic optimization of batch processes I. Characterization of the nominal solution," *Computers & Chemical Engineering*, vol. 27, no. 1, pp. 1–26, 2003.

[23] M. Schlegel, K. Stockmann, T. Binder, and W. Marquardt, "Dynamic optimization using adaptive control vector parameterization," *Computers & Chemical Engineering*, vol. 29, no. 8, pp. 1731–1751, 2005.

[24] X. Tang, Z. Liu, and Y. Hu, "New results on pseudospectral methods for optimal control," *Automatica*, vol. 65, pp. 160–163, 2016.

[25] D. A. Benson, G. T. Huntington, T. P. Thorvaldsen, and A. V. Rao, "Direct trajectory optimization and costate estimation via an orthogonal collocation method," *Journal of Guidance, Control, and Dynamics*, vol. 29, no. 6, pp. 1435–1440, 2006.

[26] G. Elnagar, M. A. Kazemi, and M. Razzaghi, "The pseudospectral Legendre method for discretizing optimal control problems," *Institute of Electrical and Electronics Engineers Transactions on Automatic Control*, vol. 40, no. 10, pp. 1793–1796, 1995.

[27] D. Garg, M. Patterson, W. W. Hager, A. V. Rao, D. A. Benson, and G. T. Huntington, "A unified framework for the numerical solution of optimal control problems using pseudospectral methods," *Automatica*, vol. 46, no. 11, pp. 1843–1851, 2010.

[28] C. L. Darby, W. W. Hager, and A. V. Rao, "Direct trajectory optimization using a variable low-order adaptive pseudospectral method," *Journal of Spacecraft and Rockets*, vol. 48, no. 3, pp. 433–445, 2011.

[29] V. V. Karelin, "Penalty functions in a control problem," *Remote Control*, vol. 65, no. 3, pp. 483–492, 2004.

[30] K. F. Bloss, L. T. Biegler, and W. E. Schiesser, "Dynamic process optimization through adjoint formulations and constraint aggregation," *Industrial & Engineering Chemistry Research*, vol. 38, no. 2, pp. 421–432, 1999.

[31] Q. Zhang, S. R. Li, X. D. Zhang, and Y. Lei, "Constraint aggregation based numerical optimal control," in *Proceedings of the 29th Chinese Control Conference (CCC)*, pp. 1560–1565, Beijing, China.

[32] R. Fonseca, B. Chen, J. D. Jansen et al., "A stochastic simplex approximate gradient (stosag) for optimization under uncertainty," *International Journal for Numerical Methods in Engineering*, vol. 109, pp. 1756–1776, 2016.

[33] B. Chen and A. C. Reynolds, "Optimal control of ICV's and well operating conditions for the water-alternating-gas injection process," *Journal of Petroleum Science and Engineering*, vol. 149, pp. 623–640, 2017.

[34] M. G. Shirangi, O. Volkov, and L. J. Durlofsky, "Joint optimization of economic project life and well controls," in *Proceedings of the SPE Reservoir Simulation Conference*, Montgomery, TX, USA, 2017.

[35] A. F. Izmailov, A. S. Kurennoy, and M. V. Solodov, "Some composite-step constrained optimization methods interpreted via the perturbed sequential quadratic programming framework," *Optimization Methods and Software*, vol. 30, no. 3, pp. 461–477, 2015.

[36] P. Sarma, W. H. Chen, L. J. Durlofsky, and K. Aziz, "Production Optimization With Adjoint Models Under Nonlinear Control-State Path Inequality Constraints," *SPE Reservoir Evaluation & Engineering*, vol. 11, no. 02, pp. 326–339, 2006.

[37] H. Bernardo, B. M. Silvana, and V. P. Carlos, "Using control cycle switching times as design variables in optimum waterflooding management," in *Proceedings of 2nd International Conference on Engineering Optimization*, pp. 1–10, Instituto Superior Tecnico, Lisbon, Portugal, 2010.

[38] L. Y. Su, *Reservoir Flooding Mechanisms*, Petroleum Industry Press, Beijing, China, 2009.

[39] C. Z. Yang, *Enhanced Oil Recovery for Chemical Flooding*, vol. 10, Petroleum Industry Press, Beijing, China, 2007.

[40] Y. L. Ge, S. R. Li, P. Chang, R. Zang, and Y. Lei, "Optimal control for an alkali/surfactant/polymer flooding system," in *Proceedings of the China Control Conference*, pp. 2631–2636, IEEE, Chengdu, China, July 2016.

[41] T. Ohsawa, "Contact geometry of the Pontryagin maximum principle," *Automatica*, vol. 55, pp. 1–5, 2015.

[42] P. Wang, C. Yang, and Z. Yuan, "The combination of adaptive pseudospectral method and structure detection procedure for solving dynamic optimization problems with discontinuous control profiles," *Industrial & Engineering Chemistry Research*, vol. 53, no. 17, pp. 7066–7078, 2014.

[43] C. G. Raspanti, J. A. Bandoni, and L. T. Biegler, "New strategies for flexibility analysis and design under uncertainty," *Computers & Chemical Engineering*, vol. 24, no. 9-10, pp. 2193–2209, 2000.

[44] Q. Zhang, S. R. Li, X. Zhang, and Y. Lei, "Constraint aggregation based numerical optimal control," in *Proceedings of the 29th Chinese Control Conference*, pp. 1560–1565, 2010.

[45] Y. L. Ge and S. R. Li, "Computation of reservoir relative permeability curve based on RBF neural network," *Journal of Chemical Engineering*, vol. 64, no. 12, pp. 4571–4577, 2013.

[46] Y. Ge, S. Li, and K. Qu, "A novel empirical equation for relative permeability in low permeability reservoirs," *Chinese Journal of Chemical Engineering*, vol. 22, no. 11, pp. 1274–1278, 2014.

2

Chaos Suppression via Euler-Lagrange Control Design for a Class of Chemical Reacting System

Ricardo Aguilar-López

Department of Biotechnology and Bioengineering, Centro de Investigación y de Estudios Avanzados del I.P.N. (CINVESTAV), Av. Instituto Politécnico Nacional, No. 2508, Colonia San Pedro Zacatenco, C.P. 07360, Ciudad de México, Mexico

Correspondence should be addressed to Ricardo Aguilar-López; raguilar@cinvestav.mx

Academic Editor: Libor Pekař

In this work the problem of chaos suppression for a class of continuous chemical reactor with chaotic dynamics is tackled via a nonlinear control strategy. The proposed controller is developed under the framework of optimal control theory, where a functional is proposed to maximize the chemical reaction rate via a proposed Lagrangian-type which contains directly the state equation of the reacting system, avoiding the problems of Lagrange, the Hamiltonian formulation, and consequently the explicit constraints to the system. This allows solving in an easier form the optimization problem in comparison with the standard methods. This procedure allows suppressing the chaotic behavior of the reacting system by stabilizing the reaction rate term by leading it to an extreme value. Numerical experiments are done in order to show the satisfactory performance of the proposed methodology.

1. Introduction

The stabilization of nonlinear systems with complex or chaotic behavior has been analyzed from several years ago. In particular, the stabilization via nonlinear controllers with regulation purposes in chemical reactors is also a classical task for processes engineers; the control of chemical reactors has been done via linear PID, adaptive, predictive, I/O linearizing, fuzzy, neural, and optimal controllers with success [1–6]. A widely employed control strategy is related to the optimal control theory, considering external or without constraints. For improved reactor operation, the need to implement optimal operational trajectories, which include maximum productivity and optimal cost, for example, leaded to the tracking trajectory control problem, where the optimal control designs have been successful [7–10]. In optimal control approach, the Hamiltonian techniques have been applied to nonlinear processes, where the corresponding Hamiltonian equations must be developed to include nonlinear constraints and construct an adequate functional which is named objective function; in order to obtain a controller for the required task, as common in optimal control theory, here the Pontryagin's maximum (or minimum) principle is used to find the best possible control for taking a dynamical system from one state to another, especially in the presence of constraints for the state variables and input controls [10]. Also, in optimal control theory the corresponding functional (objective function) to be maximizing or minimized only contains in its structure nonlinear terms of the state variables and the control input in order to generate a control law design for a specific purpose. In this work a Lagrangian based directly on the state equation of the chemical reactor with chaotic behavior is proposed, in order to develop a functional to stabilize the reactor's operation in a critical point the reaction rate, suppressing the complex oscillations in the process.

2. Control Design

The fundamental framework of the optimal control theory lies on the calculus of variations, which is related to the basic trajectory optimization problem, where a functional $\mathcal{F}(\mathcal{L}(\bullet)) := \mathbb{R}^q \longrightarrow \mathbb{R}$ is a scalar cost function or cost index or performance index, which needs to be maximizing or minimizing; this objective is reached by solving the corresponding Euler-Lagrange equation:

$$\frac{\partial \mathcal{L}}{\partial x_1} - \frac{d}{dx_2}\left(\frac{\partial \mathcal{L}}{\partial \dot{x}_1}\right) = 0 \tag{1}$$

The term $\mathscr{L}$ is known as the Lagrangian of the system under study.

In general the cost functional $\mathscr{F}$ can be represented as

$$\mathscr{F} = \psi\left(x_f, t_f\right) + \int_{t_0}^{t_f} \mathscr{L}(t, x, u)\, dt \quad (2)$$

where $\psi(x_f, t_f)$ is an algebraic term to be minimized (or maximized) at final conditions, subject to the following constraints:

(i)

$$\dot{x} := \frac{dx}{dt} = f(x) + g(x)\, u \quad (3)$$

The above equation is the state equation.

Here, $x \in \mathbb{R}^n$ is the state variable vector, $f(x) : \mathbb{R}^n \longrightarrow \mathbb{R}^n$ is a nonlinear vector field, where $f(x) \subset \Sigma \in C^{\infty}$ and Σ is a compact set, $g(x)$ is a smooth and invertible bounded function, and $u \in \mathbb{R}^m$, with $m \leq n$ is the control input.

(ii)

$$\omega\left(x_f, t_f\right) = 0 \quad (4)$$

The vector ω represents the terminal constraints.

(iii)

$$x(t_0) = x_0 \quad (5)$$

which represent the initial conditions.

The case related to the solving of (2) subject to constraints (2)–(5) is known as the Problem of Bolza [11, 12]. Particular cases, where the functional is defined as $\mathscr{F} = \psi(x_f, t_f)$ or $\mathscr{F} = \int_{t_0}^{t_f} \mathscr{L}(t, x, u) dt$ subject to the same constraints (2)-(5), are classified as the Problem of Mayer or the Problem of Lagrange, respectively [13, 14].

Note that the solutions of the above-mentioned problems can be complex given the nonlinear nature of the cost functional and the corresponding constraints; an alternative methodology consists of avoiding the nonlinear constraints by incorporating them to the cost functional via the Hamiltonian formulation [8], which is an augmented cost functional with the following structure:

$$\mathscr{F} = \psi\left(x_f, t_f\right) + \int_{t_0}^{t_f} \left(\mathscr{L}(t, x, u) + \lambda^T \left[f(x), u - \dot{x}\right]\right) dt \quad (6)$$

As mentioned above, the solution of (6) also can be complex by the highly nonlinearities of the augmented functional.

Now, let us consider the following functional form:

$$\mathscr{F}(\mathscr{L}) = \int_0^T \mathscr{L}(x, \dot{x}, u)\, dt \quad (7)$$

Now, it is needed to determine the extreme value of functional (7); therefore,

$$\delta\mathscr{F}(\mathscr{L}) = \int_0^T \delta\mathscr{L}(x, \dot{x}, u)\, dt \quad (8)$$

The differential form of the corresponding Lagrangian $\mathscr{L}$ is giving by

$$\delta\mathscr{L}(x, \dot{x}, u) = \frac{\partial \mathscr{L}}{\partial x}\delta x + \frac{\partial \mathscr{L}}{\partial \dot{x}}\delta \dot{x} + \frac{\partial \mathscr{L}}{\partial u}\delta u \quad (9)$$

Substituting (9) into (8),

$$\delta\mathscr{F}(\mathscr{L}) = \int_0^T \left(\frac{\partial \mathscr{L}}{\partial x}\delta x + \frac{\partial \mathscr{L}}{\partial \dot{x}}\delta \dot{x} + \frac{\partial \mathscr{L}}{\partial u}\delta u\right) dt \quad (10)$$

The following terms of the differential form of the Lagrangian are now represented as

$$\frac{\partial \mathscr{L}}{\partial x}\delta x = \frac{\partial \mathscr{L}}{\partial x}\delta x \frac{\delta u}{\delta u} = \frac{\partial \mathscr{L}}{\partial x}\frac{\delta u}{u'} \quad (11)$$

where

$$\frac{\delta u}{\delta x} := u' \quad (12)$$

and integrating by parts

$$\int_0^T \frac{\partial \mathscr{L}}{\partial \dot{x}}\delta \dot{x}\, dt = \frac{\partial \mathscr{L}}{\partial \dot{x}}\, \delta x|_0^T - \int_0^T \frac{d}{dt}\left(\frac{\partial \mathscr{L}}{\partial \dot{x}}\right)\delta x\, dt \quad (13)$$

Now, from (3) the following is considered:

$$f(x) = \dot{x} - g(x)\, u \quad (14)$$

The particular Lagrangian for (7) is proposed as

$$\mathscr{L}(x, \dot{x}, u) \equiv \dot{x} - g(x)\, u \quad (15)$$

Note that the proposed Lagrangian only contains information of the corresponding state equation of the system under study.

Therefore, from (15) and (10)

$$\frac{\partial \mathscr{L}}{\partial x} = -g'(x)\, u; \qquad \frac{\partial \mathscr{L}}{\partial \dot{x}} = 1; \qquad \frac{\partial \mathscr{L}}{\partial u} = -g(x) \quad (16)$$

Then

$$\int_0^T \frac{d}{dt}\left(\frac{\partial \mathscr{L}}{\partial \dot{x}}\right)\delta \dot{x}\, dt = 0 \quad (17)$$

$$\frac{\partial \mathscr{L}}{\partial \dot{x}}\, \delta x|_0^T = 0 \quad (18)$$

because $\delta x|_0 = 0$ and $\delta x|_T = 0$.

Finally, from (9), (10), and (16)-(18)

$$\frac{\delta\mathscr{F}(\mathscr{L})}{\delta u} = \int_0^T \left(\frac{u}{u'}g'(x) + g(x)\right) dt \quad (19)$$

The extreme value of functional (19) is then determined by the following restriction:

$$\frac{\delta\mathscr{F}(\mathscr{L})}{\delta u} = 0 \quad (20)$$

Equivalently,

$$\delta\mathscr{L}(x, \dot{x}, u) = \frac{u}{u'}g'(x) + g(x) = 0 \quad (21)$$

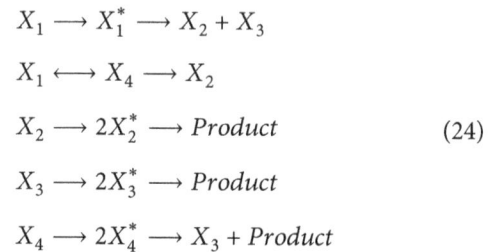

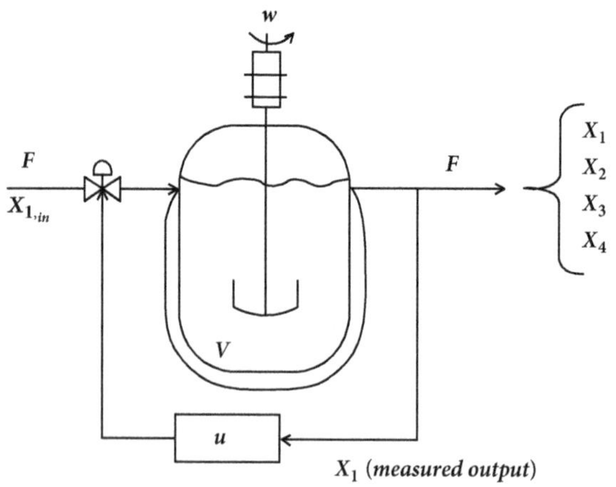

FIGURE 1: Scheme of the operation of the continuous chemical reactor.

From (21), the differential equation to describe the corresponding control law u to stabilize the system (3) in an extreme point can be obtained, without considering any explicit constraints, as follows:

$$u' = -\frac{g'(x)}{g(x)}u \tag{22}$$

Solving (22),

$$u(x) = u_0 \exp\left(\int_0^T -\frac{g'(x(t))}{g(x(t))}dt\right) \tag{23}$$

where u_0 is the initial condition of (22).

3. Application Example

From several years ago experimental evidence of chaotic behavior in chemical reacting systems exists, since the corresponding chemical species act as reactive, intermediates, and products, where reversible and autocatalytic chemical paths are generally present and hyperchaotic dynamics can undoubtedly be found [15, 16]. In particular, the chemical reactors are process equipment designed to produce high value compounds or products from the corresponding reactions via a chemical transformation, to reach this important objective, process analysis must be performed in order to determinate steady-states multiplicity, open-loop instabilities between others, to select under thermodynamic and kinetic restrictions, the optimal operating regions which maximize the reactor's productivity, process security, and operation cost, where the closed-loop operation is frequently needed to keep the reactor's variables in the selected set points. The continuous reactor operation consists of a multicomponent inlet flow F, which feeds the chemical reactions to the process equipment; the corresponding chemical reactions are carried out inside the reactor with a multicomponent output flow, where a specific set of state variables (x_1 in this case) can be measured to give feedback to the control algorithm and generate the control actions by the corresponding actuator, in this case the control valve, as can be observed in Figure 1.

If the control input u is assumed constant with a nominal value, the corresponding operation is in open-loop regimen. The main task of the control input is to compensate via the input flow manipulation, the residence time in the chemical reactor under the proposed control law (see (23)) to reach an extremum seeking of the corresponding reaction rate.

In particular, the mathematical model presented by [17] is considered, where, via mass conservation principle, a four-state dynamics system is proposed. For the kinetic model developing, several assumptions are considered; the chemical reactions are carried out in perfect homogeneous conditions in a well stirred tank reactor under isothermal operation, first-order kinetic is considered for the reactions from X_4 to X_1 and X_2 and X_1 to X_2, the reactive X_1 catalyzes the production of the compound X_3, the two reactions from X_1 are catalyzed by X_2 and X_3, the reaction from X_4 is also catalyzed by X_3, and all the chemical species are involved in autocatalytic reactions and they are mathematically modeled by Michaelis-Menten structures [17].

The following general chemical kinetic pathway is proposed:

$$\begin{aligned} &X_1 \longrightarrow X_1^* \longrightarrow X_2 + X_3 \\ &X_1 \longleftrightarrow X_4 \longrightarrow X_2 \\ &X_2 \longrightarrow 2X_2^* \longrightarrow Product \\ &X_3 \longrightarrow 2X_3^* \longrightarrow Product \\ &X_4 \longrightarrow 2X_4^* \longrightarrow X_3 + Product \end{aligned} \tag{24}$$

where X_i^* for i= 1, 2, 3, 4 are the corresponding activate chemical complexes, which are assumed in pseudo steady state, as usual.

This kinetic model is extended to continuous reactor's operation, showing complex (chaotic) oscillations. The mathematical model is represented by the following set of nonlinear ordinary differential equations and represents the mass balances for each one of the chemical compounds in terms of the corresponding mass concentrations.

Mass Balances

$$\begin{aligned} \dot{x}_1 = {} & d_0 + k_8 x_4 - k_1\frac{x_1 x_2}{x_1 + K} - k_2\frac{x_1 x_3}{x_1 + K} \\ & + (x_{1,in} - x_1)u \end{aligned} \tag{25}$$

$$\dot{x}_2 = k_3 x_1 + k_4 x_2 + k_9 x_4 - k_5\frac{x_2}{x_2 + K} + (x_{2,in} - x_2)u \tag{26}$$

$$\dot{x}_3 = d_1 + k_6 x_1 x_3 - k_7 x_3 + (x_{3,in} - x_3)u \tag{27}$$

$$\dot{x}_4 = (k_{10} - k_8 - k_9)x_4 - k_{11}\frac{x_3 x_4}{x_4 + K} + (x_{4,in} - x_4)u \tag{28}$$

Here, $x = [x_1 \ x_2 \ x_3 \ x_4]$ is the vector of mass concentrations. The set of kinetic parameters is given as follows: k_1 = 1.0; k_2 = 1.0; k_3 = 1.0; k_4 = 0.25; k_5 = 152.5; k_6 = 1.0; k_7 = 130; k_8 = 0.001; k_9 = 1.0; k_{10} = 1.051; k_{11} = 0.5; K = 0.001. The inlet concentrations to the reactor are as follows: $x_{1,in}$ = 150;

$x_{2,in} = x_{3,in} = x_{4,in} = 0$, and the nominal value of the control input is considered as $u = 0.015$. The corresponding initial conditions are $x_{10} = 129.1$; $x_{20} = 76.06$; $x_{30} = 0.5895$; and $x_{40} = 21.38$, $d_0 = 90$; $d_1 = 2.2$ are constant disturbances which are included to simulate realistic process operation. As usual for this processes, let us consider that the control input is defined as $u = F/V$, where F is the volumetric inlet flow and V is the volume of the reactor, which is assumed as constant as usual for this systems.

Now, for this application case, let us consider the state equation (21) to construct the corresponding Lagrangian:

$$\mathcal{F}(\mathcal{L}) = \int_0^T (\dot{x}_1 - (x_{1,in} - x_1)u)\,dt \tag{29}$$

Taking into account (14), note that

$$f(x) = d_0 + k_8 x_4 - k_1 \frac{x_1 x_2}{x_1 + K} - k_2 \frac{x_1 x_3}{x_1 + K} \tag{30}$$

Here:

$$g(x) = (x_{1,in} - x_1) \tag{31}$$

As observed, (31) represents the input and output mass flows, where

$$g'(x) = -1 \tag{32}$$

Applying (31)-(32) to (23), the corresponding control law is giving by

$$u(x) = (x_{1,in} - x_1)^{-u_0} \tag{33}$$

Noting the simple structure of the resulting controller, the controller's structure is properly model free and only needs online measurements of the state variable x_1 and some parameters as the inlet concentration $x_{1,in}$ and the initial condition of the control input u_0; these parameters are easily available for the standard operation of this kind of process; these characteristics would allow its real time implementation.

4. Results and Discussion

Numerical experiments were done in order to show the performance of the proposed control strategy. The numerical simulations were carried out employing ODE 23s Matlab library to solve the dynamic model of the chemical reactor, under the parametric and initial conditions described above. In this analysis a Single-Input Single Output (SISO) closed-loop operation is considered, where the state equation to be controlled corresponds to the state variable x_1. Firstly, the open-loop behavior of the bioreactor shown in Figures 2–5 observed the complex (chaotic) oscillations in the corresponding time series and the phase portraits; as previously reported in [17], the system keep positive Lyapunov indexes confirming the complexity of the oscillations. In order to show the performance of the proposed methodology, the proposed controller is activated at $t = 100$; here the effect of

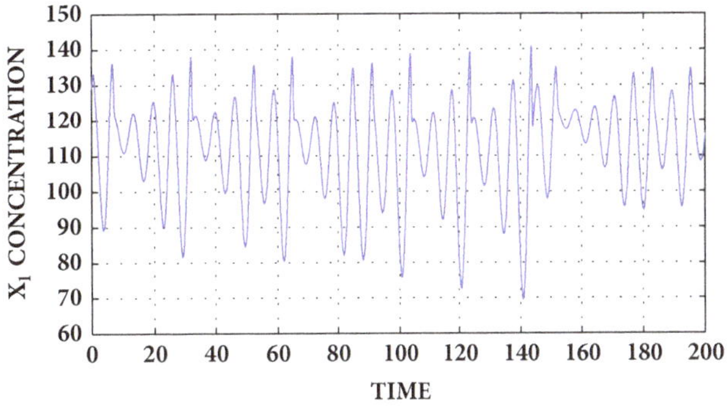

Figure 2: Open-loop dynamic behavior of the mass concentration of x_1.

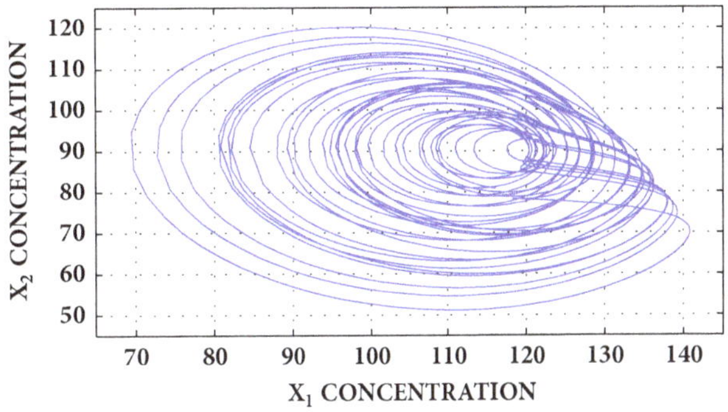

Figure 3: Open-loop 2D phase portrait.

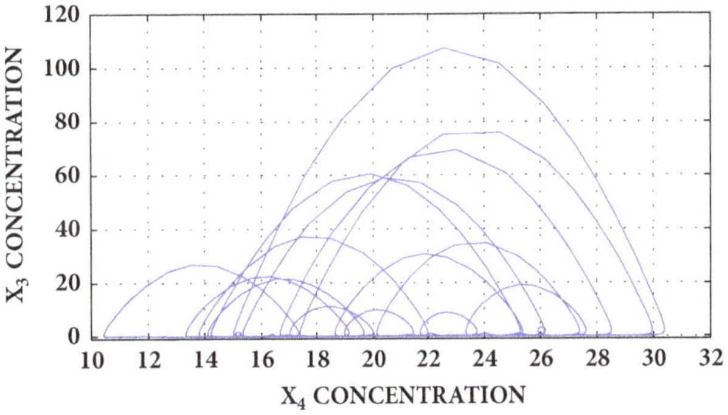

Figure 4: Open-loop 2D phase portrait.

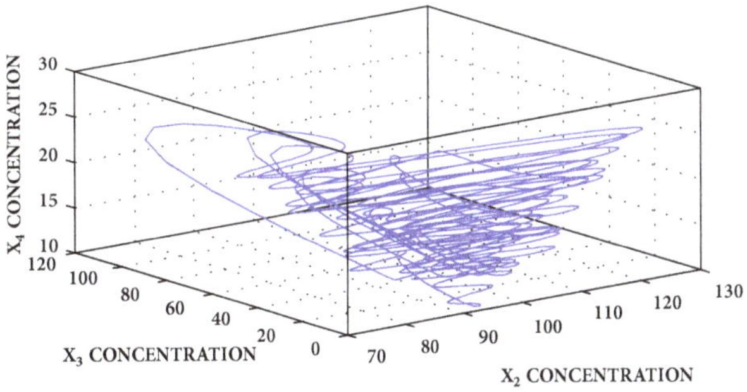

Figure 5: Open-loop 3D phase portrait.

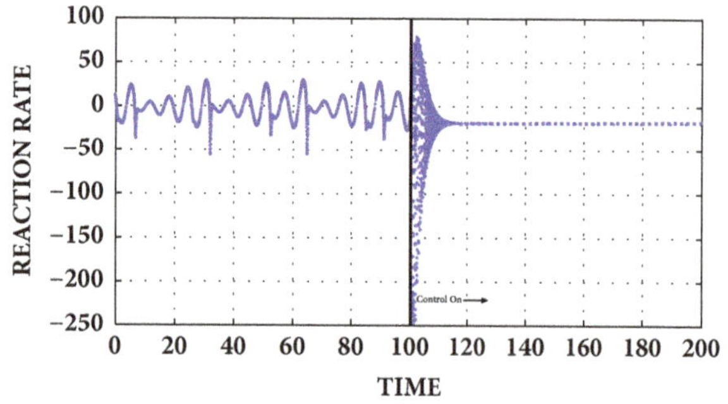

FIGURE 6: Open-loop and closed-loop dynamic behavior of the reaction rate.

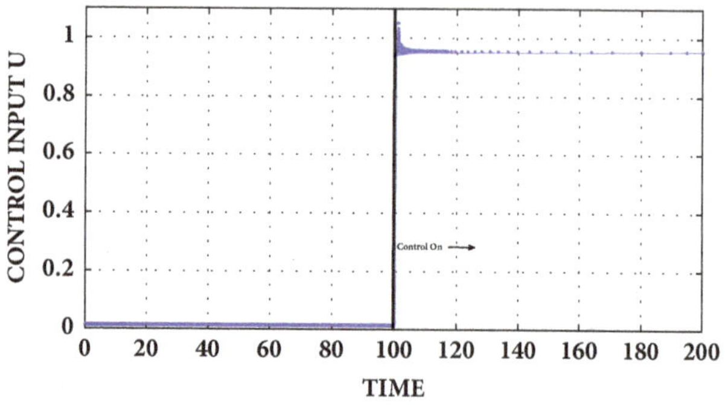

FIGURE 9: Control Input effort.

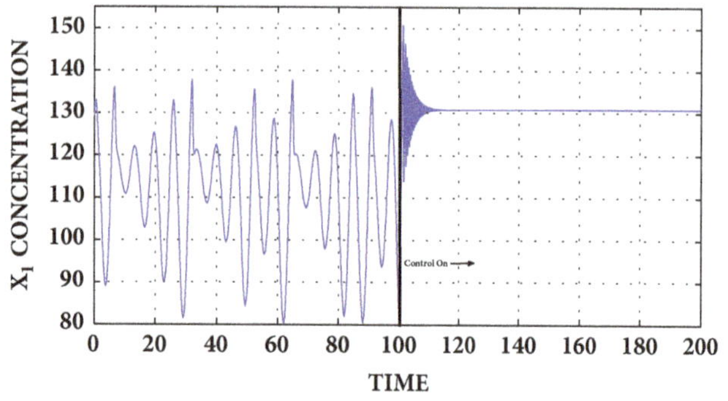

FIGURE 7: Open-loop and closed-loop dynamic behavior of the mass concentration of x_1.

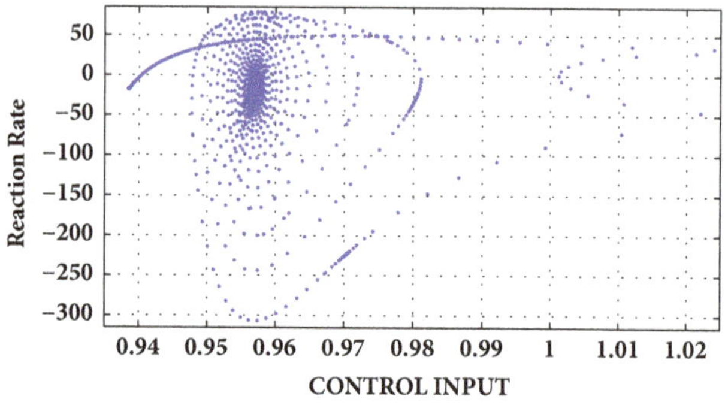

FIGURE 10: Reaction rate behavior versus control input action.

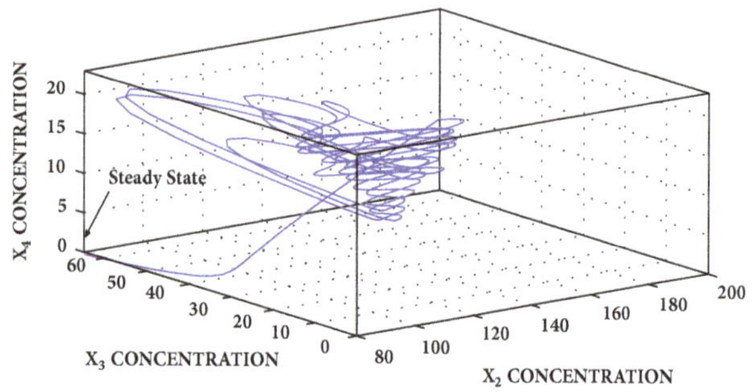

FIGURE 8: Closed-loop 3D phase portrait.

the control action in the corresponding reaction rate term is observed; Figure 6 shows that the oscillating behavior is suppressed after an overshot and a settling time of 15 time units, reaching to a stable and constant value of -20 units in the reaction rate; therefore, Figure 7 shows the corresponding effect in the mass concentration of the chemical specie x_1 which increase to a fixed value to 131.7 units; this is an important issue such that this means that the corresponding reaction rate term is maximized by the proposed controller as expected and in accordance with the corresponding design. Figure 8 is related to the closed-loop behavior of the trajectories in a 3D phase portrait, where the change in the trajectory is observed in comparison with the open-loop behavior; furthermore, in Figure 9 the control effort under the proposed controller is observed; notice that the control action acts almost immediately, with a small overshoot, increasing its value and reaching the corresponding steady state with realizable values. On the other hand, Figure 10 is related to the evolution of the reaction rate term as a function of the control input, as presented in the functional structure of (23) and (33), and it can be seen that the reaction rate reaches a stable equilibrium point $f(x_{eq})$ = -20, at u = 0.957, in accordance with the previous figures.

5. Concluding Remarks

In this work an optimal control strategy to suppress the chaotic behavior of a class of continuous chemical reactor is presented. The proposed methodology is based on a Lagrangian structure which includes directly the corresponding state equations of the system under analysis, avoiding the inclusion of implicit and nonlinear constraints as in the case of the Problem of Bolza and Hamiltonian approaches. The proposed strategy directly maximize the reaction rate term following the corresponding optimal trajectory, leading to the reactor trajectories to realizable and stable extreme values, without excessive control efforts. Numerical experiments show that the proposed control scheme is realizable keeping a satisfactory performance.

Conflicts of Interest

The authors declare no conflicts of interest.

References

[1] L. Estel, M. Poux, N. Benamara, and I. Polaert, "Continuous flow-microwave reactor: Where are we?" *Chemical Engineering and Processing: Process Intensification*, vol. 113, pp. 56–64, 2017.

[2] R. Zhang, S. Wu, and F. Gao, "State Space Model Predictive Control for Advanced Process Operation: A Review of Recent Development, New Results, and Insight," *Industrial & Engineering Chemistry Research*, vol. 56, no. 18, pp. 5360–5394, 2017.

[3] S. Vaidyanathan, "Adaptive Integral Sliding Mode Control of a Chemical Chaotic Reactor System," in *Applications of Sliding Mode Control in Science and Engineering. Studies in Computational Intelligence*, S. Vaidyanathan and C. H. Lien, Eds., vol. 709, Springer, Cham, Switzerland, 2017.

[4] M. Bahita and K. Belarbi, "Fuzzy modelling and model reference neural adaptive control of the concentration in a chemical reactor (CSTR)," *AI & Society: Knowledge, Culture and Communication*, vol. 33, no. 2, pp. 189–196, 2018.

[5] S. Baruah and L. Dewan, "A comparative study of PID based temperature control of CSTR using Genetic Algorithm and Particle Swarm Optimization," in *Proceedings of the 2017 International Conference on Emerging Trends in Computing and Communication Technologies (ICETCCT)*, pp. 1–6, Dehradun, India, November 2017.

[6] J. A. Romero-Bustamante, J. G. Moguel-Castañeda, H. Puebla, and E. Hernandez-Martinez, "Robust Cascade Control for Chemical Reactors: An Approach based on Modelling Error Compensation," *International Journal of Chemical Reactor Engineering*, vol. 15, no. 6, 2017.

[7] N. Ghadipasha, W. Zhu, J. A. Romagnoli, T. McAfee, T. Zekoski, and W. F. Reed, "Online Optimal Feedback Control of Polymerization Reactors: Application to Polymerization of Acrylamide-Water-Potassium Persulfate (KPS) System," *Industrial & Engineering Chemistry Research*, vol. 56, no. 25, pp. 7322–7335, 2017.

[8] H. Saberi Nik and S. Shateyi, "Application of Optimal HAM for Finding Feedback Control of Optimal Control Problems," *Mathematical Problems in Engineering*, vol. 2013, Article ID 914741, 10 pages, 2013.

[9] M. Yong-Quan and Q. Hong-Xing, "Integrated Multiobjective Optimal Design for Active Control System Based on Genetic Algorithm," *Mathematical Problems in Engineering*, vol. 2014, Article ID 748237, 9 pages, 2014.

[10] R. Kamalapurkar, P. Walters, J. Rosenfeld, and W. Dixon, "Optimal Control," in *Reinforcement Learning for Optimal Feedback Control. Communications and Control Engineering*, Springer, Cham, Switzerland, 2018.

[11] M. I. Krastanov and N. K. Ribarska, "A Functional Analytic Approach to a Bolza Problem," in *Control Systems and Mathematical Methods in Economics*, G. Feichtinger, R. Kovacevic, and G. Tragler, Eds., vol. 687 of *Lecture Notes in Economics and Mathematical Systems*, Springer, Cham, Switzerland, 2018.

[12] Q. Cui, L. Deng, and X. Zhang, "Second order optimality conditions for the Bolza endpoint constraint control problem evolved on a Riemannian manifold," in *Proceedings of the 2017 36th Chinese Control Conference (CCC)*, pp. 2639–2643, Dalian, China, July 2017.

[13] V. Badescu, "Generalities Concerning the Optimal Control Problems," in *Optimal Control in Thermal Engineering. Studies in Systems, Decision and Control*, vol. 93, Springer, Cham, Switzerland, 2017.

[14] I. Ekeland, "On the Euler-Lagrange equation in calculus of variations," *Vietnam Journal of Mathematics*, vol. 46, no. 2, pp. 359–363, 2018.

[15] K. Lamamra, S. Vaidyanathan, A. T. Azar, and C. Ben Salah, "Chaotic System Modelling Using a Neural Network with Optimized Structure," in *Fractional Order Control and Synchronization of Chaotic Systems. Studies in Computational Intelligence*, A. Azar, S. Vaidyanathan, and A. Ouannas, Eds., vol. 688, Springer, Cham, Switzerland, 2017.

[16] V. K. Yadav, S. Das, B. S. Bhadauria, A. K. Singh, and M. Srivastava, "Stability analysis, chaos control of a fractional order chaotic chemical reactor system and its function projective synchronization with parametric uncertainties," *Chinese Journal of Physics*, vol. 55, no. 3, pp. 594–605, 2017.

[17] H. Killory, O. E. Rössler, and J. L. Hudson, "Higher chaos in a four-variable chemical reaction model," *Physics Letters A*, vol. 122, no. 6-7, pp. 341–345, 1987.

3

Study on ADRC Parameter Optimization Using CPSO for Clamping Force Control System

Fengping Li,[1,2] **Zhengya Zhang**,[2] **Antonios Armaou**,[3] **Yao Xue**,[2] **Sijia Zhou**,[2] **and Yuqing Zhou**[2]

[1]*School of Aerospace Engineering, Xiamen University, Xiamen 361005, China*
[2]*Zhejiang Provincial Engineering Lab of Laser and Optoelectronic Intelligent Manufacturing, Wenzhou University, Wenzhou 325035, China*
[3]*Department of Chemical Engineering, Penn State University, State College, PA, USA*

Correspondence should be addressed to Fengping Li; lfp@wzu.edu.cn

Academic Editor: Carlo Cosentino

Clamping force control system is essential for clamping tasks that require high precision. In this paper, Active Disturbance Rejection Controller (ADRC) is applied for clamping force control system, aiming to achieve higher control precision. Furthermore, the CPSO-ADRC system is proposed and implemented by optimizing the critical parameters of ordinary ADRC using chaos particle swarm optimization (CPSO) algorithm. To verify the effectiveness of CPSO-ADRC, Particle Swarm Optimization- (PSO-) ADRC is introduced as a comparison. The simulation results show that the CPSO-ADRC can effectively improve the control quality with faster dynamic response and better command tracking performance compared to ordinary ADRC and PSO-ADRC.

1. Introduction

Clamping mechanism has been widely used for workpiece processing in the industrial production [1]. Therefore, clamping force should be accurately regulated to prevent damage to the workpiece. The regulation of clamping force highly relies on the quality of the clamping force controller.

Active disturbance rejection controllers (ADRCs) not only inherit the advantages of conventional PID controller, but also make up for the shortcomings, which attracts special attention. Numerous studies have been conducted on the application of ADRC. For example, Tao et al. utilized ADRC to achieve safe and accurate homing of powered parafoils [2]. Zhang and Chen applied ADRC to achieve high precision tracking control of CNC machine tool feed drives [3]. Liu et al. investigated the pitch axis control of satellite camera based on a novel active disturbance rejection controller [4]. Shen et al. tried to solve the control problem of Autonomous Underwater Vehicle (AUV) by designing an active disturbance rejection controller for diving [5].

Though ADRC has many merits compare to conventional PID controller, very limited existence of its application can be found due to its complicated parameter adjustment. To overcome this problem, key parameters of ADRC are usually adjusted in two stages. In the first stage, parameters are roughly adjusted to estimate the ballpark ranges. In the second stage, further optimization to the parameters with specific algorithm is done to determine the precise adjustment. Abundant studies have been conducted for the optimization of ADRC parameters. Observingly, among the optimization algorithms, Genetic Algorithm (GA) [6, 7] and Particle Swarm Optimization (PSO) [8, 9] have been widely used in the optimization process to determine optimal ADRC parameters. Particularly, there are improved variants of GA and PSO algorithms that have even better optimization performance. For instance, the improved GA based on time scale model identification was presented to enable ADRC parameters optimization in a large scope [10]. The adaptive genetic algorithm (AGA) was proposed to solved the adjustment problem towards ADRC systems with overabundant number of parameters [11].

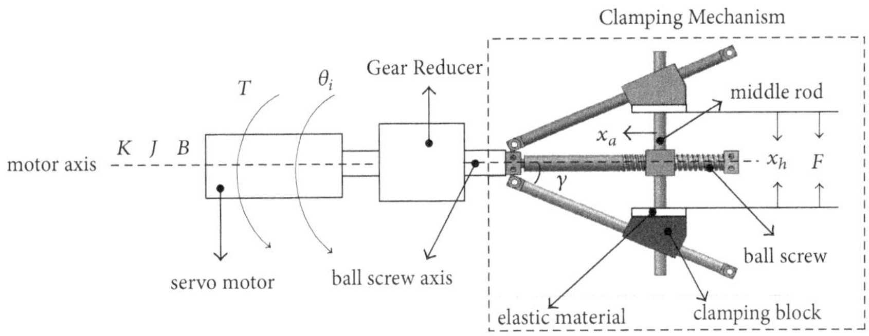

FIGURE 1: Diagram of clamping mechanism.

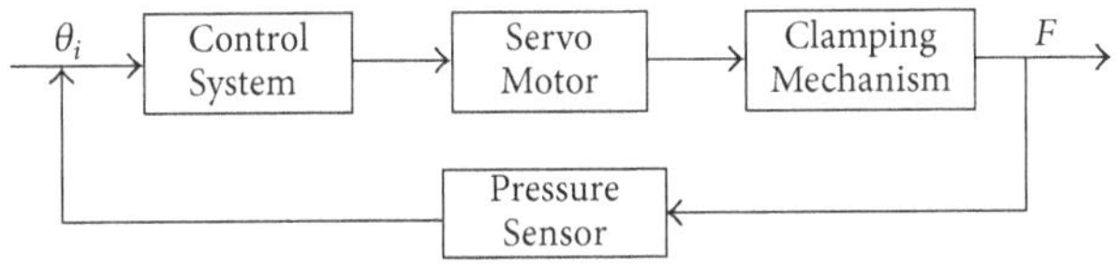

FIGURE 2: Clamping force control system structure.

Improved PSO algorithms also showed superior performances on ADRC parameter optimization. An immune binary-state particle swarm optimization algorithm (IBPSO) was proposed and successfully applied to optimize the parameters of ADRC for chaotic system [12]. A chaos particle swarm optimization (CPSO) algorithm was introduced to conduct optimization and getting optimal parameters for ADRC of an antisynchronizing different chaotic systems [13].

In this paper, ADRC is implemented to clamping force control system with consideration of its superior advantage on robustness. In order to obtain better performance for clamping force control, further parameter optimization of ADRC has been conducted. Instead of ordinary PSO algorithm, CPSO algorithm is applied to this task for its better performance that can converge easier and avoid falling into local optimum.

This paper is organized as follows: Section 2 presents the work of establishing mathematical model of clamping force control; Sections 3 and 4 describe the ADRC and CPSO-ADRC design for clamping force control, respectively; Section 5 shows simulation experiments to verify the effectiveness of CPSO-ADRC with PSO-ADRC as comparison and Section 6 proposes some concluding remarks and prospects.

2. System Control Model

As shown in Figure 1, the diagram of clamping mechanism is drawn. When the middle rod moves to the left, the two clamping blocks on both sides will move close to each other and thus clamp the workpiece. As shown in Figure 2, θ_i is the servo motor rotation angle and F is the clamping force acting on the clamped workpiece. To achieve the clamping force control, the mathematical model of clamping force control system is going to be built in the first step.

The equivalent torque balance equation for motor axis (see Figure 1) can be described as follows:

$$J\frac{d^2\theta_m(t)}{dt^2} + B\frac{d\theta_m(t)}{dt} + i \cdot T_L(t) + K\left[\theta_m(t) - \theta_i(t)\right] + T_D(t) = 0, \quad (1)$$

where J, B, K are the equivalent total moment of inertia, equivalent total viscous damping coefficient, equivalent total stiffness coefficient, respectively, i is the transmission ratio of gear reducer, $T_L(t)$ is the equivalent load torque, $\theta_i(t)$ is the input angle of servo motor, $\theta_m(t)$ is the motor shaft equivalent angle, $\theta_m(t) - \theta_i(t)$ is the shaft relative rotation angle, $T_D(t)$ is the sum of the uncertainty and disturbance torque, and θ_m is related to the middle rod displacement $x_a(t)$ as

$$\theta_m(t) = \frac{1}{i}\frac{2\pi}{L}x_a(t). \quad (2)$$

Equivalent load torque can be expressed as

$$T_L(t) = \left(\frac{L}{2\pi\mu} + \Delta k\right)F(t), \quad (3)$$

where $F(t)$ is the clamping force, L is the lead-screw pitch, μ is the reverse stroke efficiency of ball screw, ΔkF is the uncertain part of equivalent load torque, and $\Delta k < L/4\mu\pi$.

According to the geometric relationship shown in Figure 1, then (4) can be obtained as

$$\tan\gamma = \frac{x_h(t)}{2x_a(t)}, \quad (4)$$

where x_h is the total stroke length of two clamp blocks. Then clamping force is determined as

$$F(t) = \frac{kx_h(t)}{2}, \quad (5)$$

where k is *the* stiffness of elastic material.

Substituting (2), (3), (4), and (5) in (1) and developing the mathematical expression, the model is now

$$J\frac{d^2F(t)}{dt^2} + B\frac{dF(t)}{dt} + \frac{i^2Lk\tan\gamma}{2\pi} \cdot T_L(t) + KF(t) + \frac{iLk\tan\gamma}{2\pi}T_D(t) = \frac{iLk\tan\gamma}{2\pi}K\theta_i(t). \quad (6)$$

TABLE 1: Clamping force control system's parameters.

Parameters	Unit	Value
J	kg·mm^2	1.0×10^2
B	N·mm·s/rad	10
K	N·mm/rad	2.0×10^4
L	mm	5
i	/	0.32
γ	rad	$\pi/9$
k	N/mm	10
μ	/	0.85

This model can be expressed as a second-order system, where the state variable $x_1 = F$, and

$$\begin{aligned} \dot{x_1} &= x_2 \\ \dot{x_2} &= -a_1 x_1 - a_2 x_2 + g u(t) - d(t), \end{aligned} \tag{7}$$

where

$$\begin{aligned} a_1 &= \frac{K}{J} + \frac{i^2 L k \tan\gamma}{2\pi J}\left(\frac{L}{2\pi\mu} + \Delta k\right) \\ a_2 &= \frac{B}{J} \\ g &= \frac{iLk\tan\alpha}{\pi J} K \\ d(t) &= \frac{iLk\tan\alpha}{\pi J} T_D(t). \end{aligned} \tag{8}$$

As $K/J \gg (i^2 Lk\tan\gamma/2\pi J)(L/2\pi\mu + \Delta k)$, therefore $a_1 = K/J$. The parameters of clamping force control system are displayed in Table 1; then after calculation, we can obtain that $a_1 = 200$, $a_1 = 0.1$, $g = 185.37$. And to test the robustness of this system, $d(t)$ is designed with a Gaussian function (Figure 3):

$$d(t) = 100 \exp\left(-\frac{(t-a_i)^2}{2b_i^2}\right), \tag{9}$$

where $c_i = 5$, $b_i = 0.5$.

3. ADRC Design

ADRC can actively detect disturbance signal from the input and output signals of controlled object and then eliminate the disturbance before it impacts the system by adjusting the control signal. ADRC has three main components [14], tracking differential (TD), extended state observer (ESO), and nonlinear state error feedback (NLSEF). As shown in Figure 4, the ADRC structure for clamping force control system is designed.

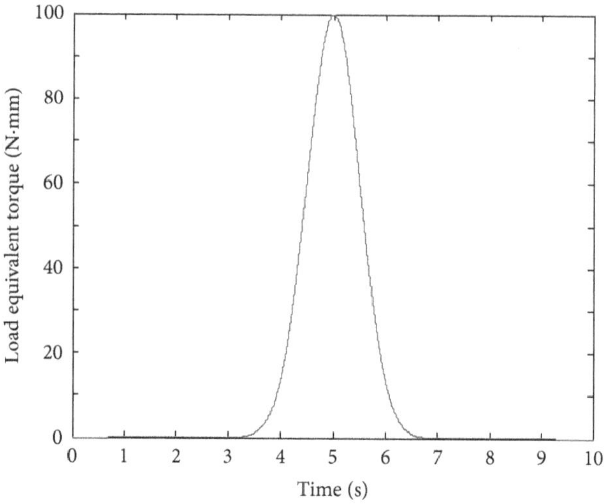

FIGURE 3: $d(t)$ varying with time.

Generally, the typical second-order ADRC can be defined as follows. TD is used for arranging the transient process, which can be expressed as

$$\begin{aligned} \dot{v_1} &= v_2 \\ \dot{v_2} &= fst(v_1 - v_0, v_2, r, h), \end{aligned} \tag{10}$$

where v_0 is the input signal, v_1 is the tracking signal of v_0, v_2 is the differential signal of v_1, r is the parameter which determines the tracking speed, and h is the sampling step length.

ESO is the key component of ADRC, which is employed to monitor the output and predict the real-time state of clamping force control system to achieve disturbance compensation; it can be expressed as

$$\begin{aligned} e &= z_1 - y \\ \dot{z_1} &= z_2 - \beta_{01} e \\ \dot{z_2} &= -\beta_{02} fal(e, \alpha_1, \eta) + bu \\ \dot{z_3} &= -h \times \beta_{03} fal(e, \alpha_2, \eta), \end{aligned} \tag{11}$$

where y is the system output, z_1 is the tracking signal of y, e is the deviation signal, z_2 is the differential signal of z_1, z_3 is the tracking signal of system disturbance, β_{01}, β_{02}, and β_{03} are gain coefficients relating to e, u is the control input, and η is the filtering factor.

Unlike traditional PID that uses linear strategy, NLSEF is a nonlinear control strategy which can enhance the precision for clamping force control system. It can be expressed as

$$\begin{aligned} e_1 &= v_1 - z_1 \\ e_2 &= v_2 - z_2 \\ u_0 &= \beta_1 fal(e_1, \alpha_{01}, \eta) + \beta_2 fal(e_2, \alpha_{02}, \eta) \\ u &= u_0 - \frac{z_3}{b}, \end{aligned} \tag{12}$$

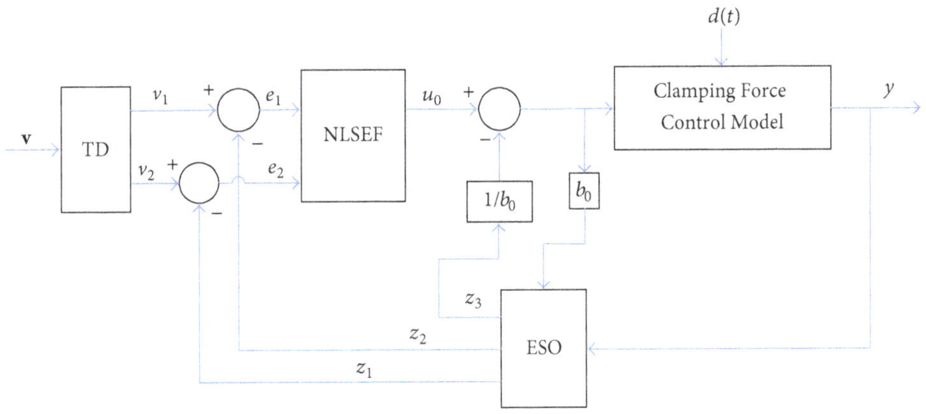

FIGURE 4: ADRC structure for clamping force control system.

TABLE 2: Common parameters.

Parameter	h	η	r	α_{01}	α_{02}	α_1	α_2	b
Value	0.01	0.0025	100	0.75	1.5	0.5	0.25	185.37

where e_1 is the system error, e_2 is the differential of system error, β_1 and β_2 are gain coefficients, and z_3/b is the total disturbance compensation of system.

The nonlinear function $fst(x_1, x_2, r, h)$ can be expressed as

$$\begin{aligned} u &= fst\left(x_1, x_2, r, h\right) \\ d &= rh, \\ d_0 &= hd \\ y &= x_1 + hx_2 \\ a_0 &= \sqrt{d^2 + 8r\left|y\right|} \\ a &= \begin{cases} x_2 + \left[\dfrac{\left(a_0 - d\right)}{2}\right] \times \operatorname{sgn}\left(y\right) & \left|y\right| > d_0 \\ x_2 + \dfrac{y}{h} & \left|y\right| \le d_0 \end{cases} \\ fst &= \begin{cases} -\dfrac{ra}{d} & \left|a\right| \le d \\ -r\operatorname{sgn}\left(a\right) & \left|a\right| > d. \end{cases} \end{aligned} \tag{13}$$

The nonlinear function $fal(e, \alpha, \eta)$ can be defined as

$$fal\left(e, \alpha, \eta\right) = \begin{cases} \left|e\right|^{\alpha} \operatorname{sgn}\left(e\right) & \left|e\right| > \eta \\ \dfrac{e}{\eta^{1-\alpha}} & \left|e\right| \le \eta \end{cases} \quad \eta > 0, \tag{14}$$

where α is the nonlinear factor. To achieve higher control precision, it is necessary to further optimize the parameters of TD, ESO, and NLSEF. There are two parameters in TD, $[h, r]$, six in ESO, $[\alpha_{01}, \alpha_{02}, \beta_{01}, \beta_{02}, \beta_{03}, b]$, and five in NLSEF, $[\alpha_1, \alpha_2, \beta_1, \beta_2, \eta]$. In this paper, these parameters are divided into two groups, which are common parameters and key parameters. The common parameters and initial key parameters are determined based on empirical experience, which can be seen in Tables 2 and 3, respectively.

TABLE 3: Initial key parameters.

Parameter	β_{01}	β_{02}	β_{03}	β_1	β_2
Value	200	700	2000	5.00	0.1

4. CPSO-ADRC Design

CPSO algorithm inherited the advantage of fast convergence from PSO algorithm, and to further avoid falling into local optimum situation, it also adopted CO (Chaos Optimization) algorithm for random traversal. Combining the two gives CPSO great advantage in practice.

As shown in Figure 5, flow chart of CPSO algorithm is drawn, where N is the number of iterations, pbest is the current optimal solution, and gbest is the global optimal solution. After current optimal solution is obtained by PSO algorithm, further search in the nearby region of current optimal solution by CO algorithm will be continually conducted.

The fitness function J is designed as follows:

$$J = \int_0^{\infty} \left(\omega_1 \left|e\left(t\right)\right| + \omega_2 u^2\left(t\right)\right) dt + \omega_3 t_u \tag{15}$$

with

$$\omega_1 + \omega_2 = 1, \tag{16}$$

where t_u is the rise time, ω_1, ω_2, and ω_3 are weighting coefficients. In this work, the weighting coefficients are given as $\omega_1 = 0.99$, $\omega_2 = 0.01$, and $\omega_3 = 100$. The CPSO-ADRC structure for clamping force control system is shown in Figure 6.

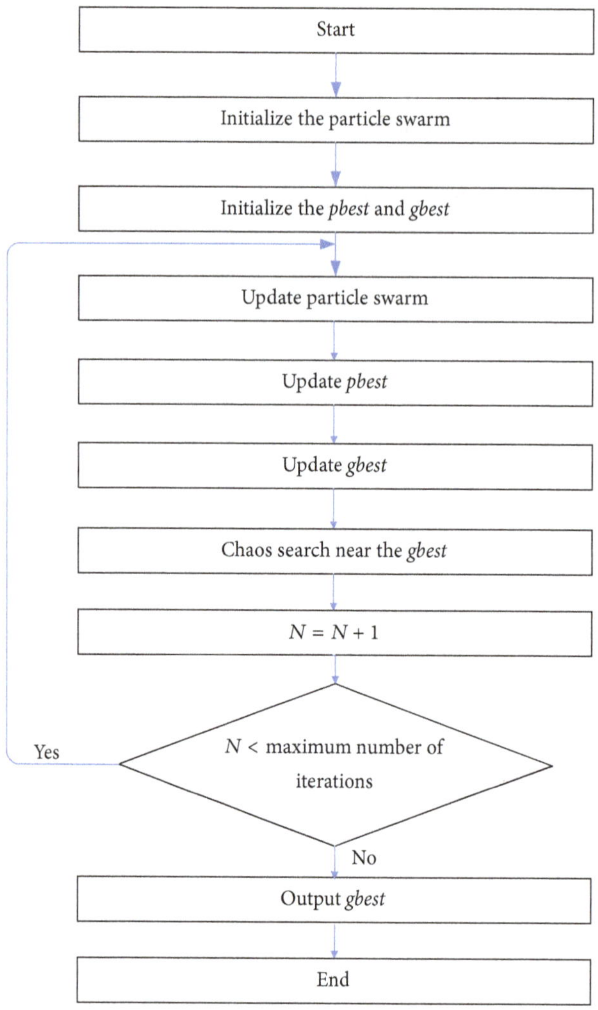

FIGURE 5: CPSO algorithm flow chart.

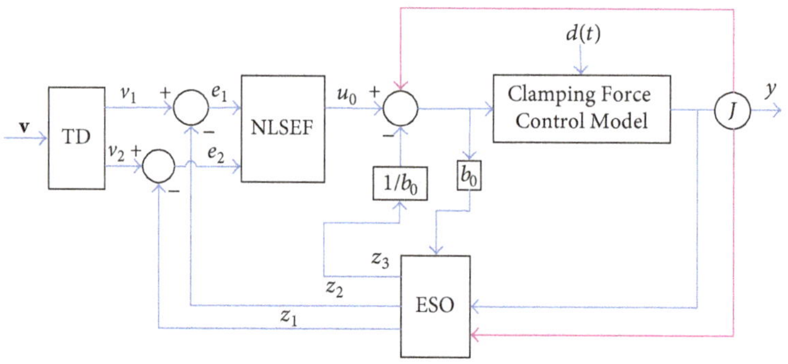

FIGURE 6: CPSO-ADRC structure for clamping force control system.

5. Simulation and Comparison Results

In order to verify the effectiveness of CPSO-ADRC, PSO-ADRC is introduced as a comparison, which uses the PSO algorithm to optimize the ordinary ADRC and then simulation experiments are conducted. In Figure 7, the number of iterations is set to 100; it shows that the fitness function J of CPSO-ADRC can reach a lower value than that of PSO-ADRC during the iterative process. As shown in Table 4, the optimized key parameters have been obtained for both PSO-ADRC and CPSO-ADRC.

In Figure 8, step responses of ordinary ADRC, PSO-ADRC and CPSO-ADRC are drawn; it reveals that CPSO-ADRC has higher response speed and stronger robustness compared with ordinary ADRC and PSO-ADRC.

To further prove the superiority of CPSO-ADRC, the input signal is changed to square wave, which is $\mathrm{rin}(k)$ =

Table 4: Optimized key parameters of ADRC.

Controller	Parameter				
	β_{01}	β_{02}	β_{03}	β_1	β_2
PSO-ADRC	230.00	878.19	2076.66	5.90	0.010
CPSO-ADRC	229.96	792.86	2499.26	6.85	0.013

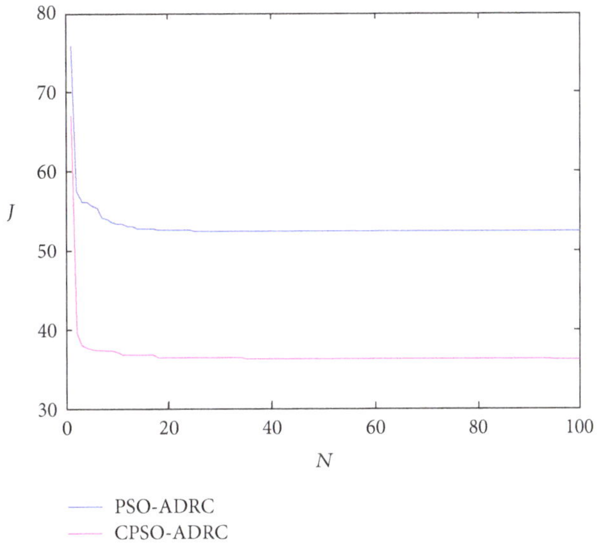

Figure 7: Comparison between the iterative processes obtained from PSO-ADRC and CPSO-ADRC.

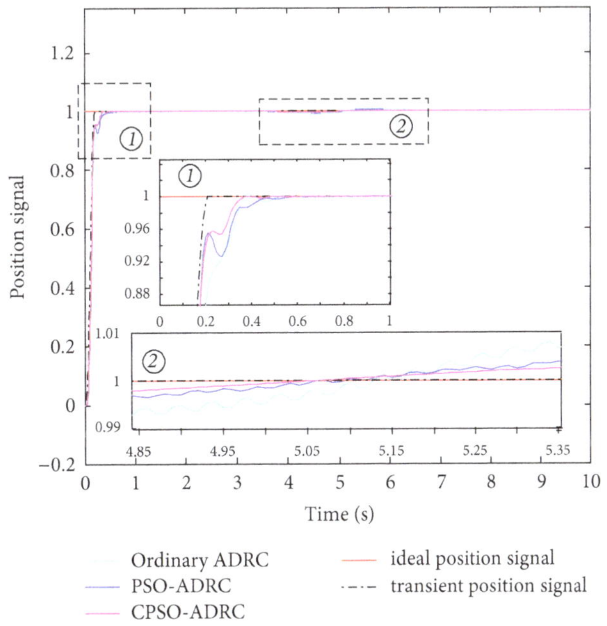

Figure 8: Comparison between step responses obtained from ordinary ADRC, PSO-ADRC, and CPSO-ADRC.

sign(sin(kh)) and the corresponding performances of three kinds of ADRC can be seen in Figures 9 and 10. It can be observed in Figure 9 that the CPSO-ADRC has better performance in responding speed, which is consistent with what Figure 8 implies. And it can also be observed from Figure 10 that CPSO-ADRC has excellent capability of command tracking.

In Figure 11, the control input performances of three controllers are compared. Ordinary ADRC produces significant chattering; PSO-ADRC also generates certain level

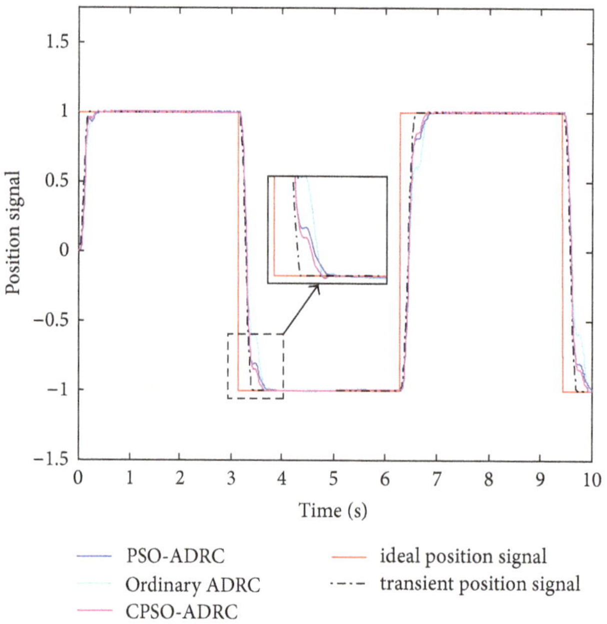

FIGURE 9: Comparison between square wave signal responses obtained from ordinary ADRC, PSO-ADRC, and CPSO-ADRC.

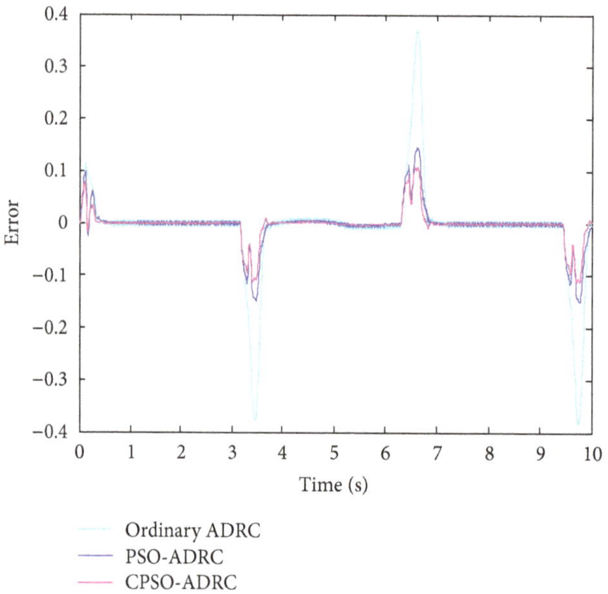

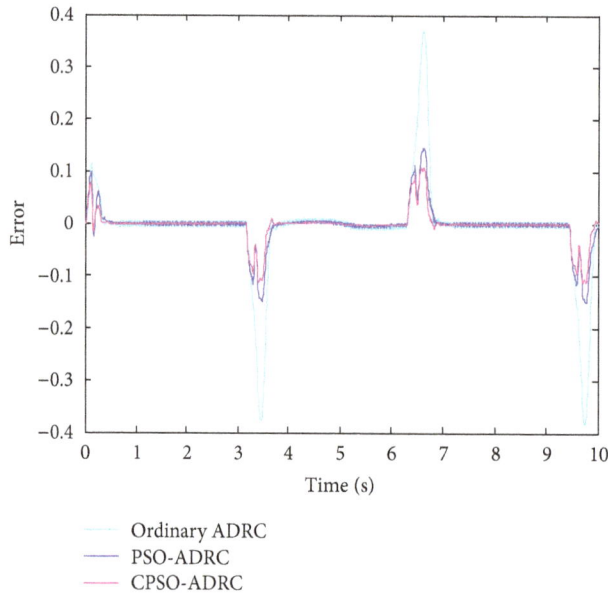

FIGURE 10: Comparison between errors of square wave signal response obtained from ordinary ADRC, PSO-ADRC, and CPSO-ADRC.

of chattering, though much milder than ordinary ADRC; CPSO-ADRC performs the best in reducing chattering; no obvious chattering can be observed from its signal.

6. Conclusion

This study focused on improving the performance of active disturbance rejection controllers for clamping force control system via online optimization. The CPSO-ADRC combination was proposed aiming to achieve faster response and better command tracking performance. Simulation studies illustrated that the CPSO-ADRC in this work has exhibited better dynamic performance. Furthermore, the disturbance signal was designed with a Gaussian function to test the robustness of CPSO-ADRC and the result indicated that CPSO-ADRC has also showed better robustness. To sum all up, CPSO-ADRC has better control performance compared to ordinary ADRC and PSO-ADRC, with similar computational burden to the PSO-ADRC.

Conflicts of Interest

The authors declare that there are no conflicts of interest regarding the publication of this paper.

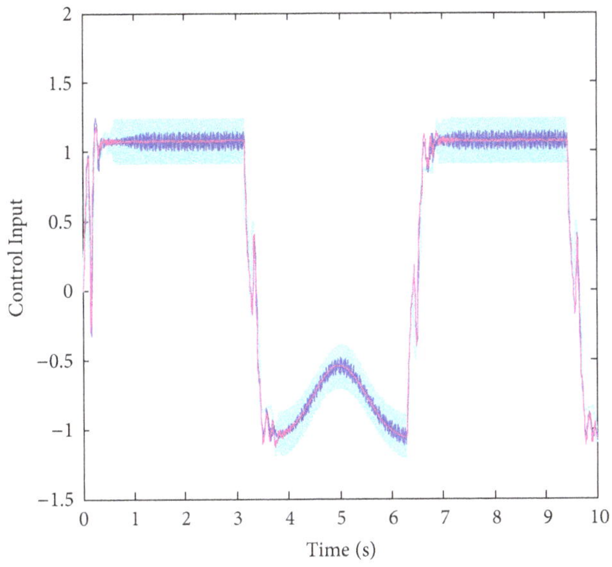

FIGURE 11: Comparison between control inputs obtained from ordinary ADRC, PSO-ADRC, and CPSO-ADRC.

Acknowledgments

The authors are grateful for the support from the Intergovernmental International Science and Technology Cooperation Foundation of China Key R&D Projects (no. 2016YFE0105900), the Zhejiang Provincial Natural Science Foundation of China (nos. LR13E050002, LQ16105005), and the Wenzhou Technologies R&D Program of China (nos. G20140047, ZG2017003).

References

[1] F. P. Li, Z. Y. Zhang, and S. J. Zhou, "Optimized design and kinematic analysis of clamping mechanism for trimming robot," *Chinese Journal of Mechanical Engineering*, vol. 27, no. 3, pp. 387–390, 2016.

[2] J. Tao, Q. Sun, P. Tan, Z. Chen, and Y. He, "Active disturbance rejection control (ADRC)-based autonomous homing control of powered parafoils," *Nonlinear Dynamics*, vol. 86, no. 3, pp. 1461–1476, 2016.

[3] C. Zhang and Y. Chen, "High-precision tracking control of machine tool feed drives based on ADRC," in *Proceedings of the ASME 2016 International Mechanical Engineering Congress and Exposition*, Phoenix, Arizona, USA, 2016.

[4] B. Liu, Y. Jin, C. Zhu, and C. Chen, "Pitching axis control for a satellite camera based on a novel active disturbance rejection controller," *Advances in Mechanical Engineering*, vol. 9, no. 2, 2017.

[5] Y. Shen, K. Shao, W. Ren, and Y. Liu, "Diving control of autonomous underwater vehicle based on improved active disturbance rejection control approach," *Neurocomputing*, vol. 173, pp. 1377–1385, 2016.

[6] H. Geng, H. Yang, Y. Zhang, and H. Chen, "Auto-disturbances-rejection controller design and it's parameter optimization for aircraft longitudinal attitude," *Journal of System Simulation*, vol. 22, no. 1, pp. 89–91, 2010.

[7] K. Hu, X.-F. Zhang, and C.-B. Liu, "Unmanned underwater vehicle depth ADRC based on genetic algorithm near surface," *Acta Armamentarii*, vol. 34, no. 2, pp. 217–222, 2013.

[8] Q.-M. Cheng, Y.-F. Wang, Y.-M. Cheng, K. Wu, and Y.-F. Bai, "Modified double hysteresis current control method for unified power quality controller," *International Transactions on Electrical Energy Systems*, vol. 25, no. 4, pp. 713–730, 2015.

[9] W. T. Pan, Y. Q. Sun, and J. Sheng, "Simulation of auto-disturbance-rejection STATCOM controller based on PSO method," *Computer Simulation*, vol. 28, no. 8, pp. 298–301, 2011.

[10] H. S. Li and X. F. Zhu, "On parameters tuning and optimization of active disturbance rejection controller," *Control Engineering of China*, vol. 11, no. 5, pp. 419–423, 2004.

[11] D. Liu, X. L. Liu, and Y. X. Yang, "Research of ADRC and its application based on AGA," *Journal of System Simulation*, vol. 18, no. 7, pp. 1909–1911, 2006.

[12] Z.-H. Liu, Y.-J. Zhang, J. Zhang, and J.-H. Wu, "Active disturbance rejection control of a chaotic system based on immune binary-state particle swarm optimization algorithm," *Acta Physica Sinica*, vol. 60, no. 1, Article ID 019501, 2011.

[13] F. C. Liu, Y. F. Jia, and L. N. Ren, "Anti-synchronizing different chaotic systems using active disturbance rejection controller based on the chaos particle swarm optimization algorithm," *Acta Physica Sinica*, vol. 62, no. 12, pp. 1-8, 2013.

[14] J. Q. Han, "From PID technique to active disturbance rejection control technique," *Control Engineering of China*, vol. 9, pp. 13–18, 2002.

4

Sensitivity Analysis of Microstructure Parameters and Mechanical Strength during Consolidation of Cemented Paste Backfill

Xuebin Qin, Pai Wang, Lang Liu, Mei Wang, and Jie Xin

School of Electrical and Control Engineering, Xi'an University of Science and Technology, 58 Yanta Rd, Xi'an, Shaanxi, China

Correspondence should be addressed to Xuebin Qin; qinxb@xust.edu.cn

Academic Editor: Giovanni Minafò

Parameter sensitivity is an important part of the quantitative model uncertainty, which helps to effectively identify the key parameters, reduce the uncertainty of the parameters, and then improve the efficiency of parameter optimization. In order to accurately and intuitively analyze the influence of the microscopic parameters and the mechanical response of consolidation process of cemented paste backfill (CPB), a method is used to characterize the geometric and morphological features of the CPB. In this paper, digital image processing technology is used to propose a method for the identification and quantitative analysis of microscopic pore images based on CPB. The pore images on the microscopic scale of CPB are obtained by the microscopic analysis of SEM images, binarization, denoising, and other operations; further, several microscopic parameters are calculated on the pore image, such as the porosity, uniformity coefficient, fractal dimension, probability entropy, and other quantitative parameters, realizing quantitative analysis of pore images. The microstructure of pore images of CPB is extracted under different curing times and then the parameter sensitivity between the microstructure parameters and the mechanical response based on the finite-difference method is analyzed. Set microstructure parameter software of consolidation process of CPB is developed based on this idea, which can be used to identify microscopic pore images and analyze the morphology quantitatively. The microcosmic parameters of CPB with strong sensitivity are uniformity coefficient, average shape coefficient, sorting coefficient, fractal dimension, average length of long axis, average pore area, weighted probability entropy, pore number, and porosity. The sensitivity of the remaining parameters is relatively low. Therefore, the CPB is preferably used in the strength testing process. The method provides a new method for the quantitative analysis of parameter sensitivity on the microscale of CPB.

1. Introduction

In recent years, with the gradual reduction of shallow mineral resources in the earth, deep mining of mineral resources already tends to normalize. Mining filling technology has become more and more important [1, 2]. With the development of cement filling technology and the need of mine environmental protection, the tailings discharged from the concentrator become the main aggregates of cementing and filling in the mine gradually. With different water-cement ratio, different curing times of CPB have a direct impact on the mechanical properties of the relationship. Many scholars use different research methods and techniques to study the factors affecting the strength of tailings CPB. Kesimal A studied the relationship between the strength of the deboning copper-lead-zinc tailings and the paste and found that the grain size distribution of the tailings sand had a great influence on the strength of the CPB [3]. Fall et al. studied the effect of curing temperature on the strength of tailings CPB [4]. G Xiu et al. experiments were carried out under different material ratios; the work reveals the microscopic chemical reaction mechanism of the tailings and studied the influence on the stability of the CPB [5]. Chun Liu et al. extracted the basic parameters of pores such as porosity, fractal dimension, and uniformity coefficient by the manual threshold on a single SEM image and analyzed the relationship between the microstructure and mechanics of rock [6, 7]. Serge Outllet et al. studied SEM images of different samples of CPB and analyzed the pore structure. This method is used to estimate three pore structure parameters: total porosity, pore size distribution, and spatial curvature of pores [8]. Many researchers have made deep research on the composition ratio, stabilization process, and mechanical strength of tailing CPB [9, 10]. It is very important to analyze the microscopic parameter sensitivity, for example, the characterization of the microscopic parameters of the CPB, how to establish the primary and secondary relationship of the parameters, the parameters of the inversion calculation, and the accurate calculation of the model. At present, there are few sensitive parameters in the microstructure of CPB. The authors investigated sensitivity analysis of landslide parameters [11], sensitivity analysis of hydrological model parameters [12], and sensitivity analysis of geotechnical parameters of slope engineering [13]. The intelligent technique is a combination of the artificial neural network and particle swarm optimization, which is used to predict the unconfined compressive strength of CPB [14]. The Curing Under Applied Pressure System (CUAPS) is used to simulate CPB placement and curing processes and then to assess the strength development of the samples under consolidated and drained conditions via the application of external pressure on the fill sample top [15, 16]. The effects of curing conditions on changes in CPB microstructure and corresponding unconfined compressive strength were assessed using both an improved CUAPS system and conventional plastic molds [17, 18]. A predictive model is proposed for unconfined compressive strength of CPB by self-weight consolidation [19, 20]. In the paper, a method of mechanical response and parameter sensitivity analysis is proposed based on microscale image processing technology of CPB.

This paper takes the CPB of different curing times as the research object; the image processing technology is used to extract the micropore image by the indoor microtest. According to measure indexes, such as total pore area, uniformity coefficient, fractal dimension, average pore area, number of pores, maximum pore area, and weighted probability entropy, geometric features and morphology of pores are characterized. The influence of the microcosmic parameters for the sensitivity of the mechanical response of CPB is analyzed. It provides an accurate and direct quantitative method for the relationship between the pore development structure and the mechanical response of the CPB. The core goal of this paper is to characterize the macro mechanical characteristics of CPB by microscopic parameters. The microcosmic parameters can be used to characterize the microscopic mechanical response and predict macroscopic mechanical characteristics of CPB.

2. Materials and Methods

2.1. Material Component. The main material for the filling test is the full tailings of the Xiang Furnace Shan Tungsten mine, the strength grade of 325# type composite Portland cement (PCC), and the urban tap water. The basic performance of the tailings is tested in the experiment and the distribution curve of the grain size is obtained. As shown in Figure 1, d_{10} indicates that the grain size of the tailings is 3.630μm. d_{50} is 52.667μm, d_{90} is 132.095μm, and

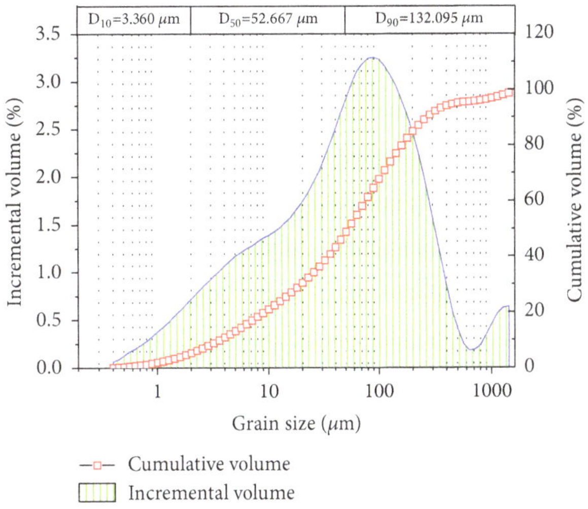

FIGURE 1: Tailing grain size distribution curve.

specific surface area is 631m2/kg, The uniformity coefficient of the grain size of tailings is 20.82. The best grade of tailings particles should be consistent with Thabo equation. Generally, it should be between 4 and 6.

It is known from the grain size curve of the tailings. The content of coarse grains in the test tailings is less, which belongs to the type of relatively missing coarse grains. The natural gradation of tailings belongs to the relative discontinuous gradation. The measurement results of the specific gravity, bulk density, porosity, and natural rest angle of the tailings are listed in Table 1. The main chemical composition and mass fraction of tailings were obtained by X-ray fluorescence spectrum analysis of tailings. It is shown in Table 2.

2.2. Test Process

2.2.1. Sample Preparation. In the experiment, four identical samples were made (one was a spare sample and three were test samples) according to the calculation of the cement sand ratio and required Portland cement weight, and then they were weighed. According to the ratios in Table 3, the tailings and cement were weighed and mixed evenly, and the sample was prepared by adding city tap water (the slurry mass fraction was 72%). A small agitator was used for more than 5 min until the CPB was stirred evenly. A layer of Vaseline was applied to the circular cast iron test mold, which had a diameter of 50 mm and a height of 100 mm. The CPB was loaded in the test mold in three layers. Each layer was then compacted by vibration. The CPB remained still for 24 h after being loaded. The surface of the test sample was smoothed by scraping, and the mold was removed. The test samples were properly labeled and placed in a constant-temperature, constant-humidity curing box at a temperature of 20 ± 1°C and humidity of 95 ± 1%.

2.2.2. Uniaxial Compression Strength Test. Each test sample was removed, and its height and diameter were measured with a Vernier caliper 3, 7, 14, 28, and 56 days after curing. A

Table 1: Test of specific gravity, loose bulk density, dense bulk density, porosity, and natural rest angle.

material	Specific gravity	Loose bulk density (t/m^3)	Dense bulk density (t/m^3)	Porosity (%)	Angle of repose (°)
Full tailings	2.992	1.392	1.955	34.659	42.997

Table 2: Full-tail grading composition.

chemical composition	TFe	SiO_2	Al_2O_3	MgO	CaO	S	WO_3	others
content (%)	9.45	48.22	5.01	2.96	12.68	2.75	0.055	18.875

Table 3: Experimental design.

No.	Slurry concentration/%	Gelling agent	cement-sand ratio
1		OPC	
2	72	OPC	1:12
3		OPC	

Figure 2: Uniaxial compressive strength test of cemented tailings backfill.

computer-controlled 20-kN pressure machine was employed to apply pressure at a constant rate of 1 mm/min until the test sample failed, as shown in Figure 2. The data were then collated to compute the uniaxial compressive strengths of the test pieces, which were then averaged to obtain the result.

2.3. Preparation of SEM Samples. In this study, SEM samples were fabricated. The method is a modern detection technology, which has the characteristics of high resolution, large magnification time, and wide field of view of three-dimensional images. Consequently, the samples had to be dried and gilded to obtain a true and clear observation.

The CPB was selected at different curing times for preparation of the SEM test samples. First, the middle part of the CPB was located, and a double-sided blade-wire-saw coated with Vaseline was used to cut a 10 mm × 10 mm × 10 mm roughcast cube with a 1.5 mm border. A sharp backfill paste steel knife was then used to cut the blank and create a fresh cross section, baring a complete natural structural surface. A 5 mm × 5 mm × 5 mm block was used for observation under the electron microscope. For this, the fresh surface had to be as flat as possible, with all disturbance particles removed by a rubber suction bulb.

3. Extraction of Micropore Image and Quantitative Analysis

3.1. Extraction of Micropore Image. Based on the method of kernel fuzzy clustering, we used the image processing technology to extract the microimage of the CPB in the SEM image obtained by the indoor microtest. First, the method of kernel fuzzy clustering was used to divide the SEM images into five clusters. The darkest category in the cluster image was used as the adaptive extracted pore image. The micropore image was treated by binarization, and the binary pore image was obtained.

3.1.1. Pore Image Extraction Based on the Kernel Fuzzy C Means Clustering Method. The kernel fuzzy C means clustering algorithm is an iterative dynamic clustering algorithm. It is applied to clinical diagnosis, pathology analysis, image compression, image recognition, and so on. In this work, a fuzzy clustering algorithm was used to extract the pore image on the microscale of the CPB. Based on many experiments, the microscale images of the CPB were divided into five types, and the darkest clustering image was used as the pore image. The principle of the algorithm is the minimization of the target function. In the experiment, the kernel function fuzzy clustering algorithm was used to obtain the pore image [21]. The target function is as follows:

$$J(U, V) = \sum_{i=1}^{N} \sum_{k=1}^{c} u_{ik}^{m} \left(1 - K\left(x_i, v_k\right)\right) \tag{1}$$

where $U = \{u_{ik}\}_{c\times N}$ is a matrix of fuzzy membership degree. $V = \{v_k\}_{c\times 1}$ represents the cluster center matrix. N represents the total number of sample points. C represents the number of clusters. And x_i $(i = 1, 2, \ldots, N)$ represents the sample set; u_{ik} indicates that x_i belongs to the membership degree of cluster k. v_k $(k = 1, 2, \ldots, c)$ represents the cluster center. Kernel function $K(x_i, v_k) = \exp(-\|x_i - v_k\|^2/\sigma)$ is a Gauss kernel function. The expression of u_{ik} is

$$u_{ik} = \frac{1}{\sum_{l=1}^{c}\left(\left(1 - K\left(x_i, v_k\right)\right) / \left(1 - K\left(x_i, v_l\right)\right)\right)^{1/(m-1)}} \tag{2}$$

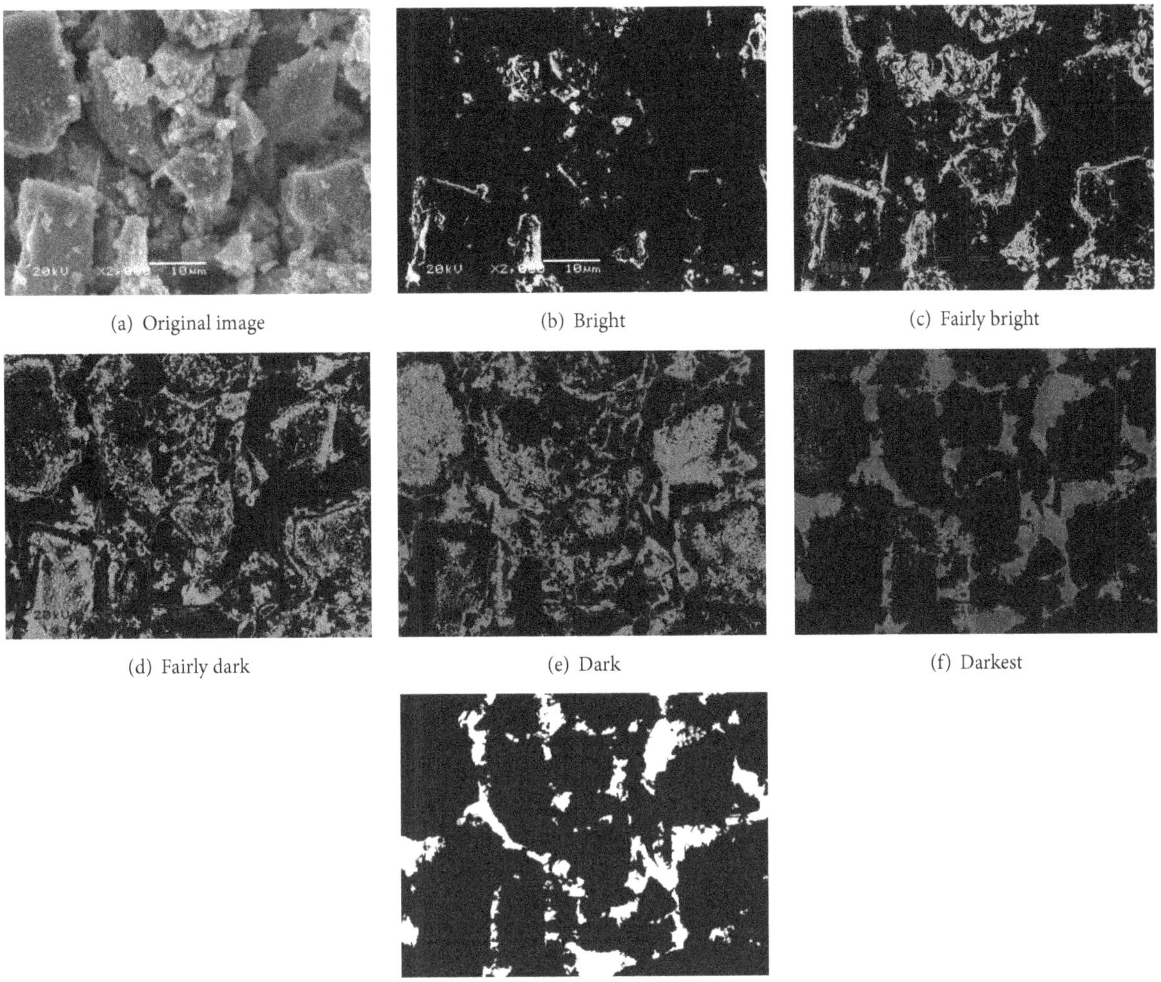

(a) Original image (b) Bright (c) Fairly bright

(d) Fairly dark (e) Dark (f) Darkest

(g) Binary image of pore

FIGURE 3: Clustering results of pore images for CPB based on the fuzzy clustering method.

The expression of cluster center v_k is

$$v_k = \frac{\sum_{i=1}^{N} u_{ik}^{m} K\left(x_i, v_k\right) x_i}{\sum_{i=1}^{N} u_{ik}^{m} K\left(x_i, v_k\right)} \tag{3}$$

The steps of the algorithm are as follows.

Step 1. Set the number of clusters $C(c >= 2)$, fuzzy exponent m, and the iterative terminating condition ε.

Step 2. Initialize cluster center $v_k^{(0)}$.

Step 3. Set cycle counter $b = 0$.

Step 4. Update membership function $u_{ik}^{(b)}$ according to formula (2).

Step 5. Update cluster center function $v_k^{(b+1)}$ according to formula (3).

Step 6. If $\max_{1\le k\le c}|v_k^{(b)} - v_k^{(b+1)}| < \varepsilon$ or when the number of iterations is greater than the threshold for a given number of iterations, then the cyclic iteration is terminated. Otherwise, set $b = b + 1$ and jump to Step 4.

In this experiment, SEM images of CPB with different curing times were used, and the curing times were 3 d, 7 d, 14 d, 28 d, and 56 d. The pore image was extracted based on the fuzzy clustering method. The SEM images were divided into five clusters: bright, fairly bright, fairly dark, dark, and darkest. The darkest image was extracted from the cluster as a pore image because the pore is in the darkest region of the image. The parameters of the clustering algorithm are as follows: $m = 2$, $c = 5$, $\delta = 2$, x_i is the gray value of the gray image, and ε is an iterative error, $\varepsilon \in (0, 1)$. As shown in Figure 3, the fuzzy clustering method is used for classification. The picture is an SEM image of the 3D (day) curing times. The picture is divided into five clusters. (a) is the original image, and (b) through (f) are five cluster images: bright, fairly bright, fairly dark, dark, and darkest. From the diagram in (f), it can be seen that this type of image objectively reflects the distribution of pores on a microscale, so this type of image is taken as a pore image. The image (g) is a binary image of the image (f).

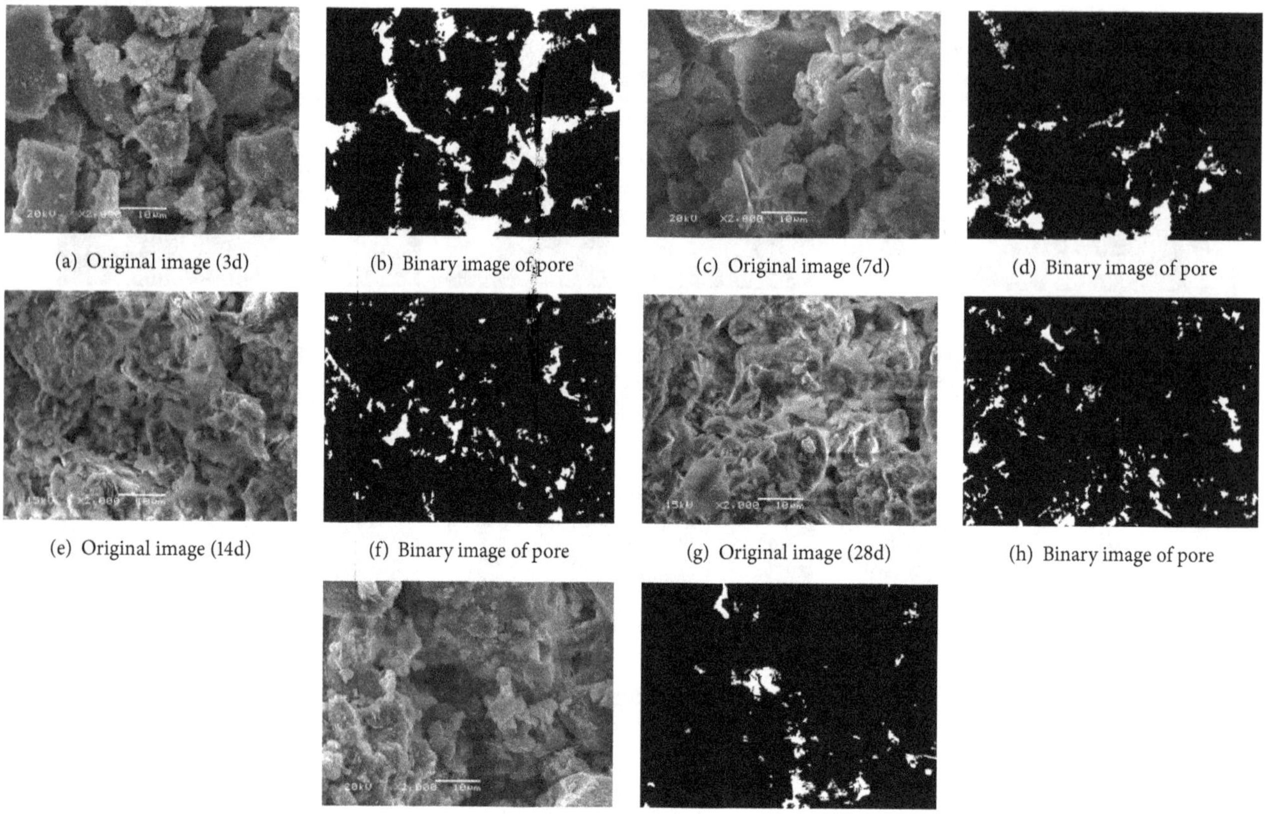

(a) Original image (3d) (b) Binary image of pore (c) Original image (7d) (d) Binary image of pore

(e) Original image (14d) (f) Binary image of pore (g) Original image (28d) (h) Binary image of pore

(i) Original image (56d) (j) Binary image of pore

FIGURE 4: SEM image and corresponding pore binary image under different curing times.

3.1.2. Accurate Extraction of Pore Images. To extract the exact pore area, there are many small heterojunction areas in Figure 3(f). In the process of software processing, areas less than 40 pixels were removed as noise, and the remaining binary image as a pore was reversed. Fine connections were removed by morphological treatment. Figure 4 shows the binary image when the miscellaneous area is removed. The curing ages of (b) through (f) are 3 d, 7 d, 14 d, 28 d, and 56 d, respectively, and the SEM scanning images and corresponding pore binary images are shown.

3.2. Quantitative Parameter Analysis. In this study, the pore image was extracted from the SEM image of CPB. To describe the microscopic pore characteristics of the CPB quantitatively, the relationship between the distribution, quantity, direction, size, and mechanical response of the pores is described quantitatively by many microscopic parameters.

The microscopic parameters included the pore number, total pore area, maximum pore area, average pore area, porosity, uniformity coefficient, sorting coefficient, curvature coefficient, roundness, fractal dimension, and weighted probability entropy.

According to the morphological and geometric characteristics of the binary pore images, the microcosmic index of the CPB was used in the quantitative analysis of the pores.

(1) **Pore number** represents the number and sizes of pores on the SEM image.

(2) **Average length** is defined as the length of a region by the Feret diameter, for all pores, which represents the accumulation of the long axis for each pore.

(3) **Total pore area** represents the sum of the total areas of all pores on the SEM image.

(4) **Pore porosity** represents the integrity of CPB pores, which is the ratio of the pore region to the total image area. The quantified index is a 2D parameter; it can indirectly reflect the changes of pore ratio in 3D space.

(5) **Uniformity coefficient**, C_u, represents the ratio of d_{60} to d_{10}:

$$C_u = \frac{d_{60}}{d_{10}} \tag{4}$$

where d_{10} is the diameter of the pore image area when the cumulative area on pore image is 10%. d_{60} is the diameter of the pore image area when the cumulative area on the image is 60%.

(6) **The curvature coefficient** represents continuity of the cumulative curve for the diameter of the pore image block.

$$C_c = \frac{d_{30}^2}{(d_{60} * d_{10})} \tag{5}$$

where the meaning of d_{10} and d_{60} are the same as uniformity coefficient C_u, and d_{30} is the diameter of the corresponding pore image block when the cumulative area is 30%.

(7) **Sorting coefficient**, S_c, is used to sort the pore image blocks in descending order by area. When the size of the pore area is uniform, the values, P25 and P75, are similar. Thus, S_c is closer to 1, and the other is larger than 1.

$$S_c = \frac{p_{25}}{p_{75}} \quad (6)$$

where d_{25} and d_{75} represent the diameter of the pore image block corresponding to the cumulative pore areas of 25% and 75%, respectively.

(8) The **fractal dimension** of porosity is the quantitative index used to describe the CPB size distribution, which directly reflects the changing pore shape. The formula is defined as follows:

$$D_c = \lim \frac{\ln N(r)}{\ln r} \quad r \longrightarrow 0, \quad (7)$$

where the pore image was divided into small square grids. In our experiment, r of each grid is 1, 3, 5, 7, etc. The maximum value of r is a quarter of the image width. N(r) denotes the number of pores in a square grid corresponding to r. A larger fractal dimension of porosity(D_c) leads to a lower level of pore homogenization and a larger difference of size among pores.

(9) **Weighted probability entropy** is a quantitative parameter reflecting the regularity of structural units. It describes the overall CPB arrangement of pores at a microscopic scale. For the distribution in each small region, the computational formula of probability entropy, h_m, is defined as follows:

$$h_{\mathrm{m}} = -\sum_{i=1}^{n} p_i \log_n p_i \quad (8)$$

where p_i is the frequency of a structural body in a certain directional region. n is the interval of the orientation angles in the arrangement direction of structural units. The value, hm, is between 0 and 1. A larger hm leads to a more disordered arrangement of pores and lower regularity, and vice versa. In the experiment, given that $n = 36$, every 10° is a sector. The midpoint of the long axis is selected as the coordinate of the original point, with a horizontal x-axis and a vertical y-axis.

Owing to the size of each area block being different across the whole region, the contribution rate is also different. Thus, a new parameter is defined as the weighted probability entropy. All regional blocks on the pore image are subject to normalization, as shown in

$$a_i = \frac{s_i}{\sum_{i=0}^{N-1} s_i} \quad (9)$$

where there are N image blocks on the pore image. s_i is the area of the ith region. a_i is the contribution rate of the ith region following normalization, which is regarded as the weighted value. The final weighted probability entropy is

$$H_m = \sum_{i=0}^{N-1} a_i h_{mi} \quad (10)$$

where h_{mi} is the probability entropy of the ith region, and H_m is the overall probability entropy of the pore image.

(10) The definition of the circular degree R is as follows:

$$R_i = \frac{4\pi S}{L^2} \quad (11)$$

where S represents the area of the block, and L represents the circumference of the block. If the error of the circular degree of a single pore is too large and is meaningless, the average circular degree is used to analyze the characteristics of the pore shape. Average shape coefficient is defined as follows:

$$R = \sum_{i=1}^{n} \frac{R_i}{n} \quad (12)$$

where n represents the number of pores. The larger R is, the more slippery the shape of the pores is, and vice versa.

4. Sensitivity Analysis of Microcosmic Parameters and Strength of Pores

4.1. Parameter Sensitivity Principle. From the SEM image of the CPB for different curing times, the pore image was extracted based on the fuzzy clustering method, and the corresponding microparameters were obtained. There are many factors affecting the mechanical properties of CPB owing to the pore microstructure: (1) the geometric parameters for the microcosmic characteristics of the CPB, such as porosity, pore number, average pore area, average pore width, pore distribution characteristics, and pore fractal dimension; (2) characteristics of pore arrangement change in CPB, such as probability entropy and rose map; (3) characteristics of pore morphological change in CPB, such as circularity and the fractal dimension of the pore; and (4) distribution characteristics, such as uniformity coefficient, curvature coefficient, and separation coefficient. The relationship between the microscopic parameters and mechanical response is blindly constructed, and the size of the influence degree cannot be accurately evaluated. Therefore, sensitivity analysis is needed; the greater the sensitivity, the higher the degree of influence of the factor. To improve the accuracy of the relationship between the microparameters and mechanical response, the principle of priority is adopted for the sensitive factors. This work uses the dimensionless sensitivity.

Based on the finite-difference method, the microcosmic parameter sensitivity and mechanical response of the cemented filling body are analyzed. The finite-difference method is a numerical method to solve differential equations. The basic idea is that the continuous solution domain is replaced by a finite discrete node, and the approximate solution process is used to replace the partial differential equation with difference. The mathematical formula is as follows:

$$\frac{dy}{dx} = \lim_{\Delta x \to 0} \frac{\Delta y}{\Delta x} = \lim_{\Delta x \to 0} \frac{f(x + \Delta x) - f(x)}{\Delta x} \quad (13)$$

where dy, dx are the differential of function and independent variable, respectively; dy/dx is the first derivative of the function to the independent variable.

TABLE 4: Microparameter calculation.

No.	Microcosmic parameters	Curing times				
		3d	7d	14d	28d	56d
1	Total area of image	288000	288000	288000	288000	288000
2	Pore number	75	84	78	51	52
3	Total pore area	27227	37935	32581	32279	52809
4	Maximum pore area	4876	6254	6869	5519	10770
5	Average pore area	363.02	451.61	417.71	632.92	1015.57
6	Average length of long axis	35.22	44.55	40.21	48.26	48.75
7	porosity	0.09	0.13	0.11	0.11	0.18
8	Uniformity coefficient	3.11	2.18	4.29	3.56	4.59
9	Sorting coefficient	3.20	1.77	2.77	2.27	2.59
10	curvature coefficient	0.82	1.19	0.91	1.09	0.92
11	Average shape coefficient	0.26	0.20	0.26	0.25	0.29
12	fractal dimension	0.94	0.95	0.95	0.91	0.93
13	Weighted probability entropy	0.86	0.85	0.84	0.86	0.89

TABLE 5: Sensitive values and contribution rate of 12 microscopic parameters on the pore image.

No.	Microcosmic parameters	Sensitive value	Contribution rate (%)
1	Pore number	4.70	7.44
2	Total pore area	3.06	4.84
3	Maximum pore area	1.00	1.58
4	Average pore area	5.54	8.77
5	Average length of long axis	6.31	9.99
6	porosity	3.06	4.84
7	Uniformity coefficient	11.00	17.46
8	Sorting coefficient	7.58	12.00
9	curvature coefficient	1.33	2.11
10	Average shape coefficient	8.19	12.97
11	fractal dimension	6.35	10.05
12	Weighted probability entropy	5.02	7.95

4.2. Calculation of Microparameter. The steps of calculating the sensitivity between the microscopic parameters and the mechanical response are as follows:

(1) The digital image processing software was used to analyze the SEM image of the CPB, and the related microparameters, as in Table 4, were obtained.

(2) The same microscopic parameters in each continuous curing times in Table 4 were subtracted, and four differences were taken for each parameter. The dimensionless processing was carried out, mapping the values to 0.1~1.1, as shown in the following formula:

$$x_t = \frac{x - x_{\min}}{x_{\max} - x_{\min}} + 0.1 \tag{14}$$

(3) According to formula (13), the sensitivity and contribution rate of the influencing factors are calculated, as shown in Table 5.

4.3. Parameter Sensitivity Analysis. To analyze the sensitivity of the parameters, the finite-difference method was used to analyze the relationship between the multiple microscopic parameters and mechanical strength. The dimensionless treatment was used in the experiment because of the different dimensions between parameters. The parameter values of five different curing ages were obtained by the difference method, resulting in four different Δx_i. Then, the four difference values were mapped to 0.1~1.1. The minimum difference was 0.1, the maximum difference was 1.1, and the other difference was mapped. In the same way, the strength parameters of the mechanical response for five different curing times were calculated by differential operation; that is, four Δy_i were obtained. Figure 5 is a mechanical strength map corresponding to five SEM images. To increase the recognition of the sensitive value mapping Δy_i to 1~11, the minimum difference was 1, and the maximum difference was 11; the sensitivity of each parameter was defined as follows:

$$s = \sum_{i=1}^{N} \left(\frac{\Delta y_i}{\Delta x_i} \right)^2 \tag{15}$$

where, in the experiment, $N = 4$.

The size of the sensitivity of the microscopic parameters can be obtained from Table 5. The order of specific sensitivity from large to small is uniformity coefficient, average shape

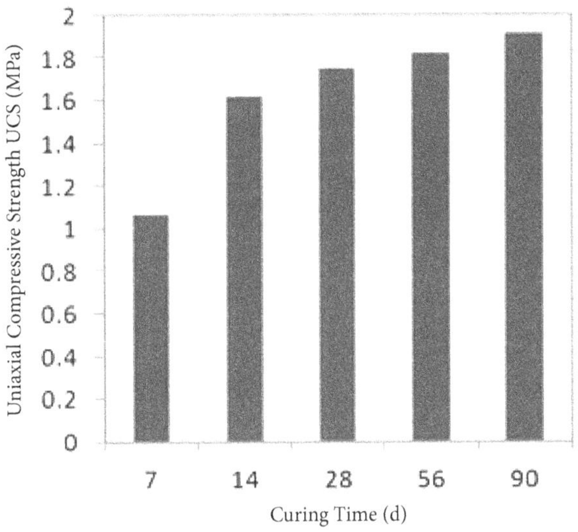

Figure 5: Uniaxial compressive strength of CPB at different curing times.

coefficient, sorting coefficient, fractal dimension, average length of long axis, average pore area, weighted probability entropy, pore number, and porosity. The cumulative contribution rate of the first nine microparameters' sensitivity was 91.11%.

Therefore, when constructing the relationship between microscopic parameters and mechanical response, the nine parameters uniformity coefficient, average shape coefficient, sorting coefficient, fractal dimension, average length of long axis, average pore area, weighted probability entropy, pore number, and porosity are preferentially used.

5. Conclusion

(1) A computer intelligent recognition method for microscopic pore images of CPB was established. For the geometric and morphological structure of the CPB, the quantitative description of the pore number, total pore area, maximum pore area, average pore area, long-axis average length, porosity, uniformity coefficient, sorting coefficient, curvature coefficient, roundness, fractal dimension, and weighted probability entropy index were proposed. This method can reduce human error and objectively reflect the microstructure of the CPB.

(2) The results of sensitivity analysis showed that the contribution of the microparameters to the compressive strength of the CPB is different. The first nine parameters of the contribution rate are arranged in order from large to small: uniformity coefficient, average shape coefficient, sorting coefficient, fractal dimension, average length of long axis, average pore area, weighted probability entropy, pore number, and porosity. Therefore, in the process of strength testing, these parameters are preferred.

(3) To express the direction of the pores effectively, the concept of weighted probability entropy was proposed, and the contribution rate of the image region was different.

(4) Digital image processing technology provided an effective method for the quantitative analysis of cemented filling pores. It obtained good pore-quantizing results. When the three-dimensional image analysis technology of CPB is developed, the analysis of CPB can be realized more accurately and fully.

(5) This method is simple and easy, and its objective is to describe the microstructure of the CPB.

Conflicts of Interest

The authors declare that there are no conflicts of interest regarding the publication of this paper.

Authors' Contributions

Xuebin Qin and Pai Wang conceived and designed the experiments; Lang Liu and Mei Wang performed the experiments; Jie Xin analyzed the data; Lang Liu contributed SEM image; Xuebin Qin wrote the paper.

Acknowledgments

This work was supported by the National Natural Science Foundation of China (no. 51704229 and no. 51405381) and Key Scientific and Technological Project of Shaanxi Province (no. 2016GY-040).

References

[1] X. Dongsheng, "Discussion on reducing pressure technology of filling pipeline transport system in deep well," *Express Information of Mining Industry*, vol. 10, no. 2, pp. 25–28, 2007.

[2] Q. Zhang, Q. Liu, J.-W. Zhao, and J.-G. Liu, "Pipeline transportation characteristics of filling paste-like slurry pipeline in deep mine," *The Chinese Journal of Nonferrous Metals*, vol. 25, no. 11, pp. 3190–3195, 2015.

[3] A. Kesimal, B. Ercikdi, and E. Yilmaz, "The effect of desliming by sedimentation on paste backfill performance," *Minerals Engineering*, vol. 16, no. 10, pp. 1009–1011, 2003.

[4] M. Fall, J. C. Célestin, M. Pokharel, and M. Touré, "A contribution to understanding the effects of curing temperature on the mechanical properties of mine cemented tailings backfill," *Engineering Geology*, vol. 4, no. 10, pp. 397–413, 2010.

[5] G. Xiu, W. Dang, and Z. Liu, "Microstructure test and macro size effect on the stability of cemented tailings backfill," *International Journal of Digital Content Technology and Its Applications*, vol. 6, no. 14, pp. 387–397, 2012.

[6] C. Liu, B. Shi, J. Zhou, and C. Tang, "Quantification and characterization of microporosity by image processing, geometric measurement and statistical methods: Application on SEM images of clay materials," *Applied Clay Science*, vol. 54, no. 1, pp. 97–106, 2011.

[7] C. Liu, C.-S. Tang, B. Shi, and W.-B. Suo, "Automatic quantification of crack patterns by image processing," *Computers & Geosciences*, vol. 57, pp. 77–80, 2013.

[8] S. Ouellet and M Benzaazoua, "Characterization of cemented paste backfill pore structure using SEM and IA analysis," *Bulletin of Engineering Geology and the Environment*, vol. 67, no. 2, pp. 139–152, 2008.

[9] B. Ercikdi, H. Baki, and M. Izki, "Effect of desliming of sulphide-rich mill tailings on the long-term strength of cemented paste backfill," *Journal of Environmental Management*, vol. 115, pp. 5–13, 2013.

[10] A. Ghirian and M. Fall, "Coupled thermo-hydro-mechanical-chemical behaviour of cemented paste backfill in column experiments," *Engineering Geology*, vol. 164, no. 3, pp. 195–207, 2013.

[11] S. Jiang and B.-P. Wen, "Analysis of sensitivity of landslide sliding mechanics parameters based on different methods," *Journal of Engineering Geology*, vol. 23, no. 6, pp. 1153–1162, 2015.

[12] X.-m. Song, F.-Z. Kong, and C.-S. Zhan, "Analysis of sensitivity of hydrological model parameter based on statistical theory," *Journal of Advances in Water Science*, vol. 23, no. 5, pp. 642–649, 2017.

[13] J.-j. Ge, "Sensitivity analysis of geotechnical parameter of railway slope engineering," *Journal of Railway Construction*, vol. 46, no. 11, pp. 107–110, 2014.

[14] C. Qi, A. Fourie, and Q. Chen, "Neural network and particle swarm optimization for predicting the unconfined compressive strength of cemented paste backfill," *Construction and Building Materials*, vol. 159, pp. 473–478, 2018.

[15] D. Landriault, "Paste backfill mix design for Canadian underground hard rock mining," in *Proceedings of the 97th Annual General Meeting of the CIM Rock Mechanics and Strata Control Session*, pp. 13–22, Halifax, Nova Scotia, Canada, 1995.

[16] M. Benzaazoua, T. Belem, and E. Yilmaz, "Novel lab tool for paste backfill," *Canadian Mining Journal*, vol. 127, no. 3, p. 31, 2006.

[17] E. Yilmaz, T. Belem, M. Benzaazoua, and B. Bussière, "Assessment of the modified CUAPS apparatus to estimate in situ properties of cemented paste backfill," *Geotechnical Testing Journal*, vol. 33, no. 5, 2010.

[18] E. Yilmaz, T. Belem, B. Bussière, and M. Benzaazoua, "Relationships between microstructural properties and compressive strength of consolidated and unconsolidated cemented paste backfill," *Cement and Concrete Composites*, vol. 33, pp. 702–715, 2011.

[19] T. Belem, E. Yilmaz, and M. Benzaazoua, "Predictive models for unconfined compressive strength of cemented paste backfills taking into account self-weight consolidation," in *Proceedings of the 68th Canadian Geotechnical Conference*, pp. 1–8, Quebec, Canada, 2015.

[20] E. Yilmaz, T. Belem, B. Bussière, M. Mbonimpa, and M. Benzaazoua, "Curing time effect on consolidation behaviour of cemented paste backfill containing different cement types and contents," *Construction and Building Materials*, vol. 75, pp. 99–111, 2015.

[21] Z.-d. Wu, W.-x. Xie, and J.-p. Yu, "Sensitivity analysis of geotechnical parameter of railway slope engineering fuzzy c-means clustering algorithm based on kernel method," in *Proceedings of the 15th International Conference on Computational Intelligence and Multimedia Applications*, pp. 49–54, 2003.

5

Optimal Design of Submarine Pipelines by a Genetic Algorithm with Embedded On-Bottom Stability Criteria

Juliana Souza Baioco,[1,2] Mauro Henrique Alves de Lima Jr.,[1] Carl Horst Albrecht,[1] Beatriz Souza Leite Pires de Lima,[1] Breno Pinheiro Jacob,[1] and Djalene Maria Rocha[3]

[1]*Laboratory of Computer Methods and Offshore Systems (LAMCSO), PEC/COPPE/UFRJ, Civil Engineering Department, Post-Graduate Institute of the Federal University of Rio de Janeiro, Avenida Pedro Calmon, S/N, Cidade Universitária, Ilha do Fundão, 21941-596 Rio de Janeiro, RJ, Brazil*
[2]*Chemical and Petroleum Engineering Department (TEQ), UFF, Rua Passo da Pátria No. 156, São Domingos, 24210-240 Niterói, RJ, Brazil*
[3]*Petróleo Brasileiro S.A. (Petrobras), Research & Development Center (CENPES), Avenida Horacio Macedo 950, Cidade Universitária, Ilha do Fundão, 21941-915 Rio de Janeiro, RJ, Brazil*

Correspondence should be addressed to Breno Pinheiro Jacob; breno@lamcso.coppe.ufrj.br

Academic Editor: Maurizio Brocchini

This work describes a computational tool, based on an evolutionary algorithm, for the synthesis and optimization of submarine pipeline routes considering the incorporation of on-bottom stability criteria (OBS). This comprises a breakthrough in the traditional pipeline design methodology, where the definition of a route and the stability calculations had been performed independently: firstly, the route is defined according to geographical-topographical issues (including manual/visual inspection of seabed bathymetry and obstacles); afterwards, stability is verified, and mitigating procedures (such as ballast weight) are specified. This might require several design spirals until a final configuration is reached, or (most commonly) has led to excessive costs for the mitigation of instability problems. The optimization tool evaluates each candidate route by incorporating, as *soft* and *hard* constraints, several criteria usually considered in the manual design (pipeline length, bathymetry data, obstacles); also, with the incorporation of OBS criteria into the objective function, stability becomes an integral part of the optimization process, simultaneously handling minimization of length and cost of mitigating procedures. Case studies representative of actual applications are presented. The results show that OBS criteria significantly influences the best route, indicating that the tool can reduce the design time of a pipeline and minimize installation/operational costs.

1. Introduction

Submarine pipeline systems have been extensively used in the oil and gas industry, to transport the production between offshore platforms and/or to onshore processing facilities. Being one of the higher-cost items of the subsea layout, pipeline systems significantly affect the feasibility of an offshore project, thus demanding detailed studies to obtain efficient and low-cost designs, comprising an iterative and very complex process governed by several variables, following design recommendations addressed by codes such as DNV-OS-F101 [1].

In this context, perhaps the most crucial step in the design of a submarine pipeline is the selection of its route. Traditionally, this task has been manually performed by experienced engineers, by inspecting the seabed bathymetry and available information regarding obstacles (including subsea equipment, flowlines, and other preexistent pipelines). Many environmental, commercial, regulatory, or even geopolitical issues may determine specific regions that should be avoided, for instance, corals, geotechnical hazards, or fields allotted to another oil company. There are many other variables that govern the selection of a route; thus the process has been treated almost as an "art," being highly dependent on the expertise of the engineer.

Previous works have already recognized [6–8] that the task of selecting a route with good performance and low cost could be automated by devising its formal description as a synthesis and optimization problem and building a computational tool based on evolutionary algorithms (EAs). Such algorithms have been successfully used in many complex engineering problems and have been shown to be useful for the optimization of offshore engineering problems [9–14].

In [15, 16] we have presented results of preliminary studies related to the development and implementation of a computational tool for the synthesis and optimization of submarine pipeline routes, based on Genetic Algorithms (GAs) [17, 18]. The modeling of the optimization problem as presented in [15, 16] included only the basic geographical-topographical issues associated with the route geometry and with the seabed bathymetry and obstacles. Those issues were considered for the representation of a candidate route in the context of the GA and for its evaluation in terms of criteria incorporated into the objective function and constraints. These criteria were defined considering only the main aspects already involved in the manual selection of a route, including basically the total pipeline length, interference with obstacles, minimum radius of curvature, and declivity. In [16] special focus was dedicated to the study and assessment of different constraint-handling techniques, including the *ε-constrained* method [19, 20]. Therefore, the developments presented in [15, 16] could be seen as leading to a computational tool that, although innovative in itself, merely automated the process of selecting an optimal route, following the traditional design guidelines.

Here, the focus is on incorporating technical/engineering criteria into the route optimization tool, related to the structural behavior of the pipe under hydrostatic and environmental loadings: specifically, the implementation of on-bottom stability (OBS) criteria. The stability of a pipe is reflected by its ability to remain within its outline in the installed position (taking into account allowable tolerances) under environmental conditions of wave and current. The DNV-RP-F109 code [21] presents criteria to check if a given pipe is stable or else to define minimal values of submerged weight to reach stability. This idea of incorporating OBS criteria into the route optimization tool comprises a breakthrough in the traditional pipeline design methodology and has originally been proposed in [22] where a preliminary implementation was sketched with the OBS criteria incorporated as constraints handled by the classic static penalty technique. This way, taking predefined values for the pipe submerged weight, the "optimal" route would seek areas and trajectories where the stability of the pipe is favored, according to the predominant direction of the environmental loadings.

However, that approach would possibly lead to longer routes and require a "trial-and-error" process, where the designer should perform successive runs of the optimization tool increasing the pipe weight in order to obtain smaller routes. Now, this work describes an improved approach to incorporate the OBS criteria into the optimization tool, where an additional term is introduced into the objective function, representing the ballast weight as an additional variable to be optimized. Thus, the weight required for stability at each pipe segment is directly provided as a result of an optimization run.

In the optimization procedure, candidate routes are represented by specific geometric parameterization and are evaluated in terms of several criteria incorporated in the *objective function* and in the *constraint functions*. These comprise the core of the route optimization model, which will be described in Section 2 that focuses on the route parameterization and encoding in the context of the GA; Section 3 presents the final form of the objective function and describes specifically the incorporation of the OBS criteria; Section 4 presents all remaining criteria that comprise the constraint functions (including their distinction between *soft* and *hard* constraints according to the consequences of their violation). Case studies are presented in Section 5 to illustrate the use of the optimization tool and assess the influence of the stability criteria on the definition of the optimal pipeline route. Lastly, final remarks and conclusions are presented in Section 6.

2. Modeling and Solving the Route Optimization Problem

2.1. Objective Function. A general constrained optimization problem is formally defined in an n-dimensional search space S comprised by a vector of design variables $\mathbf{x} = (x_1, x_2, x_3, \ldots, x_n)$. The following expression mathematically defines the problem:

$$\begin{aligned} &\text{minimize} \quad f(\mathbf{x}) \\ &\text{subject to} \quad g_i(\mathbf{x}) \leq 0, \quad i = 1, \ldots, m \\ &\qquad\qquad h_j(\mathbf{x}) = 0, \quad j = 1, \ldots, p \\ &\qquad\qquad l_k \leq x_k \leq u_k, \quad k = 1, \ldots, q. \end{aligned} \tag{1}$$

The goal is to minimize an *objective function* $f(\mathbf{x})$, considering inequality and equality constraints (resp., $g_j(\mathbf{x}) \leq 0$ and $h_j(\mathbf{x}) = 0$) that define the feasible region. The components x_i may have lower and upper bounds $[l_k, u_k]$. In engineering problems, the constraints are defined in terms of appropriate design criteria. As will be seen later, in the case of the pipeline route optimization problem these criteria may be expressed as inequality constraints $g_j(\mathbf{x})$ only, so equality constraints are not considered in this particular engineering application.

Here it is assumed that the outer diameter and wall thickness of the pipe segments have already been selected from previous design steps. Usually, the outer diameter is dictated by the amount of oil or gas to be transported, according to the yield of the well; and the wall thickness is dictated by strength constraints related to collapse under external hydrostatic pressure [23]. Thus, amongst the relevant

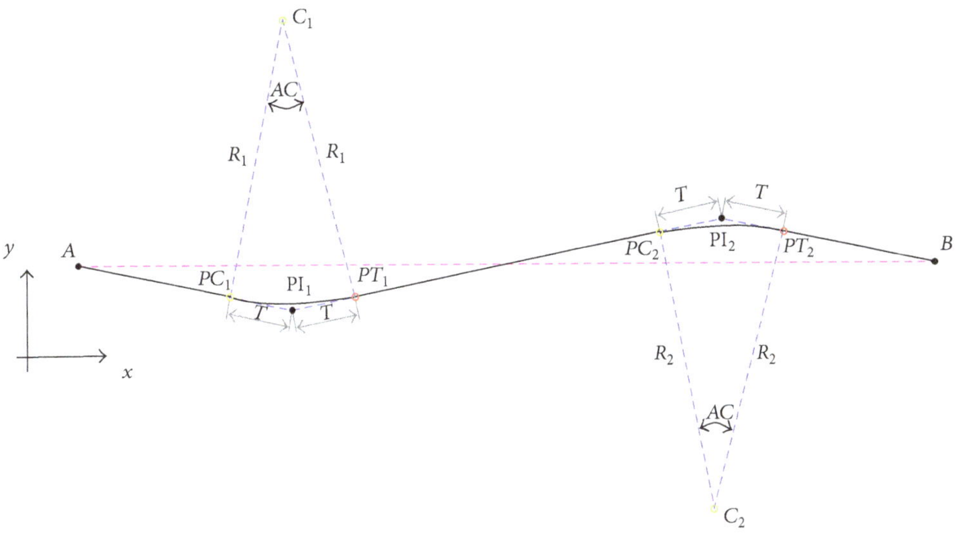

FIGURE 1: Planar representation of a route.

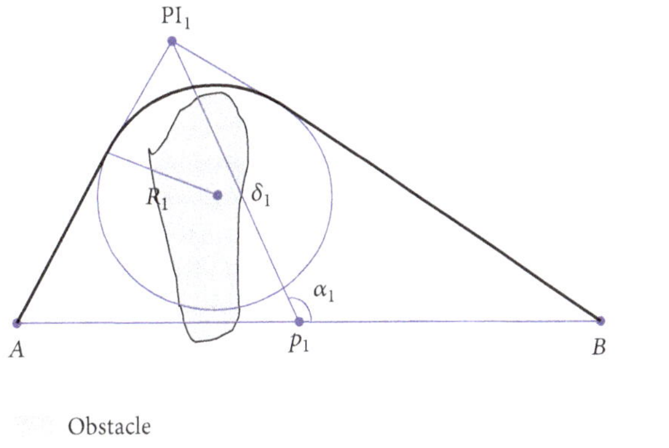

FIGURE 2: Primary parameters: curvature radius (R_i); coordinates of PI (δ_i, α_i).

variables that should be considered in the design of a pipeline route, the first factor that comes to mind is the total length, which should be minimized to reduce material and installation costs. Therefore, as presented in [16], the objective function f could be defined simply as the ratio between the lengths of a given candidate route (L_{route}) and of the straight line connecting endpoints A and B (L_{AB}):

$$f(\mathbf{x}) = \frac{L_{Route}}{L_{AB}}. \quad (2)$$

Considering the focus of the present paper which is to incorporate OBS criteria into the optimization procedure, later in Section 3 we will introduce an additional term into this objective function to take into account the weight of ballast that would be required to assure the stability of the pipeline. Now, before describing the other design criteria that comprise the constraints, in the remainder of this section we will describe the geometric representation of each candidate route in the context of the GA.

2.2. Geometric Parameterization of a Route. The modeling of the route optimization problem requires geometric parameterization to allow the representation of each individual candidate solution. The full formulation for this parametrization has already been detailed in [16]; here we will present only a brief overview. Figure 1 illustrates the planar representation of a route between endpoints A and B, as a horizontal line defined in the xy-plane and comprised by a sequence of straight lines and curves; the curved segments are defined as circular arcs. Figure 2 illustrates the three primary parameters associated with each curve that, as demonstrated in [16], completely describe this representation:

(a) the curvature radius (R_i) of each curve;

(b) the radial (δ) and angular (α) polar coordinates of *Points of Intersection* (PI$_i$)—at which the prolongation of the straight lines (before and after each curve) intersects. These coordinates are relative to *base points* (p_i) uniformly distributed along the straight line AB, as illustrated in Figure 2 showing the position of one PI$_i$ associated with its corresponding base point p_i.

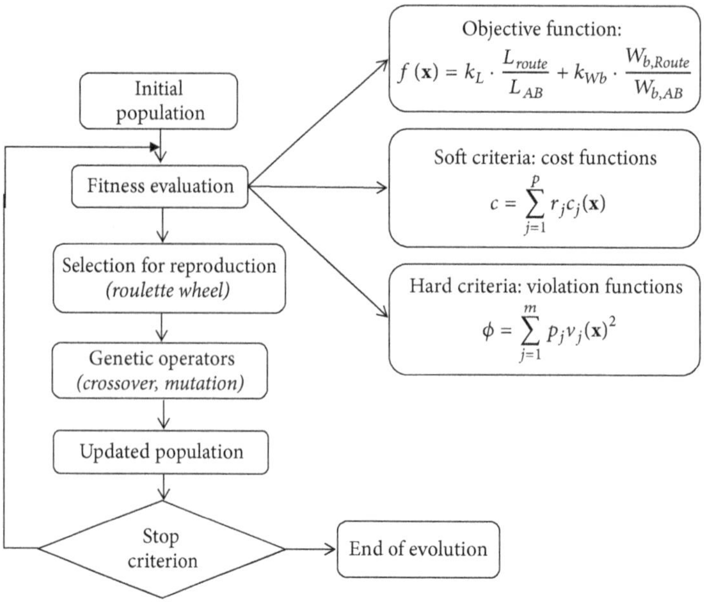

FIGURE 3: Schematic view of the main algorithm.

To obtain the complete three-dimensional representation of the route, in terms of the (x, y, z) coordinates of a series of nodal points along each candidate route, the optimization tool incorporates facilities to import bathymetric data for a given subsea region. Such data are generated by specialized oceanographic vessels equipped with side-scan sonars and are usually available from design databases maintained by oil companies. They are provided as contour maps of isobathymetric curves, or *isobaths*. Having imported the isobaths, a standard gridding technique is then employed to generate a wireframe of quadrangular elements. Along the optimization process, the planar representation of each candidate route is divided into N equal-length segments connecting $N + 1$ nodal points. The (x, y) coordinates are determined using the formulation mentioned above and described in detail in [16]. From these (x, y) coordinates, the corresponding vertical z-coordinate is then obtained by interpolating on the quadrangular wireframe mesh.

2.3. Encoding the Routes into the GA. The pipeline route optimization tool described in this work employs canonical GA implementation. Each individual candidate route is encoded with real value representation in a *chromosome* with n sets of *genes*. Each set is associated with a curve and comprises four genes. The last three genes correspond to the primary route parameters described above: radial coordinate δ_i, angular coordinate α_i, radius R_i; the first gene is an "activation key" A_i that will be described shortly. The full codification of a chromosome can then be written as

$$x^r = \left(A_1\delta_1\alpha_1R_1A_2\delta_2\alpha_2R_2\cdots A_n\delta_n\alpha_nR_n\right), \qquad (3)$$

where $A_1\delta_1\alpha_1R_1$ are the genes corresponding to the first curve and so on.

A population is represented by a set of N individuals; in general, an initial population $P_0 = \{x_1^1, x_2^1, \ldots, x_N^1\}$ is randomly created, where x_i^r is the ith individual in the rth generation. Following Darwin's evolution theory, the fittest individuals have higher probability of surviving and reproducing, and their descendants keep the good genetic material in the species; thus, GAs involve mechanisms of natural selection, genetic recombination, and mutation. The "fitness" is provided by the evaluation of each individual via the objective and constraint functions. The individuals are selected for mating and reproduction by selection operators that generally follow probabilistic rules; here the fitness-proportional *roulette wheel* method is employed.

Mating is performed with *crossover*, combining genes from different parents to produce offspring and generate a new population. Here we consider single-point crossover, with the breakpoint on the parents' chromosomes randomly set according to a probability value equal to 0.6 (following usual guidelines). Thus the offspring inherit features from each of the parents and may be submitted to *mutation*, which confers innovative characteristics to the individual and provide a better exploration of the search space. Mutation alters each bit randomly with a relatively small probability; in the present implementation the mutation rate is set to 0.2.

The implementation of the GA follows a generational approach, where the population is updated by replacing all parents by their offspring, which are made to compete with each other. Also, here we consider one-individual elitism; that is, the fittest individual from the previous population is directly injected into the new population. The process ends when a predefined stopping criterion is reached, and the individual with the best fitness is then defined as the solution of the optimization problem.

Figure 3 presents a schematic view of the basic steps of the GA, including the main expressions that will be employed

FIGURE 4: Procedures to stabilize a pipeline [2–4].

to evaluate each individual candidate route, that is, the objective, cost, and violation functions. Those expressions will be defined later in Sections 3.1 and 4.2.

2.4. Selective Activation of Curves. The number of curves needed to adequately represent a route may vary, depending on its length and the complexity of the scenario (in terms of seabed bathymetry and obstacles). Therefore, besides the three parameters for each curve described above (radius R_i; coordinates of PI: δ_i, α_i), the encoding of the route into the GA incorporates a fourth parameter: the "activation key" A_i. With this parameter, the number of curves actually employed to define each candidate route varies along the evolution process—from zero, corresponding to the trivial straight line AB, up to a user-defined maximum value $n_{\max}$. That is, the number of curves and their associated PI is also a variable of the optimization process.

During the generation of the first population (with individuals randomly created), a real value for the activation key A_i corresponding to each curve in a route is randomly generated within the range [0, 1]. This value indicates whether the corresponding curve is active or not, associated with a given user-defined activation threshold A_t that defines the probability of activation of the curve as $(1.0 - A_t) \times 100$. Then, these A_i values are compared with the activation threshold A_t. If $A_i > A_t$, the corresponding curve is generated by the other three parameters: its radius and the polar coordinates of the PI relative to the closest base point. Otherwise, the curve is inactive and the corresponding section of the route is straight.

For instance, if the user feels that a given scenario is relatively simple, the user may define a higher value for the threshold parameter; for instance, $A_t = 0.6$; this means that the probability of a given curve to be generated is only 40%, and therefore the generated candidate routes will tend to have fewer curves. On the other hand, the user may force all $n_{\max}$ curves to be permanently active for all routes, simply by defining $A_t = 0.0$. As the evolution of the GA proceeds, the selection/reproduction operators of the GA propagate the values of the activation keys A_i, favoring routes with either fewer or more curves depending on the complexity of the scenario; also, the mutation operator may introduce new random values for these genes.

3. On-Bottom Stability Criteria

3.1. Objective Function considering OBS. Now, to incorporate the OBS criteria into the optimization procedure, an additional term is introduced into the objective function of (2). The goal is to guide the optimization process towards an optimal solution that complies with the on-bottom stability with lower intervention costs. This term is defined as the weight of ballast required to stabilize a candidate route $W_{b,Route}$, normalized by the ballast $W_{b,AB}$ required to stabilize the straight route L_{AB}. With this additional term, the objective function now reads as follows:

$$f(\mathbf{x}) = k_L \cdot \frac{L_{Route}}{L_{AB}} + k_{Wb} \cdot \frac{W_{b,Route}}{W_{b,AB}}. \quad (4)$$

Usually those two objectives (minimizing length and ballast weight) may be conflicting: routes with shorter lengths may require more ballast weight, and vice versa. Thus, weighting factors k_L and k_{Wb} are inserted so that the user can specify a relative measure of the importance of minimizing, respectively, length and ballast. As will be seen later in the results of the case studies, this feature presents a crucial role in the definition of an optimal route—recalling that there are different alternative procedures to provide ballast weight (including concrete mattress, trenching, burying, or rock dumping as illustrated in Figure 4); in a preliminary design stage (which is the focus of this optimization tool), actual data regarding the costs of those different procedures to provide ballast may not be easy to obtain. This way, the weighting factors allow the user to adjust the relative importance of each term of the objective function (length and ballast), favoring the minimization of either the pipeline length or the ballast weight, or even obtaining an intermediate solution. This allows the decision-maker to obtain different optimal routes that satisfy the conflicting objectives of the optimization. This feature will be illustrated later in the case studies.

The remainder of this section will begin by briefly summarizing, in Section 3.2, the main concepts behind the OBS criteria as defined in [21], focusing on the two design methods that will be incorporated into the optimization tool. It will be seen that such methods depend on several parameters related to not only the pipe segments themselves, but also the soil properties and environmental loads. All those parameters may vary along each candidate route and along the spatial domain it occupies on the seabed; thus, to consider these variations, each route is discretized into a given number of nodes (*nNodes*) and segments (*nSegm*). In Section 3.3, the OBS expressions of [21] will be rearranged to provide, for each segment, the ballast weight (per unit length) $w_{b,pipe}$ required to stabilize the pipe (either directly, or using iterative procedures to comply with the selected safety factors). The expressions will also be extended to consider slopes in irregular seabed. Then, the total ballast weight for the entire route $W_{b,Route}$ (to be incorporated into (4)) may be obtained by a summation for all pipe segments:

$$W_{b,Route} = \sum_{i=1}^{nSegm} w_{b,pipe}(i) \cdot L_{segm}(i). \tag{5}$$

3.2. Approaches for the Analysis and Verification of OBS. The Recommended Practice DNV-RP-F109 [21] describes three different methodologies for analysis and verification of the lateral stability of a given pipeline configuration: (1) *"absolute lateral static stability"* with zero displacement, that is, ensuring that the hydrodynamic loads acting on the pipe are less than the soil resistance and that the vertical lift load is lower than the submerged weight; (2) *"generalized lateral stability method,"* ensuring *"no break-out"* for a "virtually stable" pipe, allowing small displacements (less than about one-half diameter), taking advantage of the passive resistance of the soil and ensuring that the pipe does not move out of its cavity, with maximum displacements independent of time; and (3) *"dynamic lateral stability analysis,"* allowing "accumulated displacements," with the pipe able to break out of (and return to) its cavity, and the soil resistance is dependent on the time-history of pipe displacements.

These approaches may be associated with analytical expressions, precalibrated curves, or dynamic FE analyses. In the case of the *absolute lateral static stability* method, static analytical expressions provide *safety factors* associated with the ratio between hydrodynamic loads and horizontal soil resistance; in the case of the *generalized lateral stability* method, precalibrated curves provide the minimum required weight for a given maximum allowable displacement.

Finally, the more complex *dynamic lateral stability analysis method* is based on the generation of numerical models and the execution of dynamic analyses under environmental loadings of current and irregular wave, for each specific pipe configuration, taking into account the appropriate soil resistance forces. Differently from the previous methods, it does not directly provide specific design values (in terms of safety factors or required weight); rather, it provides results regarding the behavior of the pipeline, in terms of motions and stresses. These results should then be compared with the limit values of the respective design criteria, in order to obtain the corresponding safety factors.

In this work, we will study the implementation of the first two methodologies (absolute static stability and generalized stability) into the route optimization tool. The third approach is more adequate for more advanced stages of the design of the pipeline; its implementation would not be feasible in the context of the optimization tool, since it would require the dynamic analysis of each candidate route by a complete model in a finite element simulation program, demanding excessively high CPU costs.

3.2.1. Absolute Lateral Static Stability. According to [21], "absolute lateral static stability" is assured when both of the following conditions are met, respectively, for the lateral and vertical directions, considering a given safety factor γ_{SC}:

$$\gamma_{SC} \cdot \frac{F_Y^* + \mu \cdot F_Z^*}{\mu \cdot w_s + F_R} \le 1.0,$$
$$\gamma_{SC} \cdot \frac{F_Z^*}{w_s} \le 1.0. \tag{6}$$

The first condition corresponds to the static lateral equilibrium, where the ratio between the horizontal components of the hydrodynamic loads and the soil resistance is checked against a safety factor γ_{SC}. Hydrodynamic loads are incorporated in the horizontal force F_Y^* as well as the vertical lift force F_Z^*, all calculated by the Morison formula taking the current velocity at the level of the pipe and the water velocity obtained by a linear wave theory considering a single regular wave component. In [21] specific considerations are presented for the use of the Morison formula, in terms of current velocity reduction factors to take into account the soil rugosity z_0 and load reduction factors to take into account the pipe-soil interaction (in terms of the soil permeability, pipe penetration, and trenching). The soil resistance is calculated in two parts: one proportional to the normal force acting on the soil, being determined by a friction coefficient μ affecting both the lift force F_Z^* and the pipe submerged weight w_s, and another that corresponds to the passive resistance F_R due to an initial penetration of the pipe.

The second condition, which corresponds to the static vertical equilibrium, simply checks the ratio between the lift force F_Z^* and the pipe weight w_s against the same safety factor γ_{SC}.

3.2.2. Generalized Lateral Stability. The *generalized lateral stability method* described in Section 3.5 of [21] is a relatively more complex and less conservative method. Differently from the *absolute lateral static stability method*, it now allows some pipe displacement, under the action of a design spectrum of oscillatory wave-induced velocities U_s at the pipeline level.

This method depends on the level of allowable displacement, specified as a value that does not result in excessive pipe deformations or stresses. Two levels of allowable displacements are suggested: (a) up to one-half pipe diameter (corresponding to the "virtually stable pipe") and (b) up to ten diameters. According to the desired level, the method provides values for the weight required for global stability of the pipeline, in terms of a *significant weight parameter* L: that is, L_{stable} and L_{10}, respectively. Intermediate displacement criteria can be established by defining the required weight L_Y

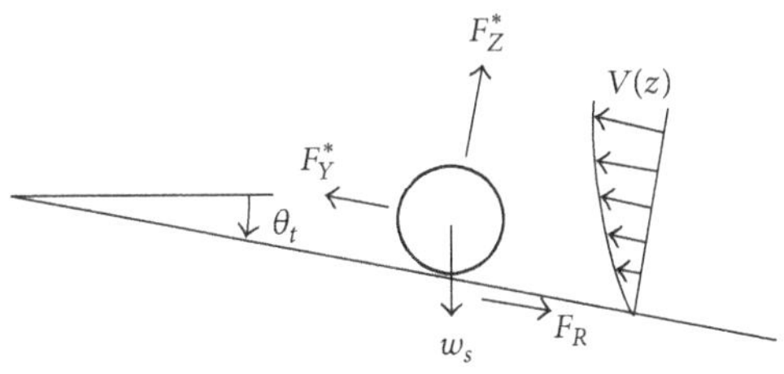

(a) Up slope velocity

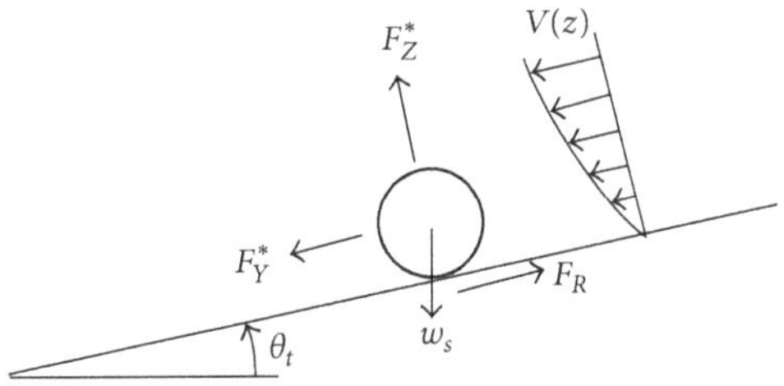

(b) Down slope velocity

FIGURE 5: Hydrodynamic Forces considering slopes (adapted from [5]).

for an allowable displacement Y according to the following formula:

$$\log(L_Y) = \log(L_{\text{stable}}) + \frac{\log(L_{\text{stable}}/L_{10})}{\log(0.5/(0.01\cdot\tau))} \cdot \log\left(\frac{Y}{0.5}\right), \tag{7}$$

where τ is the number of oscillations in the design bottom velocity $= T/T_u$ (with T being the wave period and T_u the spectrally derived mean zero up-crossing period).

Values for the parameters L_{stable}, L_{10} are obtained through empirical expressions and design curves, calibrated from a database of results for a large number of dynamic analyses. A detailed description can be found in [21]; in summary, they depend on a set of dimensionless parameters, including parameters related to the type of soil (clay or sand) and to the wave and current loadings (spectrally derived values of the oscillatory fluid flow due to waves and steady current velocity). To check if a pipe is stable, the required value for the significant weight parameter (L_{stable}, L_{10}, or L_Y) given by the design curves of [21] may be compared with the L parameter calculated for the pipe by the following expression [21]:

$$L = \frac{w_s}{0.5\cdot\rho_w\cdot D\cdot U_s^{\,2}}, \tag{8}$$

where ρ_w is the mass density of water, D is the external pipe diameter, and U_s is the spectrally derived oscillatory velocity (significant amplitude) for design spectrum, perpendicular to the pipeline. If the value calculated by expression (8) is higher than the required value (L_{stable}, L_{10}, or L_Y), then the pipe is stable.

3.3. Calculating the Ballast Terms of the Objective Function. The expressions presented in Sections 3.2.1 and 3.2.2 allow the verification of the OBS for a given pipeline configuration; in the context of a conventional design procedure, the engineer should then select a set of values for the pipe design parameters in order to obtain a minimum submerged weight that complies with those expressions. On the other hand, in the context of the optimization procedure, these expressions may be rearranged to provide safety factor values for each pipe segment into which a candidate pipeline route has been discretized; in some cases, they may provide directly the ballast weight required to stabilize the pipe.

The "absolute lateral static stability" expressions (6) may be rearranged to provide values for the lateral and vertical safety factors γ_{SC_y} and γ_{SC_z}, respectively, as follows:

$$\gamma_{SC_y} = \frac{\mu\cdot w_s + F_R}{\left|F_Y^* + \mu\cdot F_Z^*\right|}, \tag{9}$$

$$\gamma_{SC_z} = \frac{w_s}{F_Z^*}. \tag{10}$$

Similarly, the "generalized lateral stability" expressions that define the different "significant weight parameters" L_{stable}, L_{10}, or L_Y may be combined to provide a safety factor γ_{SC_y} as the ratio between the significant weight parameter L of the pipeline and the significant weight parameter L_{stable}, L_{10}, or L_Y required to obtain the stability; in the case of L_{10}, for instance, we have

$$\gamma_{SC_y} = \frac{L}{L_{10}}. \tag{11}$$

The OBS expressions described above were originally derived in DNV-RP-F109 [21] considering only horizontal seabed. However, since the optimization tool considers irregular seabed defined by actual bathymetric data, it is desirable to employ expressions that take into account the declivity of the bathymetric floor transversal to the pipe axis (Figure 5). This allows the optimization process to favor relatively leveled trajectories (perpendicular to the pipe), to avoid the sliding of the pipe on regions with high transversal slopes.

To reach this goal, (9) that provides the lateral safety factor γ_{SC_y} is altered to take into account the transversal declivity angle θ_t as follows:

$$\gamma_{SC_y} = \frac{\mu\cdot w_s\cdot\cos(\theta_t) + F_R}{\left|F_Y^* + w_s\cdot\sin(\theta_t) + \mu\cdot F_Z^*\right|}. \tag{12}$$

A similar expression for the vertical safety factor γ_{SC_z} may be derived from (10):

$$\gamma_{SC_z} = \frac{w_s\cdot\cos(\theta_t)}{F_Z^*}. \tag{13}$$

Regarding the generalized stability method, the following expression can be defined to provide a safety factor considering slopes:

$$\gamma_{SC_y} = \frac{L \cdot \cos(\theta_t)}{L_{10}}. \quad (14)$$

Taking these expressions for the safety factors, a simple iterative procedure has been devised to provide the ballast weight per unit length $w_{b,pipe}$ required to stabilize each pipe segment. Basically, this procedure gradually increases an initial estimate for $w_{b,pipe}$ until it provides a safety factor that meets a given minimum safety factor γ_{SC}, such as those listed in [21]. Then, the corresponding values for all *nSegm* pipe segments are summed to obtain the total ballast weight for the entire route $W_{b,Route}$, as indicated in (5). The ballast $W_{b,AB}$ required to stabilize the straight route L_{AB} (that nondimensionalizes the OBS term of (4)) is also calculated following a similar procedure.

Considering particularly the "generalized lateral stability" method, the expressions that define the different "significant weight parameters" (L_{stable}, L_{10}, or L_Y) might provide directly the submerged weight of the pipe (per unit length) required for stability—simply by equating (for instance) $L_{10} = L$ and then rearranging (8) to obtain the total required submerged weight, from which the desired ballast weight per unit length may be obtained by subtracting the original pipe weight. However, strictly speaking this would apply only for procedures that are not influenced by the calculation of hydrodynamic loads, which depend on the outer diameter D of the pipe; that is, concrete mattress, trenching, burying, or rock dumping as illustrated in Figure 4. For the most usual procedures to apply ballast weight to a pipe, which are based on installing an additional layer of concrete around the pipe itself, two terms of (8) would be simultaneously affected (D and w_s); to solve this issue and provide the desired value for $w_{b,pipe}$, an iterative procedure similar to that described above is also applied.

4. Other Design Criteria: Constraints

4.1. Design Criteria. Besides the total length and the OBS criteria that have been incorporated into the objective function as described above, the route optimization tool incorporates several other design criteria related to geographical/topographical aspects associated with the seabed bathymetry (such as slopes and interference with obstacles). These criteria and their respective violation functions had already been described in detail in [16]; in the remainder of this section we will present only a brief summary.

4.1.1. Interference with Obstacles. Besides automatically importing isobathymetric curves from design databases to generate the three-dimensional representation of the route (as described in Section 2.2), the tool also incorporates facilities to selectively or fully import data describing the positioning and specification of subsea obstacles that should be avoided. Such data is gathered by specialized vessels equipped with a ROV (remotely operated vehicle), including also geophysical and geotechnical data obtained from the bathymetry and sonography, with information about facies and geotechnical hazards that identify critical areas such as geological faults, rocky outcrops, natural reefs.

Thus, different types of obstacles may be considered: subsea equipment, wellheads, flowlines, other preexistent pipelines, regions with corals or geohazards that should be avoided. To each one the user may associate a level of severity, according to the consequences of the possible interference: *Level 0*, tolerated, with low severity and easily mitigated; *Level 1*, conditionally allowed, with moderate severity introducing relatively higher mitigation costs; *Level 2*, not allowed, with severity being high; *Level 3*, not allowed, with severity being critical. During the optimization process the number intersections for each severity level is counted and stored in variables *nInterS* and *nInterH* (according to the classification of the interference as *soft* and *hard*, as will be seen later); interference is identified by verifying the intersections between the segments or volumes that define the obstacles, against nodal points and segments along each candidate route.

4.1.2. Self-Crossing. The geometric representation described in Section 2.2 might eventually lead the optimization process to generate routes in which the pipeline passes over itself. To identify such configurations with self-crossings or loops, the optimization tool incorporates an algorithm that spans all nodes and segments of the route and counts the number of self-crossings (*nSelfCross*).

4.1.3. Minimum Length of Straight Sections. Pipeline launching operations require straight sections between two consecutive curves, to allow proper space for the maneuvering of the launching ship. Straight sections are already incorporated in the geometric parameterization described before, but if a given minimum length $L_{\min}$ is not respected, the operation may be infeasible. To identify such situations, the length L_i of all straight sections (*nStraight*, including the first and last at the beginning/end of the route) is checked against this limit $L_{\min}$.

4.1.4. Minimum Radius of Curvature. During the pipeline launching operation, the top connection device at the launching ship applies a given tension to the pipe. After the completion of the operation and the accommodation of the pipeline on the seabed, the pipe segments still maintain a certain level of residual tension. It is the balance between this residual tension and the lateral friction forces from the soil that maintains the curved sections of the pipeline over the predefined route. On curves with smaller radius the friction forces may not be sufficient, and the pipe can slide sideways. Therefore, to assure the feasibility of a given route, the curved sections should present radius of curvature larger than a given value $R_{\min}$, function of the pipe-soil friction coefficient μ, pipe weight w_s, and residual pipe tension T_{residual} at its equilibrium configuration after installation.

To identify such situations, the optimization tool spans each node of the route along the curved sections, calculates the required limit $R_{\min}$, and checks this limit against the radius of the curve R_i. Being a function of the soil and pipe parameters that are also variable, $R_{\min}$ may vary from curve to curve, or even along each curve.

4.1.5. Declivity. The slope of the regions where the pipeline is set also comprises a design criterion for the definition of the route. Values for the longitudinal declivity θ_L^i along the nodes of a candidate route (*nNodes*) can be calculated taking the z-coordinate of two consecutive nodes, which in turn are determined from the grid of quadrangular elements interpolated from the isobathymetric curves. To implement this criterion, the optimization tool incorporates an algorithm that check those declivities θ_L^i against a user-defined limit θ_{L_Lim}.

4.1.6. Attractors. During the design of a pipeline route, it may be desirable to keep the route near some regions of interest, referred here as "attractors," for instance, a manifold connecting other pipelines or flowlines, or even production platforms, allowing the pipeline to collect the production of adjacent oil fields. To identify such configurations, the tool incorporates an algorithm that, for a total number of *nAttr* attractors, calculates the distance between the route segments closest to a given attractor ($minDist^i$) and checks if this value is smaller than a user-defined limit expressed as the radius of the "attractor circle" $R_{Attr}{}^i$.

4.2. Soft and Hard Criteria: Handling of Constraints. All the aforementioned design criteria should be incorporated as constraints into the optimization process. Since GAs and other evolutionary algorithms were originally designed to deal with unconstrained search spaces, specific constraint-handling techniques are required to guide the evolutionary search process to feasible regions.

Following the approach presented in [16], according to the consequences of their violation the criteria described above are classified as *soft* and *hard*, which will be handled separately using different techniques. The violation of the so-called *soft* criteria would not indicate that the corresponding solution is infeasible, since mitigation procedures could be adopted. This is the case, for instance, of interference with Level 0 and Level 1 obstacles, such as preexistent pipelines; in this case there are specific installation procedures that allow the new pipe to pass safely over the existing one. Solutions that violate these *soft* criteria would still be feasible, although requiring additional costs; therefore they do not formally characterize constraints in the search space.

On the other hand, violation of *hard* criteria cannot be mitigated and should be avoided. Their violation will mark the solution as infeasible, and therefore they should be formally considered as constraints, to be treated by a particular constraint-handling technique. Amongst the criteria listed in Section 4.1, the ones that are classified as *hard* are interference with Level 2 and Level 3 obstacles (such as subsea equipment or mooring lines), self-crossing, minimum length of straight sections, minimum radius of curvature.

4.2.1. Soft Criteria. Besides interference with Level 0 and Level 1 obstacles, the longitudinal declivity and attractors may be considered as *soft* criteria; in the former case, mitigation procedures can be adopted for route segments with declivity exceeding the specified limit. Attractors can also be considered a *soft* criterion if one assumes that routes passing far from such regions would not be infeasible but would require higher costs.

The *soft* criteria are handled by adding, to the objective function of (4), dimensionless *cost functions* $c_j(\mathbf{x})$ associated with the violation of the jth *soft* criterion, weighted by positive constant values r_j that provide a relative measure of the importance of the criterion. This is equivalent to the classic static penalty technique and leads to the following expanded expression for the objective function f:

$$f(\mathbf{x}) = k_L \cdot \frac{L_{route}}{L_{AB}} + k_{Wb} \cdot \frac{W_{b,Route}}{W_{b,AB}} + \sum_{j=1}^{p} r_j c_j(\mathbf{x})\Big|_{SoftCriteria}. \tag{15}$$

This objective function corresponds to a minimization problem, where the goal is to reduce the value of L_{route}, the weight of ballast $W_{b,Route}$, and the cost terms related to the *soft* criteria. For the ideal route one would trivially have $L_{route} = L_{AB}$, $W_{b,Route} = 0$, and $f = 1$. Solutions that violate one of the *soft* criteria are not considered as infeasible; adding the cost terms only represents a handicap, to favor the other possible solutions that do not violate them. Mathematical expressions for the cost functions $c_j(\mathbf{x})$ associated with those *soft* criteria will be provided shortly in Section 4.2.3.

4.2.2. Hard Criteria: The ε-Constrained Method. The *hard* criteria that formally characterize constraints are handled by the *ε-constrained* method [19, 20]. In this technique, the individuals are ranked based on a lexicographic order of sets of values (f, ϕ) assigned to each individual, where ϕ is a function that indicates the level of constraint violation and $v_j(\mathbf{x})$ are the *violation functions* associated with the jth *hard* criterion, weighted by positive constant values p_j that provide a relative measure of the importance of each criterion:

$$\phi = \sum_{j=1}^{m} p_j v_j(\mathbf{x})^2,$$
$$v_j(\mathbf{x}) > 0 \text{ if constraint is violated}, \quad v_j(\mathbf{x}) = 0 \text{ otherwise}. \tag{16}$$

Considering two individuals 1 and 2, their ranking is performed as follows:

$$(f_1, \phi_1) <_\varepsilon (f_2, \phi_2) \Updownarrow \begin{cases} f_1 < f_2, & \text{if } \phi_1, \phi_2 \le \varepsilon, \\ f_1 < f_2, & \text{if } \phi_1 = \phi_2, \\ \phi_1 < \phi_2, & \text{otherwise.} \end{cases} \tag{17}$$

TABLE 1: *Soft* criteria and cost functions.

Soft criteria	Cost function
Interference with Level 0 and Level 1 obstacles *Severity*: weight factor assigned by the specialist to each level	$c_{obst} = \sum_{i=1}^{nInterS} Severity_i$ (19)
Longitudinal declivity Activated for the nodes of the route with declivity θ_L^i greater than the limit θ_{L_Lim}	$c_{decliv} = \frac{\sum_{i=1}^{nNodes} \theta_L^i - \theta_{L\_Lim}}{nNodes}$ (20)
Attractor Encourage routes near regions of interest; activated whenever route does not pass into the "attractor circle." w_i: weight factor indicating relative importance of each attractor	$c_{attr} = \sum_{i=1}^{nAttr} w_i \frac{minDist^i}{L_{AB}}$ (21)

TABLE 2: *Hard* criteria and violation functions.

Hard criteria	Violation function
Interference with Level 2 and Level 3 obstacles *Severity*: weight factor assigned to each level	$v_{obst} = \sum_{i=1}^{nInterH} Severity_i$ (22)
Self-Crossing	$v_{SelfCross} = nSelfCross$ (23)
Minimum length of straight sections Activated whenever one or more straight sections have length L_i lower than $L_{\min}$	$v_{LminStr} = \sum_{i=1}^{nStraight} \frac{L_{\min} - L_i}{L_{\min}}$ (24)
Minimum radius of curvature Activated whenever a curve has radius lower than $R_{\min}$	$v_{R\min} = \sum_{i=1}^{nNodes} \frac{(R_{\min} - R_i)}{R_{\min}}$ (25) $R_{\min} = \frac{T_{\text{residual}}}{\mu * w_s}$ (26)

In this expression ε is a specified value that represents tolerance related to the sum of constraint violation. The first option means that if both solutions in the pairwise comparison are feasible or slightly infeasible, as determined by the ε value, they are compared using their objective function values f. The second and third options refer to infeasible solutions: the second option means that two infeasible solutions presenting the same sum of constraint violation ϕ will also be compared using their objective function values; otherwise, in the third option they will be compared based on their sum of constraint violation.

The objective function f is given by $f(\mathbf{x})$ of (15) that incorporates the cost functions for the *soft* criteria. Also, assuming that the feasibility of a solution is more important than the minimization of its objective function, we take $\varepsilon = 0$; in this case, the lexicographical ordering considers that the minimization of the sum of constraint violation ϕ precedes the minimization of the objective function f. Feasible solutions are ranked based on their unconstrained objective function values f; infeasible solutions are compared based on their sum of constraint violation ϕ. If one of the solutions is infeasible ($\phi > 0$) and the other is feasible ($\phi = 0$), the latter will be selected.

4.2.3. Summary of the Soft and Hard Criteria. Tables 1 and 2 summarize, respectively, the *soft* and *hard* criteria along with their associated cost and violation functions ($c_j(\mathbf{x})$, $v_j(\mathbf{x})$). Although the expressions for these functions may seem simple, their evaluation is not trivial since it involves calculations performed for possibly hundreds or thousands of points along each candidate route, using the three-dimensional representation that combines the planar formulation with the vertical z-coordinates determined by interpolating the seafloor bathymetric data as described in Section 2.2. Later in the presentation of the case studies, the values assigned for each parameter of these functions will be shown.

5. Case Studies

5.1. Description of the Scenarios. To assess the influence of the stability criteria on the definition of the optimal pipeline route, two different scenarios are considered. The first, illustrated in Figure 6, corresponds to a relatively simpler subsea layout where a pipeline is designed to connect two neighboring platforms, while the second (Figure 7) is associated with a complex layout, including more obstacles and also attractors. The straight line AB connecting the points of the route to be optimized is depicted in pink. Levels 0, 1, 2, and 3 obstacles are indicated, respectively, in green, yellow, red, and black.

For scenario 1 (Figure 6) only Level 2 obstacles are defined, the mooring lines of the floating platforms, indicated in red. For scenario 2 (Figure 7) the mooring lines are treated as Level 4 obstacles, indicated in black; Levels 0 and 1 obstacles are also included (preexistent pipelines and flowlines); and another Level 4 obstacle is added: the black polygon representing an area allotted to another oil company.

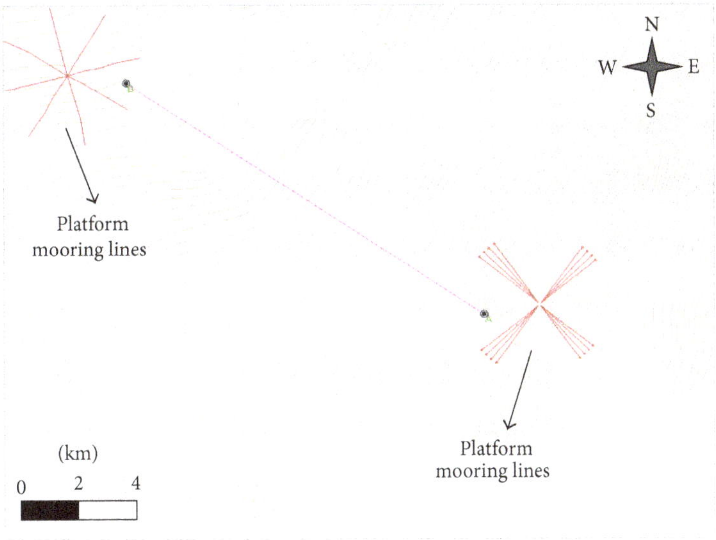

FIGURE 6: Scenario 1: simple layout.

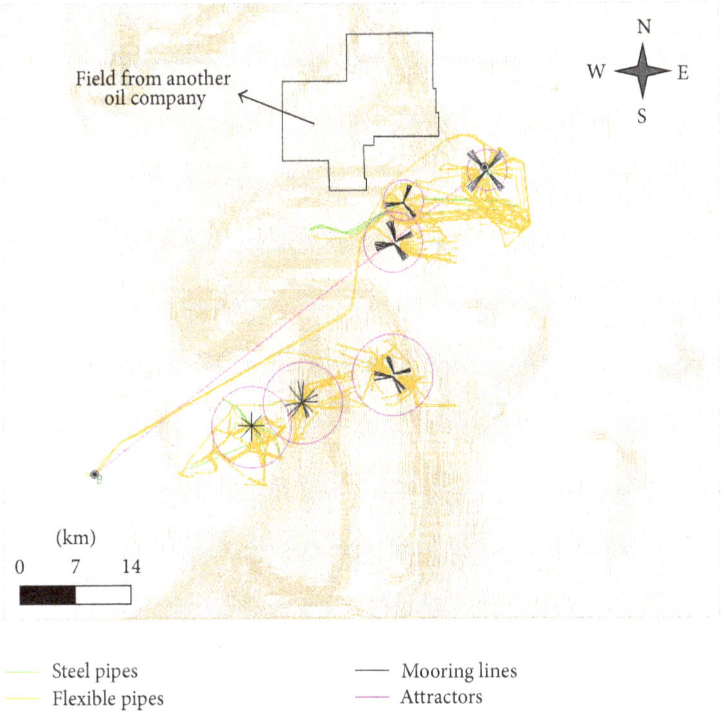

FIGURE 7: Scenario 2: complex layout with obstacles and attractors.

The "attractor" regions close to other platforms are indicated as purple circles.

Table 3 presents the main geometric parameters defining the endpoints of the routes for these scenarios, while Table 4 describes the properties of the pipeline for each scenario. Table 5 presents the values for the soil parameters required to evaluate the on-bottom stability criteria and also the minimum radius of curvature (see (26)). One can observe that scenario 2 comprises two different soil types, defined according to the water depth. The pipeline properties also vary with the water depth.

5.2. Objective Function and Constraints. Table 6 summarizes the main parameters of the optimization algorithm employed for each scenario described above: those that define the search space of the problem (the maximum number of curves $n_{\max}$ and the threshold of the activating factor A_t as described in Section 2.4), along with the number of individuals in each

TABLE 3: Main geometric parameters.

Scenario	L_{AB} (km)	Depth of endpoint B (m)	Depth of endpoint A (m)
1	12.359	14	529
2	49.393	96	1782

TABLE 4: Pipeline properties.

Parameter		Scenario 1	Scenario 2	
Outside diameter (OD)		323.85 mm/$12\frac{3}{4}$ in	457 mm/18 in	
Wall thickness t_s		19.05 mm/$\frac{3}{4}$ in	Depth (m)	Thickness
			0–300	17.5 mm/$\frac{2}{3}$ in
			300–800	19.1 mm/$\frac{3}{4}$ in
			800–1190	22.2 mm/$\frac{7}{8}$ in
			1190–1535	25.4 mm/1 in
			1535–1800	28.6 mm/$1\frac{1}{8}$ in
Steel specific weight ρ_s			77000 N/m^3	
Corrosion coating			Thickness	Spec. weight (N/m^3)
	Thickness (t_p)	76 mm/3 in	0.15 mm/0.006 in	14126.4
			0.20 mm/0.008 in	8829
	Spec. weight (ρ_p)	8826 N/m^3	3.65 mm/$\frac{1}{7}$ in	8829
			48 mm/$1\frac{8}{9}$ in	8829

TABLE 5: Soil properties.

Parameter	Scenario 1	Scenario 2	
		0–300 m	300–1800 m
Type	Sand	Sand	Clay
Pipe-soil friction coefficient (μ)	0.7	0.6	0.2
Dry unit soil weight (γ_s)	-	-	18000 N/m^3
Submerged unit soil weight (γ_s')	13500 N/m^3	13500 N/m^3	-
Undrained clay shear strength (s_u)	-	-	10000 N/m^2

TABLE 6: Parameters of the optimization algorithm.

	N_{pop}	*MaxGen*	$n_{\max}$	A_t
Scenario 1	60	200	4	0.15
Scenario 2	100	300	6	0.15

population N_{pop} and the maximum number of generations *MaxGen* comprising the termination criterion.

To provide a relative measure of the importance of minimizing, respectively, total length and ballast weight, different sets of optimization runs are performed varying the weighting factors k_L and k_{Wb} employed in the objective function ((4) and (15)). Table 7 presents three sets of values: the first and third favors, respectively, the minimization of the pipeline length (possibly requiring more ballast weight) and vice versa, and the second corresponds to an intermediate value for the relative pipeline length weighting factor (k_L).

The general form of the objective function is represented by (15), recalling that this expression incorporates the cost functions $c_j(\mathbf{x})$ for the *soft* criteria whose violation can be mitigated. The cost functions themselves were already presented in Table 1; now, Table 8 presents the cost factors r_j and the other user-assigned values for the parameters associated with these *soft* criteria that will be employed in the

TABLE 7: Weighting factors associated with the objective function.

	Weighting factors Eq. (15)	Scenario 1	Scenario 2
Favoring length	k_L	50	10
	k_{Wb}	0.01	0.01
Intermediate	k_L	----	5
	k_{Wb}	----	0.01
Favoring ballast	k_L	1	1
	k_{Wb}	0.01	0.01

TABLE 8: Parameters of the cost functions associated with the soft criteria.

Criterion	Cost factors r_j, (15)		Parameter	Values	
	Scenario 1	Scenario 2		Scenario 1	Scenario 2
Interference with obstacles	1	1	Severity values, (19)		
			Level 0 (green)	0.2	
			Level 1 (yellow)	0.5	
Declivity	5	1	θ_{L_Lim}, (20)	5°	8°
Attractor	--	1	Radii of the three attractor groups ($R_{Attr}{}^i$), (21)	--	2000 m/3000 m/4000 m

TABLE 9: Parameters of the violation functions associated with *hard* criteria/constraints.

Criterion	Violation factors p_j, (16)		Parameter	Values	
	Scenario 1	Scenario 2		Scenario 1	Scenario 2
Interference with obstacles	5	1	Severity values, (22)		
			Level 2 (red)	1	
			Level 3 (black)	100	
Self-crossing	1	10	--	--	--
Min. length of straight sections	1	1	$L_{\min}$, (24)	100 m	500 m
Min. radius of curvature	1	1	$R_{\min}$, (26)		
			Friction coefficient (μ)	According to Table 5	
			Residual tension (T_{residual})	500 kN	
			Pipe weight w_s	Calculated from pipeline properties (Table 4)	

case studies. Those values were tuned through preliminary studies, considering the characteristics of the respective scenario and the previous experience in the design of submarine pipeline routes.

The objective function of (15) does not incorporate the *hard* criteria that formally characterize design constraints; as shown in Section 4.2.2, these criteria are evaluated separately by the ε-constrained method, being incorporated into ϕ (function of the sum of violations $v_j(\mathbf{x})$, (16)). Table 9 presents the user-assigned values for the parameters associated with these *hard* criteria/design constraints, also tuned through preliminary studies considering the previous experience of the designer.

It is interesting to recall that the optimization tool must perform checks for the whole series of nodes and segments, into which each candidate route is discretized: not only the OBS criteria (as mentioned in Section 3.1), but also all other criteria described in Section 4 whose parameters vary along the spatial domain depending on the water depth, seabed bathymetry and obstacles, and the pipe and soil properties as seen in Tables 4 and 5.

5.3. Additional Parameters for the OBS Criteria. Besides the pipe and soil properties, evaluation of the OBS criteria also requires the definition of a set of environmental loading cases. According to [21], two extreme loading combinations of wave and current should be considered: (1) 100-year return wave combined with 10-year current and (2) 10-year wave with 10-year current. Wave and current are taken as aligned along eight directions (N, NE, E, SE, S, SW, W, and NW). Table 10 presents the JONSWAP spectral wave parameters (significant height Hs, peak period Tp); Table 11 presents the extreme near-bottom current velocities v_c, taken from metocean data for the two scenarios. Figure 8 presents polar graphs depicting the main current and wave parameters, indicating that the predominant environmental loadings are acting along the NE-SW direction.

Table 10: Wave parameters.

Direction	10-year H_s (m)	10-year T_P (s)	100-year H_s (m)	100-year T_P (s)
N	4.74	9.2	5.01	9.56
NE	4.88	9.47	5.17	9.7
E	4.34	9.9	4.87	10.4
SE	5.72	10.28	6.53	11.63
S	6.19	13.54	7.1	14.35
SW	7.16	14.78	7.84	15.55
W	3.57	8.22	3.88	8.51
NW	3.57	8.22	3.88	8.51

Table 11: Near bottom current velocities (m/s).

Direction	All scenarios Current 1		Scenario 2 Current 2		Current 3		Current 4	
	10-yr	100-yr	10-yr	100-yr	10-yr	100-yr	10-yr	100-yr
N	0.54	0.65	1.45	1.92	0.57	0.71	0.36	0.45
NE	0.75	0.96	1.13	1.43	0.52	0.63	0.44	0.56
E	0.62	0.74	1.51	2.06	0.52	0.65	0.39	0.5
SE	0.65	0.77	1.6	2.03	0.5	0.64	0.29	0.4
S	0.96	1.18	1.05	1.2	0.57	0.74	0.31	0.41
SW	1.09	1.41	0.76	0.85	0.48	0.62	0.34	0.45
W	0.63	0.75	0.92	1.08	0.38	0.47	0.56	0.77
NW	0.55	0.61	1.11	1.37	0.41	0.5	0.42	0.54

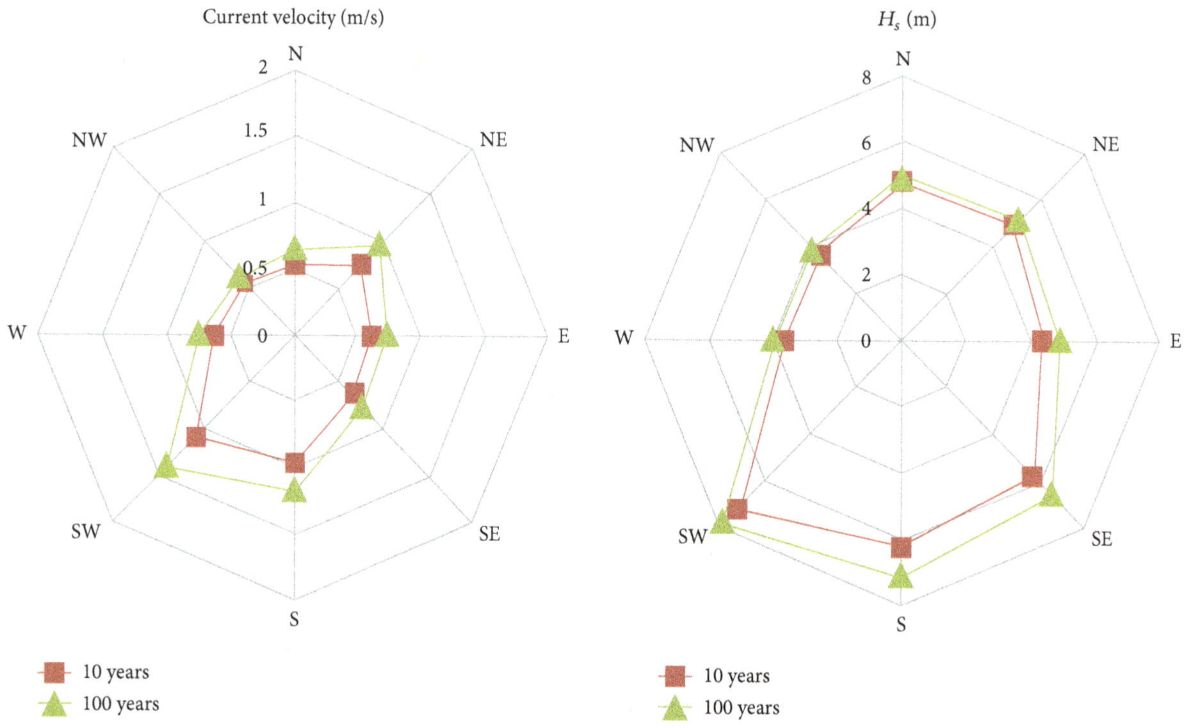

Figure 8: Environmental loadings.

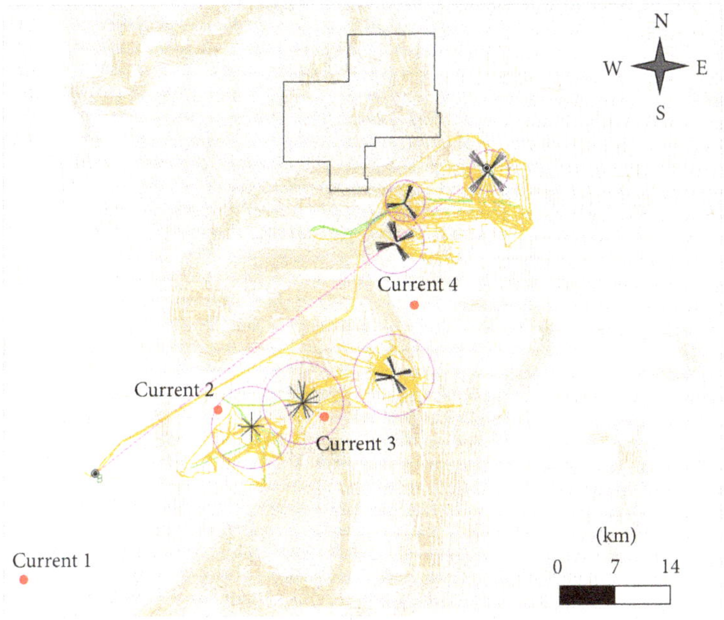

FIGURE 9: Position of the current measurements.

As seen in Table 11 for the more complex scenario 2, the metocean data indicate that the current values vary along the spatial domain according to four points of measurement depicted in Figure 9. Thus, the optimization tool must evaluate the current loading along each node/segment of a route, by taking the values corresponding to the measurement point that is nearest to the node.

The near-bottom current velocities perpendicular to the pipeline are calculated for each node of each candidate route by applying a factor to take into account the directionality and also the effect of the bottom boundary layer. For this purpose the following expression [21] is considered, where the directionality is taken into account by the angle between current direction and axis of the pipe segment θ_c; and the bottom boundary layer is a function of the elevation above sea bed z, bottom roughness parameter z_0 (whose values are defined in [21]), and reference measurement height over seabed z_r:

$$V(z) = V(z_r) \cdot \frac{\ln(z + z_0) - \ln(z_0)}{\ln(z_r + z_0) - \ln(z_0)} \sin(\theta_c). \quad (18)$$

This expression is employed for all directions of current, with their respective values of velocity and relative angle θ_c. Taking all resulting perpendicular velocities, the respective ballast weights per unit length for the node/segment are calculated following the procedures described in Section 3.3 that also takes into account the corresponding values of transversal declivity; the required ballast weight is then taken as the higher value amongst the obtained results (eventually, smaller velocity values associated with higher declivities might require more ballast).

The near-bottom wave-induced water velocities perpendicular to the pipeline must also be evaluated for each node, since they depend on the water depth. Moreover, recalling that wave loadings are relevant only for pipelines in shallower waters, the optimization tool automatically disregards wave effects for pipe segments resting in water depths larger than a user-defined limit (for instance, 100 m). Thus, as can be seen in Table 3, wave loadings are relevant only for the routes of scenario 1, where a significant part of the pipeline rests on shallow waters.

Moreover, according to [21] the "generalized stability" criterion is more adequate for shallower waters, so a depth limit may also be taken to allow the selection of this criterion. Thus, while both the "absolute" and the "generalized" criteria are employed and compared for the shallower segments of scenario 1 (the latter considering the allowable displacement as 10 times the pipe diameter), only the absolute stability criterion is activated for the deeper segments of scenario 1 and for the routes of scenario 2.

5.4. Results. The results for the case studies are compared in terms of figures that show the geometry of the optimal route, indicating the ballast weight required for stability by a color gradation scheme (where green, yellow, and red indicate successively higher weight values), and in tables that present the corresponding values for the main parameters of the objective function, for example, length L_{Route}, total ballast weight $W_{b,Route}$, and ballast weight per unit length $w_{b,pipe}$, according to (4) and (5).

5.4.1. Scenario 1. For this simpler scenario two sets of optimization runs were performed, employing, respectively, the "favoring length" and "favoring ballast" weighting factors k_L and k_{Wb} of Table 7. Each set consists of two runs: the

TABLE 12: Scenario 1: parameters of the optimal routes.

	Absolute stability		Absolute + generalized	
	Favoring length	Favoring ballast	Favoring length	Favoring ballast
Route length (m)	12,400	15,060	12,403	14,792
Total ballast weight (kN)	5,208	3,042	4,812	2,647
Ballast weight per unit length (N/m)				
Average	420	202	388	179
Min.	100	0	100	0
Max.	4,500	2,900	4,500	2,500

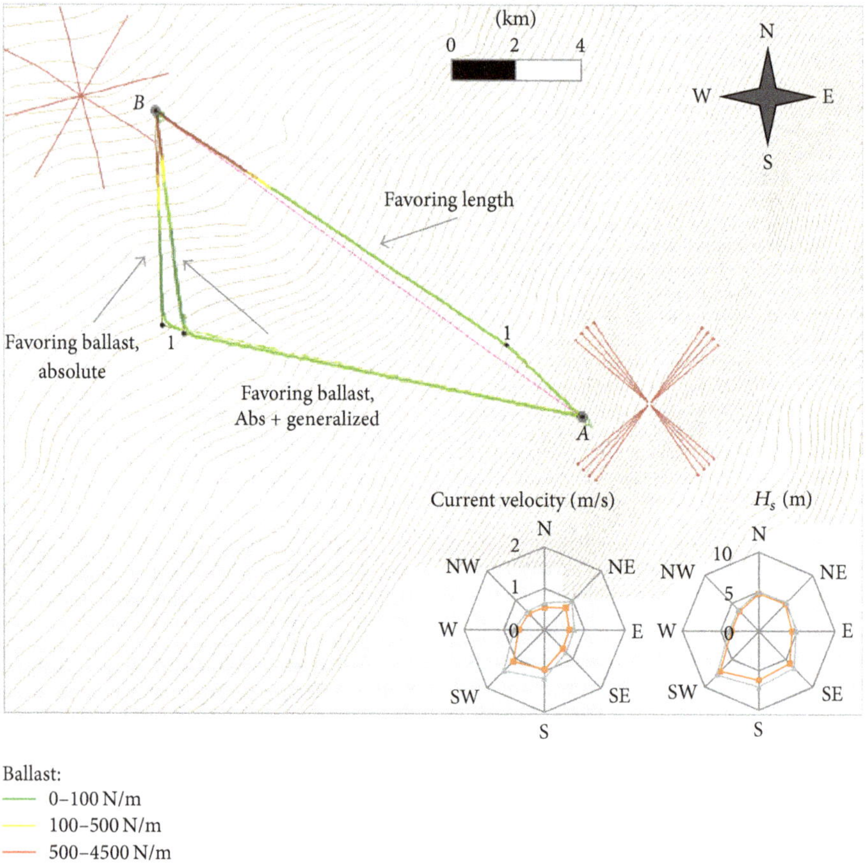

FIGURE 10: Scenario 1: comparison of optimal routes.

first using only the "absolute stability" OBS criterion and the second set considering both the absolute and generalized criteria, automatically selected according to the water depth.

Figure 10 shows the resulting optimal routes, while Table 12 compares the routes in terms of the main parameters of the objective function, that is, route length and ballast weight. As expected, the optimization runs with the "favoring length" weighting factors k_L and k_{Wb} of Table 7 generated shorter optimal routes, closer to the straight line between endpoints AB; the OBS criteria did not affect their geometries, so they are practically coincident and cannot be distinguished in Figure 10. However, they required considerably more ballast weight, mostly for the pipe segments closer to endpoint B (located in shallower waters), as can be seen from the lengthier segments depicted in red and yellow in Figure 10.

On the other hand, the routes generated with the "favoring ballast" weighting factors have lengths that are not dramatically greater (20%, approximately), while requiring almost half of the total ballast compared with the "favoring length" routes; in this sense, perhaps this choice of favoring ballast minimization may lead to a more efficient design. This is mostly due to the straight section $\overline{B1}$ located in shallower depths ranging from 20 to 150 m, where wave loads are more significant; in Figure 10 it can be seen that, to reduce the resultant loads normal to the pipeline axis, the optimization algorithm tries to align the route with the most severe directions of environmental loadings (NE-SW and

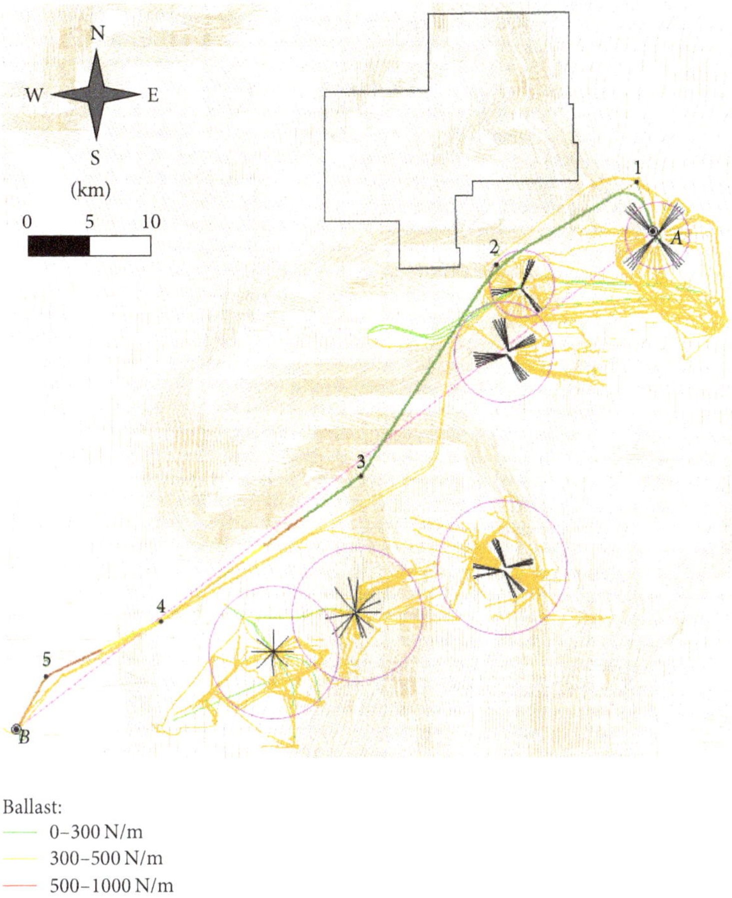

FIGURE 11: Scenario 2: "favoring length" route.

TABLE 13: Scenario 2: parameters of the optimal routes.

	Favoring length	Intermediate	Favoring ballast
Route length (m)	53,923	55,966	56,870
Total ballast weight (kN)	14,508	13,564	13,206
Average ballast weight per unit length (N/m)	269	242	232

N-S directions), just enough to avoid interference with the mooring line close to endpoint *B*.

Differently from the "favoring length" routes, the "favoring ballast" routes provided by the different stability criteria can be clearly identified in Figure 10 (indicated as *"favoring ballast, absolute"* and *"favoring ballast, Abs + generalized,"* respectively). Table 12 confirms that the "generalized stability" criterion is indeed less conservative, not only leading to shorter routes, but also requiring less ballast. This latter aspect is also indicated by the shorter length of the segments that require more ballast (depicted in red in Figure 10).

5.4.2. Scenario 2. The optimal routes obtained for scenario 2, corresponding to the three different sets of values for the weighting factors of Table 7, are shown next: favoring length (Figure 11), intermediate (Figure 12), and favoring ballast (Figure 13). Figure 14 groups those three routes for a better visual assessment.

Table 13 presents the values for the main parameters of the objective function: length and ballast weight. Again, as expected, increasing the weighting factor k_L leads to the reduction of the route length and conversely to the increase in ballast weight. Thus the "favoring length" route requires more ballast, especially in the shallower region near endpoint *B* and also in the stretch depicted in red between points 3 and 4 indicated in Figure 11. This stretch extends from the depth of 300 m that sets the limit where the soil changes from sand to clay, thus affecting the OBS calculations; see Table 5 and Figure 15 where sand and clay regions are depicted in gray and green, respectively.

The other routes ("intermediate" and "favoring ballast") require slightly less ballast in this stretch. Overall, the

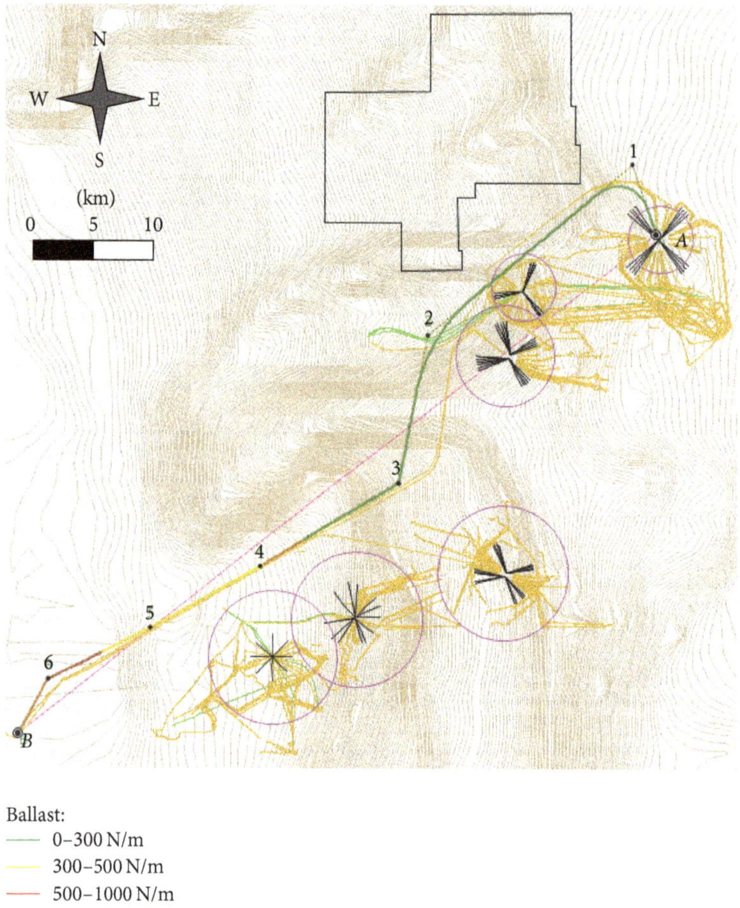

FIGURE 12: Scenario 2: "intermediate" route.

"favoring ballast" route is longer but requires less ballast, while the "intermediate" route is around 900 m shorter but requires slightly more ballast (4%). Another advantage of the "favoring ballast" route is that it deviates from the flowlines (depicted in light green, close to point 3, in Figure 13), thus minimizing the requirements for intervention to avoid interference with obstacles.

In summary, for this considerably complex scenario with a greater distance between the endpoints (approx. 50 km) and irregular seabed bathymetry (including canyons), it can be seen that the optimal routes successfully complied with the main objectives and constraints of the optimization procedure. Excessive declivities associated with the crossing of canyons have been minimized; the obstacle indicated by a black polygon (representing a *hard* constraint) has been avoided; and the proximity to the other platforms (corresponding to the "attractors" depicted as purple circles) has been attained as close as possible.

6. Final Remarks and Conclusions

It is well known that the costs of a given pipeline project are dictated not only by the material of the pipe segments themselves (that derives from the route length), but also by installation and intervention procedures that mitigate problems related to the structural behavior of the pipe under hydrostatic and environmental loadings. Considering specifically on-bottom stability, several methods to provide ballast and stabilize the pipe (including those described at the end of Section 3.3) introduce substantial costs to the installation of the pipeline along the chosen route. In the traditional pipeline design methodology, the definition of a route and the stability calculations had been performed independently: firstly, the route is defined according to geographical-topographical issues, including seabed bathymetry and obstacles; then stability is verified and the required mitigating procedures are specified.

Now, with the incorporation of OBS criteria into the route optimization tool following the approaches described in this work, stability becomes an integral part of the route optimization process. This may be seen as a breakthrough in the traditional design methodology, allowing the designer to simultaneously handle the minimization of route length and the cost of the mitigating procedures.

The optimization tool provides the decision-maker with profitable information regarding the optimal routes, considering different aspects of the design and the costs associated with the different procedures that provide ballast to stabilize

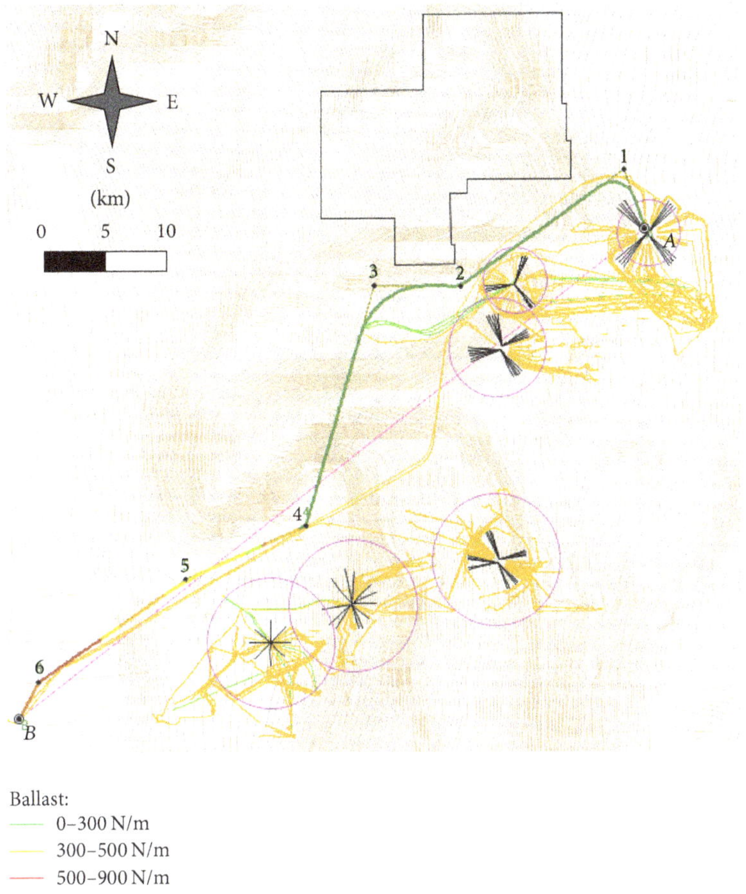

FIGURE 13: Scenario 2: "favoring ballast" route.

a pipeline (mattress, trenching, burying, concrete coating, or even employing a heavier pipe with additional steel wall thickness). For instance, relatively high values for the weight of concrete might hinder the installation of the pipeline (due to limitations of the launching vessel and/or excessive stresses on the pipe itself). In such cases, the designer might then opt for a hybrid solution combining less concrete coating with a heavier pipe (with increased wall thickness). Taking the required ballast weights as indicated in Table 12 or Table 13, different combinations of steel wall and concrete coating thickness may be provided by the iterative procedure described at the end of Section 3.3.

Overall, the computational costs associated with the stability calculations by the different OBS methods ("absolute stability" and "generalized stability") are relatively low, corroborating the feasibility of their incorporation into the optimization tool. The results of the first case study confirmed that the latter is more adequate for shallower waters with increased influence of the environmental loads on the pipeline: it leads to smaller routes and/or less ballast when compared with those obtained by the absolute method. Thus, also observing that the computational costs associated with the "generalized" criterion are not significantly higher than those for the "absolute" criterion, in the optimization tool the former may be set as the default procedure for pipe segments below a given user-specified water depth value.

Several new developments related to the route optimization tool are currently underway, including procedures to incorporate other technical/engineering criteria into the objective function (for instance, VIV-induced fatigue on free spans and multiphase flow). Also, here the multiobjective problem of optimizing the route length and ballast weight has been dealt with by combining the respective terms into a single objective function. As observed in the results of the case studies, those may be conflicting objectives since usually the optimal routes with shorter lengths have more ballast weight and vice versa. While the present approach effectively provides different optimal routes by adjusting the weighting factors k_L and k_{Wb} along different optimization runs, another promising approach would be to formally model the optimization problem with a multiobjective procedure using specialized algorithms based on the Pareto front concept, thus obtaining a set of optimal routes from a single optimization run. Preliminary studies related to those topics have been briefly outlined in conference papers [24, 25]; more detailed and conclusive results will be presented in subsequent works.

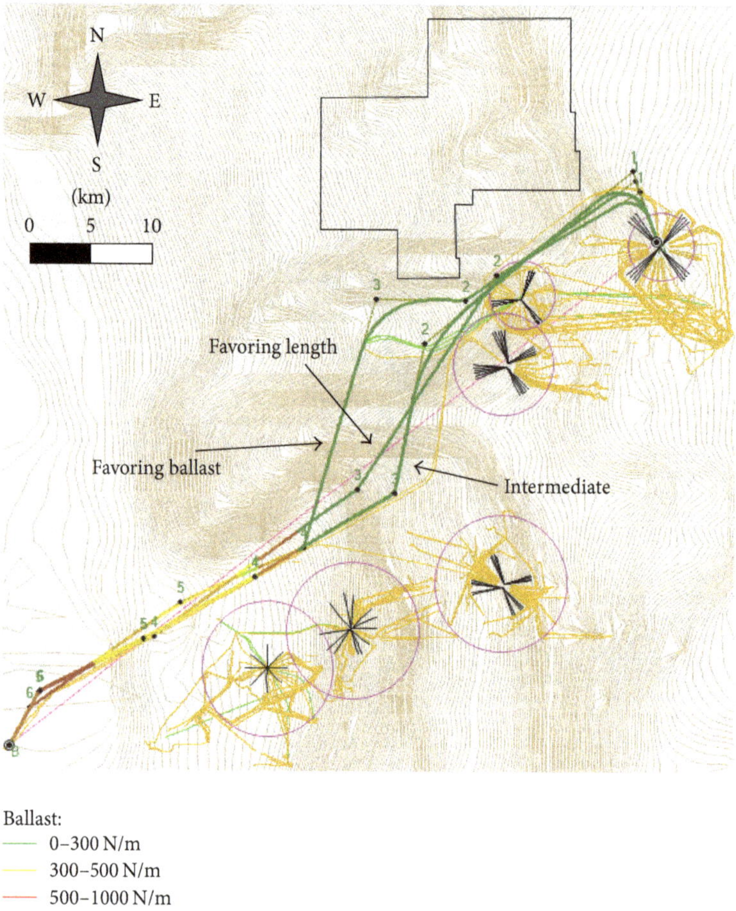

FIGURE 14: Scenario 2: comparison of optimal routes.

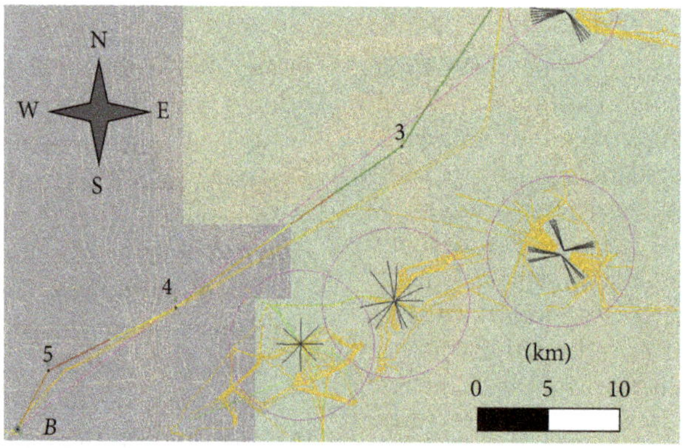

FIGURE 15: Scenario 2: "favoring length" route, detail.

To summarize, the main feature of the developments presented in this work consists of anticipating, to the stage of route definition, engineering checks and calculations traditionally related to subsequent design stages. Thus the optimization tool may reduce the design time of a given pipeline, not only for the route definition, but also for the other stages involving the detailed verification of those engineering criteria; more important, the tool can minimize installation and operational costs of the pipeline.

Acknowledgments

The first and second authors would like to acknowledge the Brazilian research funding agencies CNPq (Conselho

Nacional de Desenvolvimento Científico e Tecnológico) and FAPERJ (Fundação Carlos Chagas Filho de Amparo à Pesquisa do Estado do Rio de Janeiro) for the financial support of their doctoral studies at PEC/COPPE/UFRJ. The fourth and fifth authors acknowledge the support of CNPq and FAPERJ corresponding to the research Grants nos. 304850/2012-8, 306104/2013-0, and E-26/201.184/2014, respectively. Finally, all authors acknowledge Petrobras S.A. for supporting and allowing the publication of this work.

References

[1] Det Norske Veritas, *Offshore Standard DNV-OS-F101 Submarine Pipeline Systems*, Det Norske Veritas, Oslo, Norway, 2012.

[2] https://stoprust.com/products-and-services/retromat/.

[3] http://allseas.com/activities/pipelines-and-subsea/pipeline-protection/.

[4] http://offshore-fleet.com/data/rock-dumping-vessel.htm.

[5] W. T. Jones, "On-Bottom Pipeline Stability in Steady Water Currents," *Journal of Petroleum Technology*, vol. 30, no. 3, pp. 475–484, 1978.

[6] H. Meisingset, J. Hove, and G. Olsen, "Olsen Optimization of Pipeline Routes," in *in the Proceedings of the Fourteenth International Offshore and Polar Engineering Conference*, Toulon, France, 2004.

[7] D. H. Fernandes, B. P. Jacob, B. S. L. P. de Lima, and C. H. Albrecht, "Offshore Pipeline Route Assessment via Evolutionary Algorithms," in *in the Proceedings of the XXIX Iberian-Latin-American Congress on Computational Methods in Engineering, CILAMCE*, Maceió, AL, Brazil, 2008.

[8] D. H. Fernandes, B. P. Jacob, B. S. L. P. de Lima, A. R. Medeiros, and C. H. Albrecht, "A Proposal of Multi-Objective Function for Submarine Rigid Pipelines Route Optimization via Evolutionary Algorithms," in *in the processdings of Rio Pipeline Conference & Exposition 2009 Annals*, Rio de Janeiro, Brazil, 2009.

[9] S. Percival, D. Hendrix, and F. Noblesse, "Hydrodynamic optimization of ship hull forms," *Applied Ocean Research*, vol. 23, no. 6, pp. 337–355, 2001.

[10] B. D. S. L. P. de lima, B. P. Jacob, and N. F. F. Ebecken, "A hybrid fuzzy/genetic algorithm for the design of offshore oil production risers," *International Journal for Numerical Methods in Engineering*, vol. 64, no. 11, pp. 1459–1482, 2005.

[11] R. L. Tanaka and C. D. A. Martins, "Parallel dynamic optimization of steel risers," *Journal of Offshore Mechanics and Arctic Engineering*, vol. 133, no. 1, Article ID 011302, 2010.

[12] A. A. de Pina, C. H. Albrecht, B. S. de Lima, and B. P. Jacob, "Tailoring the particle swarm optimization algorithm for the design of offshore oil production risers," *Optimization and Engineering*, vol. 12, no. 1-2, pp. 215–235, 2011.

[13] I. N. Vieira, B. S. de Lima, and B. P. Jacob, "Bio-inspired algorithms for the optimization of offshore oil production systems," *International Journal for Numerical Methods in Engineering*, vol. 91, no. 10, pp. 1023–1044, 2012.

[14] M. A. L. Martins, E. N. Lages, and E. S. S. Silveira, "Compliant vertical access riser assessment: DOE analysis and dynamic response optimization," *Applied Ocean Research*, vol. 41, pp. 28–40, 2013.

[15] I. N. Vieira, C. H. Albrecht, B. S. L. P. de Lima, B. P. Jacob, D. M. Rocha, and C. O. Cardoso, "Towards a Computational Tool for the Synthesis and Optimization of Submarine Pipeline Routes," in *in the proceedings of the Twentieth International Offshore and Polar Engineering Conference - ISOPE*, Beijing, China, 2010.

[16] R. R. De Lucena, J. S. Baioco, B. S. L. P. De Lima, C. H. Albrecht, and B. P. Jacob, "Optimal design of submarine pipeline routes by genetic algorithm with different constraint handling techniques," *Advances in Engineering Software*, vol. 76, pp. 110–124, 2014.

[17] D. E. Goldberg, *Genetic Algorithms in Search, Optimization and Machine Learning*, Addison-Wesley, Boston, Massachusetts, USA, 1989.

[18] K. F. Man, K. S. Tang, and S. Kwong, "Genetic Algorithms: Concepts and Applications," *IEEE Transactions on Industrial Eletronics*, vol. 43, no. 5, pp. 519–534, 1996.

[19] E. Mezura-Montes and C. A. Coello Coello, "Constraint-handling in nature-inspired numerical optimization: past, present and future," *Swarm and Evolutionary Computation*, vol. 1, no. 4, pp. 173–194, 2011.

[20] T. Takahama, S. Sakai, and N. Iwane, "Constrained optimization by the epsilon constrained hybrid algorithm of particle swarm optimization and genetic algorithm," in *AI 2005: Advances in Artificial Intelligence*, Lecture Notes in Artificial Intelligence, pp. 389–400, Springer-Verlag, Berlin, Germany, 2005.

[21] Det Norske Veritas, *On Bottom Stability Design of Submarine Pipelines, Recommended Practice DNV-RP-F109*, Det Norske Veritas, Oslo, Norway, 2008.

[22] M. H. A. De Lima Jr., J. S. Baioco, C. H. Albrecht et al., "Synthesis and optimization of submarine pipeline routes considering On-Bottom Stability criteria," in *Proceedings of the ASME 2011 30th International Conference on Ocean, Offshore and Arctic Engineering, OMAE2011*, pp. 307–318, Netherlands, June 2011.

[23] B. Guo, S. Song, J. Chacko, and A. Ghalambor, *Offshore Pipelines*, Gulf Professional Publishing, Elsevier, Houston, Texas, USA, 2005.

[24] J. S. Baioco, P. Stape, M. Granja, C. H. Albrecht, B. S. L. P. De Lima, and B. P. Jacob, "Incorporating engineering criteria to the synthesis and optimization of submarine pipeline routes: On-bottom stability, viv-induced fatigue and multiphase flow," in *Proceedings of the ASME 2014 33rd International Conference on Ocean, Offshore and Arctic Engineering, OMAE*, San Francisco, California, USA, June 2014.

[25] J. S. Baioco, C. H. Albrecht, B. P. Jacob, and D. M. Rocha, "Multi-Objective Optimization of Submarine Pipeline Routes Considering On-Bottom Stability, VIV-Induced Fatigue and Multiphase Flow," in *Proceedings of the Twenty-Fifth International Ocean and Polar Engineering Conference - ISOPE*, Kona, Hawaii, USA, 2015.

6

Mathematical Model for the Fluid-Gas Spontaneous Displacement in Nanoscale Porous Media considering the Slippage and Temperature

Kang Liu,[1] Zhongyue Lin,[2,3] Daiyong Cao,[1] and Yingchun Wei[1]

[1] *College of Geoscience and Surveying Engineering, China University of Mining & Technology, Beijing 100083, China*
[2] *Beijing Dadi Gaoke Coalbed Methane Engineering Technology Research Institute, Beijing 100040, China*
[3] *National Administration of Coal Geology in China, Beijing 100038, China*

Correspondence should be addressed to Zhongyue Lin; 53674914@qq.com

Academic Editor: Sandro Longo

The fracturing fluid-gas spontaneous displacement during the fracturing process is important to investigate the shale gas production and formation damage. Temperature and slippage are the major mechanisms underlying fluid transport in the micro-/nanomatrix in shale, as reported in the previous studies. We built a fracturing fluid-gas spontaneous displacement model for the porous media with micro-/nanopores, considering two major mechanisms. Then, our spontaneous displacement model was verified by the experimental result of the typical shale samples and fracturing fluids. Finally, the influences of temperature, slip length, and pore size distribution on the spontaneous imbibition process were discussed. Slippage and temperature significantly influenced the imbibition process. Lower viscosity, higher temperature, and longer slip length increased the imbibition speed. Ignoring the temperature change and slippage will lead to significant underestimation of the imbibition process.

1. Introduction

During hydraulic fracturing process in the unconventional gas formation, a relatively large volume of fracturing fluid is pumped into formation, which can greatly stimulate the gas production [1, 2]. In this process, water will be imbibed into matrix in the fractured reservoir by many influences, including capillary pressure [3–5], chemical osmotic pressure [6], pore network [7–9], and clay mineral [10], which is called spontaneous imbibition. Also, leakage, lost circulation, and induced fracture in drilling process will lead to fracturing fluid being pumped into formation [11–13], which will also cause imbibition process and may change the stress field near wellbore [14] and drilling state [15, 16]. The spontaneous imbibition is the dominant mechanism of the water transport into the formation because of the high capillary pressure by nanopores [17–19]. Recently studies have shown that the spontaneous imbibition of fracturing fluid can be a driving force to enhance the gas recovery for shale gas reservoir [20, 21]. Thus, analysis of the spontaneous imbibition of shale and the potential effect on gas recovery needs urgent attention.

Shale has an ultralow permeability and porosity with abundant nanopores. Liquid flow mechanism in shale is much more complex than conventional formations. Slip flow is a major mechanism of liquid transport in nanotube [22, 23]. More studies have shown that the slip flow is significantly different from the no-slip boundary condition in the nanotubes [24, 25]. In addition, these studies reported that conventional flow equations, such as Darcy's law, may not be valid for shale systems because of the difference in the controlling physics of liquid flow.

Fracturing fluid commonly includes hydrochloric acid, friction reducers, guar gum, and biocides. Viscosity varies with temperature significantly [26]. Reservoir temperature is also one of the key controlling factors in spontaneous imbibition. The influence of temperature on gas production is usually ignored because the temperature change is not severe. However, for spontaneous imbibition in the hydraulic fracturing process, the temperature of fracturing liquid is quite different from formation. Thus, ignoring the variation in temperature will lead to inaccurate results and errors.

2. Mathematical Model

2.1. Spontaneous Imbibition considering Slip Effect in a Single Capillary. Spontaneous imbibition occurs in the capillaries of shale as the wetting fluid imbibed in capillaries is driven by capillary force automatically. Liquid slip in nanoscale capillaries is especially not negligible because the slip length is the same scale with the diameter. In this section, considering liquid slip effect, we established a spontaneous imbibition model. To focus on the effect of liquid slip, we have made some simplifications as follows: (1) the cross section of tube is circular; (2) liquid is the wetting phase, while gas is the nonwetting phase; (3) liquid is the Newton liquid with laminar flow, and inertial forces have been ignored; (4) the driving force of spontaneous imbibition is the capillary force; (5) slip occurs at tube wall; and (6) gravity is ignored. On the basis of the Hagen–Poiseuille equation, the fluid flux considering the liquid slip can be calculated as follows:

$$Q = \frac{\pi \Delta p}{32\mu L_f}\left[\frac{1}{2}(\lambda + 2L_s)^2\lambda^2 - \frac{1}{4}\lambda^4\right], \quad (1)$$

where Q is the fluid flux in tube, Δp is the pressure difference on fluid, μ is the dynamic viscosity, L_f refers to the length of fluid path line, λ is the tube's equivalent diameter, and L_s is the slip length. L_s can be expressed in a dimensionless form, $L_{sD} = L_s/\lambda$.

Imbibition velocity can be determined as follows:

$$v_f = \frac{4Q}{\pi\lambda^2} = \frac{dL_f}{dt}. \quad (2)$$

The real capillary in shale is tortuous. Tortuous fractal dimension is introduced to express the tortuous capillaries, according to Yu and Cheng [27] and Cai et al. [28].

$$L_f = \lambda^{1-D_T}L_0^{D_T},$$
$$v_f = \frac{dL_f}{dt} = D_T\lambda^{1-D_T}L_0^{D_T-1}v_0, \quad (3)$$

where L_0 is the distance between meniscus and liquid intake and D_T refers to the fractal dimension of a tortuous capillary.

If (1)–(3) are rearranged, we have the following equation:

$$v_0 = \frac{dL_0}{dt} = \frac{\left[(\lambda + 2L_s)^2/2 - \lambda^2/4\right]}{8\mu D_T\lambda^{2-2D_T}L_0^{2D_T-1}}\Delta p. \quad (4)$$

The driving pressure of spontaneous imbibition is the capillary force. Thus, we have the following equation:

$$\Delta p = p_{\text{out}} - p_{\text{in}} = p_c = \frac{4\sigma\cos\theta}{\lambda}, \quad (5)$$

where θ is the contact angle between liquid and tube wall and σ is the interfacial tension. During the imbibition process, the imbibition length is increasing with the movement of meniscus. For the tortuous tube, using the initial condition $L_0|_{t=0} = 0$, the relationship between imbibition length and time can be derived as follows:

$$L_0 = \left\{\frac{\left[(\lambda + 2L_s)^2/2 - \lambda^2/4\right]}{4\mu\lambda^{2-2D_T}}\left(\frac{4\sigma\cos\theta}{\lambda}\right)\right\}^{1/2D_T} t^{1/2D_T}. \quad (6)$$

2.2. Pore Size Distribution. Pore size distribution is also important. According to the statistical data by Diamond and Dolch [29] and Hwang and Powers [30], the pore size distribution of the porous media can be simulated by lognormal distribution function. This function is a good way to represent the pore size distribution of the porous media. Thus, the pore space is the generalized lognormal distribution, as follows [29]:

$$f(r) = \frac{1}{\sqrt{2\pi}\ln\sigma_0}\exp\left[-\left(\frac{\ln\left(\lambda'/\overline{\lambda}'\right)}{\sqrt{2}\ln\sigma_0}\right)^2\right], \quad (7)$$

where $\lambda' = (\lambda - \lambda_{\min})(\lambda_{\max} - \lambda_{\min})/(\lambda_{\max} - \lambda)$, λ is the equivalent diameter of pores, σ_0 and $\overline{\lambda}'$ are the distribution parameters characterizing the distribution properties of λ', $\overline{\lambda}'$ is the mean or expectation of the distribution, and sigma σ_0 is the standard deviation. $f(\lambda)$ is the percent volume of voids in diameter λ. The cumulative distribution function can be expressed as follows:

$$F(r) = \frac{1}{\sqrt{2\pi}\ln\sigma_0} \cdot \int_{\lambda/2}^{\lambda_{\max}/2}\exp\left[-\left(\frac{\ln\left(\lambda'(x)/2\right) - \ln\left(\left(\overline{\lambda}'\right)/2\right)}{\sqrt{2}\ln\sigma_0}\right)^2\right] \cdot \frac{(\lambda_{\max} - \lambda_{\min})}{(\lambda_{\max} - x)(x - \lambda_{\min})}dx, \quad (8)$$

where $F(\lambda)$ is the percent volume of voids in the diameters larger than λ. According to Diamond and Dolch [29], as σ_0 decreases, pores are more concentrated on the mean or expectation of the distribution $\overline{\lambda}'$; in addition, the peak of the curve is higher. For the capillary bundle model used in this work, the imbibition volume V can be expressed as follows:

$$V = \sum_{\lambda=\lambda_{\min}}^{\lambda_{\max}} f(\lambda)A_r L_f(\lambda), \quad (9)$$

where A_r is the area of cross section.

2.3. Temperature Influences. Viscosity of the fracturing fluid varies with temperature. Normally, viscosity decreases with the increase in the temperature. For simplicity, many empirical or semiempirical equations (correlations) are proposed

Table 1: Basic property of the samples.

Sample	Mass (g)	Diameter (mm)	Length (mm)	Porosity (%)	Dry sample permeability (10^{-6} μm^2)
S1	16.13	25	15	3.49	3.1
S2	15.93	25	15	3.37	2.2
S3	16.34	25	15	4.22	1.9

to describe the temperature dependence of the fluid viscosity [31–33]. According to the observations on the experimental data, $\ln \mu$ is a linear function of the reciprocal absolute temperature $1/T$ in the low temperature range. For the typical fracturing fluid, the relationship between viscosity and temperature follows the Arrhenius equation [31–33]:

$$\mu = A_s \exp \frac{E_a}{RT}, \tag{10}$$

where E_a is the activation energy (Arrhenius energy) of the viscous flow, T is the temperature of liquid, and μ is the apparent viscosity, as is the preexponential factor, and R is the universal gas constant. The constants can be derived by experiment. Equation (10) can be rewritten as follows:

$$\ln \mu = A + \frac{B}{T}, \tag{11}$$

where $A = \ln A_s$ and $B = T^* = E_a/R$ are the Arrhenius activation temperature [33]. Taking (3), (6), (7), and (10) into (9), the spontaneous imbibition model considering the reservoir temperature and slippage effect can be derived.

3. Experiment and Validation

We used three samples from the Longmaxi Marine Shale Formation of Lower Silurian in Sichuan Basin to validate our imbibition model. The basic properties of the samples are shown in Table 1. Sample permeabilities are tested by pulse-decay method on an ultralow permeability measurement instrument. Related introductions have been attached in Appendix. Permeability results are listed in Table 1.

In addition, the core imbibition characteristics are tested in the spontaneous imbibition experiment. This experiment can record the relationship between imbibition volume and imbibition time. We used cocurrent imbibition this time, which can eliminate the influence of mineral dissolution from samples to the imbibed water. The experiments were performed at room pressure (0.1 MPa) and temperature (298 K). Detailed introductions are listed in Appendix.

Mercury intrusion experiment was conducted to measure the pore size distribution. Because mercury intrusion has irreversible negative effect on samples, the mercury intrusion should be arranged after permeability tests and imbibition tests. The schematic diagram of mercury intrusion experiment and detailed procedure of experiment are listed in Appendix.

During the mercury intrusion experiment, as pressure increases, liquid mercury is first injected in the largest pores

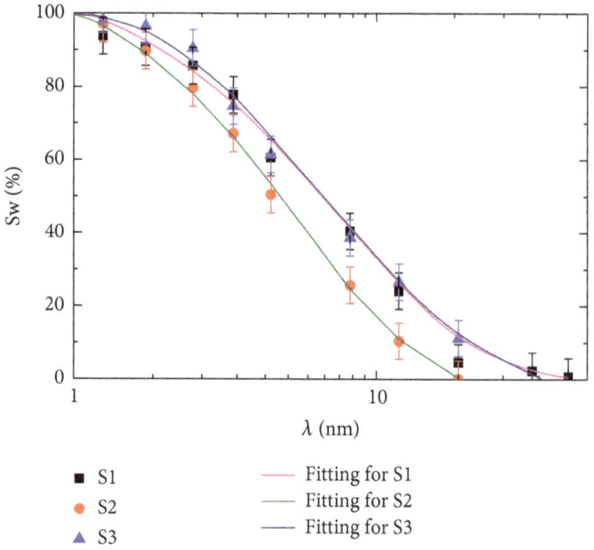

Figure 1: The simulating process for cumulative distributions (S1, S2, and S3).

and was gradually submerged to smaller pores. After the calculation of the mercury injection curve, the results of the pore size distribution for S1, S2, and S3 are shown in Figure 1. In our model, by adjusting the parameters in (8), the pore size distribution for each core sample can be rebuilt when the simulated cumulative distributions follow the experiment curves. Simulated cumulative distributions are also shown in Figure 1.

To derive the rheological parameters, we used an electronic rheometer (NDJ-5K) to test the viscosity-temperature curve. The schematic of instrument is shown in Appendix. The ingredients are shown in Table 2. Their viscosity-temperature relationship data are rearranged in $\ln \mu \sim 1/T$ form to calculate the parameters in the Arrhenius equation. The linear relationship is evident, and the parameters A and B can easily be derived in Figure 2. The results from linear fitting (see (10) and (11)) are shown in Table 2. Fluid viscosity μ reflects the ability of resistance of fluid to gradual deformation by shear stress or tensile stress. Based on the theory of Arrhenius equation, E_a is the activation energy, which reflects the intrinsic property for a liquid. E_a not only indicates the resistant of flow but also indicates the sensibility of temperature for a liquid. We can derive that L2 has higher activation energy, which means the average molecule chain of L2 is rigid or polar. During imbibition process, L2 may not easy to be imbibed in samples. The imbibition result from Figure 4 will give further explanation. Also, according to Haj-Kacem et al. [33], A_s can be closely related to the viscosity of the pure system in vapor state at the same studied pressure [34–37]. From the results of experiment, As_1 and As_2 can be derived, so As_1 and As_2 are approximately the viscosity of L1 and L2 systems at vapor state under the working pressure, if any some molecules pass in vapor state and mix with the gas fluid.

To examine the validation of this model, we draw the spontaneous imbibition curves from the experimental and simulated results for the 3 samples under the same coordinates (Figure 3). The environment parameters in the model

TABLE 2: Ingredients of the fracturing fluid we used: A is intercept in equation (11); B is Arrhenius activation temperature in equation (11); A_s can be closely related to the viscosity of the pure system in vapor state at the same studied pressure; E_s is the activation energy.

Fracturing fluid	Ingredients	A	B	A_s (mPa·s)	E_s (kl·mol^{-1})
L1	Slick water 100%	0.68	890	1.98	7.40
L2	Slick water 90% + guar gum 10%	−0.31	1553	0.73	12.91

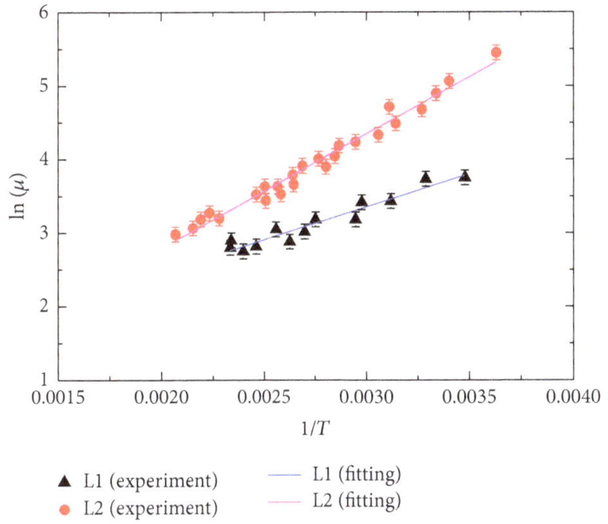

FIGURE 2: Experimental data for the relationship between viscosity and temperature for L1 and L2.

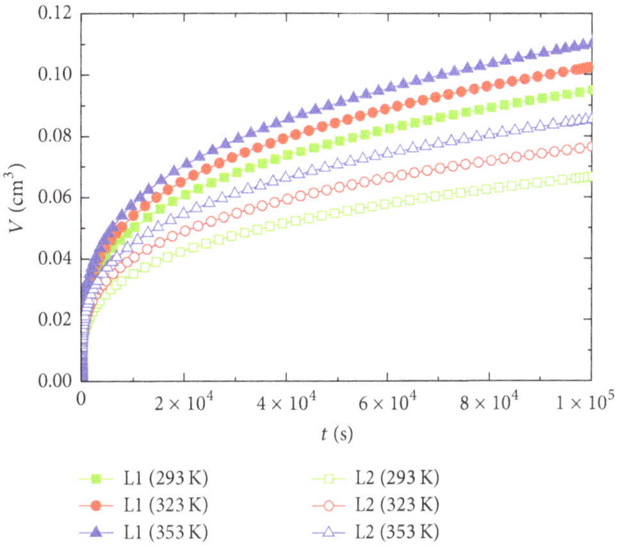

FIGURE 4: Relationship between imbibition volume and imbibition time at different temperatures for two kinds of fracturing fluids.

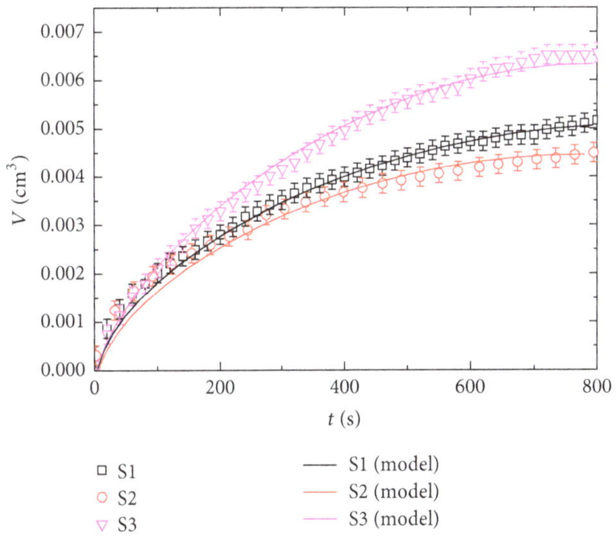

FIGURE 3: Model verification process for the spontaneous imbibition experimental results and simulated results.

were set the same as those in the experiments. Slick water was used as the imbibition fluid. Under the same working conditions, the result shows that our model coordinates well with the experimental data. Thus, the imbibition function considering slippage was verified. Moreover, the length of our samples is small, and influence of gravity on our model can be neglected. Thus, our model does not include a gravity item.

4. Results and Discussion

4.1. Temperature. On the basis of the derived spontaneous imbibition model, the effects of temperature, slip length, and pore size distribution have significant influence on the imbibition process. When temperature variation is considered, the rheological parameters of the fracturing fluid vary with the temperature. The imbibition curves for the porous media with the same pore size distribution have been calculated at 3 different temperatures, namely, 20, 50, and 80°C (or 293 K, 323 K, and 353 K), with the nondimensional slip length at 0.1 for two kinds of fracturing fluids (L1 and L2). Figure 4 shows the imbibition curve variation at different temperatures. With the same initial condition, the imbibition volume is 0 at $t = 0$. As temperature increases, the imbibition volume also increases. Under the same temperature, fracturing fluid L2 has higher viscosity than L1; however, its imbibition volume is lower. On the other hand, for the same fracturing fluid, the imbibition volume increases with the increase in temperature. This phenomenon is caused by the following: in environments with lower temperature, fracturing fluid viscosity is higher, making fracturing fluid harder to be imbibed. L2 contains guar gum; thus, its viscosity is higher than that of L1 at the same temperature, which also makes the fracturing fluid harder to be imbibed.

4.2. Slip Length. Considering the scale of micro-/nanopores and silts, the slip effect has a significant influence on the spontaneous imbibition process. The relationship curves of

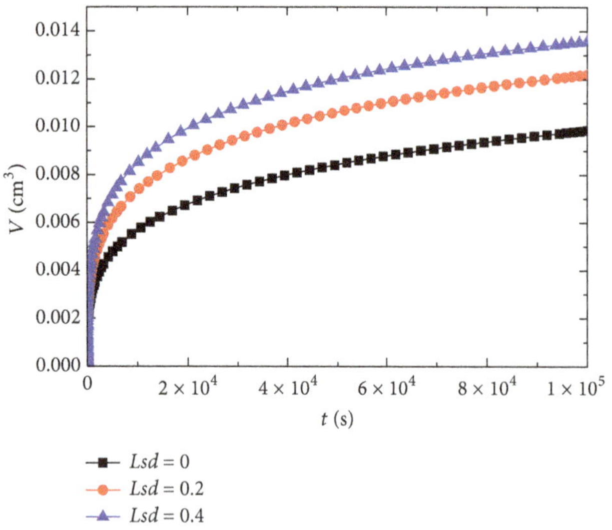

FIGURE 5: Relationship between imbibition volume and imbibition time at different slip lengths.

the imbibition volume and imbibition time under different dimensionless slip length are shown in Figure 5. In this figure, when the temperature is certain, the imbibition volume is gradually increased with increasing dimensionless slip length. Slip effect is much evident in micro-/nanopores in shale. When the dimensionless slip length reaches 0.4, which is about half of the diameter, the imbibition volume increases up to about 30% at the same time compared with the no-slip boundary. Also, as dimensionless slip length increases, the influence on spontaneous imbibition will decrease. Because as dimensionless slip length increases from 0 to 0.2, imbibition volume at $t = 10^5$ increases from 0.009 to 0.012 (with increment 0.003). While dimensionless slip length increases from 0.2 to 0.4, the increment is 0.0015 (from 0.012 to 0.0135).

Slip length mainly exists in the nanoscale pores and silts [22]. In the imbibition analysis for shale, imbibed pore spaces are mainly in the nanoscale, which has the same scale with slip length. Thus, slip length can easily reach 0.4 or higher. Ignorance of the slip effect will lead to significant underestimation of the imbibition speed and volume. When analyzing the spontaneous imbibition process, overlooking slippage will considerably underestimate the imbibition flux, which is special compared with sandstones.

4.3. Pore Size Distribution. Pore size also has significant effects on the imbibition characteristics. The distribution effect can be derived by adjusting the pore size distribution parameters in the spontaneous imbibition model. A total of 3 different kinds of distributions have been discussed, as shown in Figure 6(a). Cases 1 and 2 share the same range of diameter with different peaks, so the mean or expectation of the distributions $\overline{\lambda}'$ is the same, while the standard deviation σ_0 for Case 1 is smaller than that for Case 2. By comparing Cases 1 and 2, the influence of the concentration for pore size on spontaneous imbibition can be analyzed. On the other hand, Cases 2 and 3 share the same peak height with different average diameters, but different average diameter. To analyze the influence of porosity on spontaneous imbibition, the standard deviations σ_0 for Cases 2 and 3 are the same, while the mean or expectation of the distributions $\overline{\lambda}'$ is different. By comparing Cases 2 and 3, we can derive the influence of average pore size on spontaneous imbibition. The simulated imbibition curves are shown in Figure 6(b). Comparing Cases 2 and 3, for the cases with smaller average pores, the imbibition speed is smaller too. From (1), we can derive that flowing flux is proportional to diameter's biquadrate. Smaller pores will gain much more flowing resistance. Meanwhile, for sample with higher peak frequency, the imbibition speed is higher, which means pore space with more concentrate pore size distribution will gain higher imbibition volume.

5. Conclusions

(1) Temperature has influence on the fluid viscosity. Thus, the spontaneous imbibition process can be influenced by temperature. When the slip length is kept stable, the imbibition volume increases with the rise in the temperature at the same imbibition time. Meanwhile, when the temperature is kept unchanged, the imbibition volume increases with the rise in the slip length at the same imbibition time. Additionally, the imbibition rate decreases as imbibition time increases.

(2) Pore size also has significant effects on the imbibition process. With the same peak height of pore size distribution, the imbibition volume decreases with the decline in the average pores at the same imbibition time. On the other hand, with the same average pore diameter, higher peak of pore size distribution than that of the others will gain larger imbibition volume at the same imbibition time.

(3) Temperature and slippage are not negligible in the study of the fracturing fluid-gas spontaneous displacement in shale. Ignoring these two parameters will lead to significant underestimation on the imbibition speed and volume.

Appendix

We have used a series of experiments in this work. Detailed structures and procedures are listed in this Appendix.

The procedures of permeability test are as follows: (a) all samples are dried at temperature 338 K for 12 h before the experiment until the mass remained unchanged. (b) Sample permeability are tested by pulse-decay method on an ultralow permeability measurement instrument (Figure 7). The dried cores are placed in core chamber with confining pressure 5 MPa. The environment is set at 298 K and pore pressure is 2 MPa. Helium is used for sample tests. (c) Repeating the preceding steps 3 times, average permeability for each sample was derived.

In spontaneous imbibition experiment, the sample contacts the liquid at surface. The schematic is presented in Figure 8. The imbibed water volume was measured on line by a balance (METTLER LE204E) connected to a computer, with accuracy of 0.0001 g. The method to obtain the imbibed water volume is shown in Figure 8. The procedures of imbibition test are as follows.

(a) Keep the experimental environment stable at 0.1 MPa and 298 K. The fluctuation of temperature is controlled within

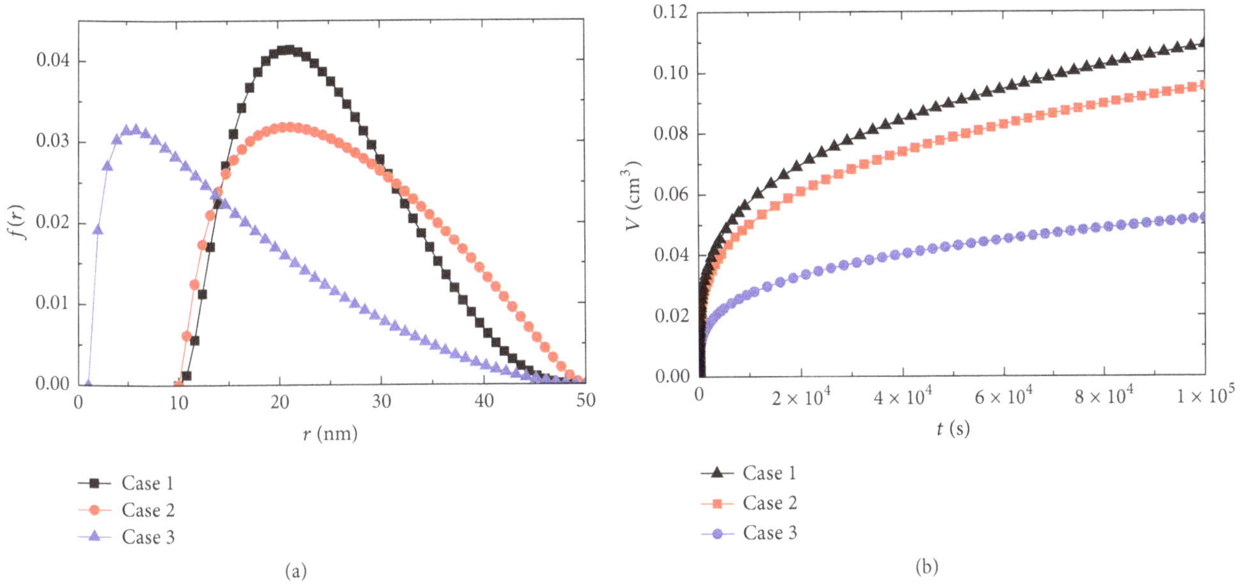

FIGURE 6: Relationship between imbibition volume and imbibition time in different pore size distribution: (a) different kinds of distributions for three cases; (b) simulated imbibition curves for three cases.

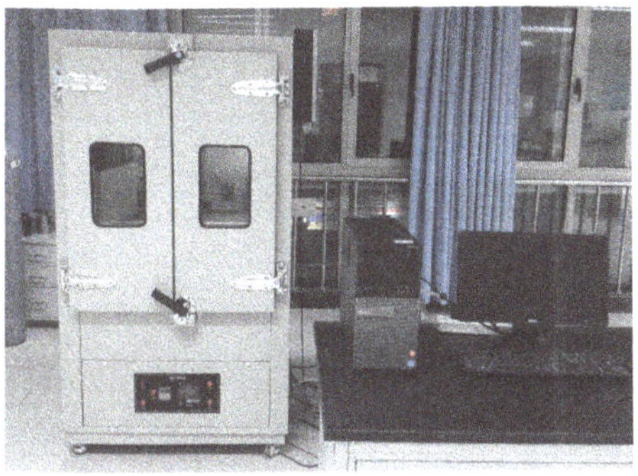

FIGURE 7: The photo of the pulse-decay permeability instrument.

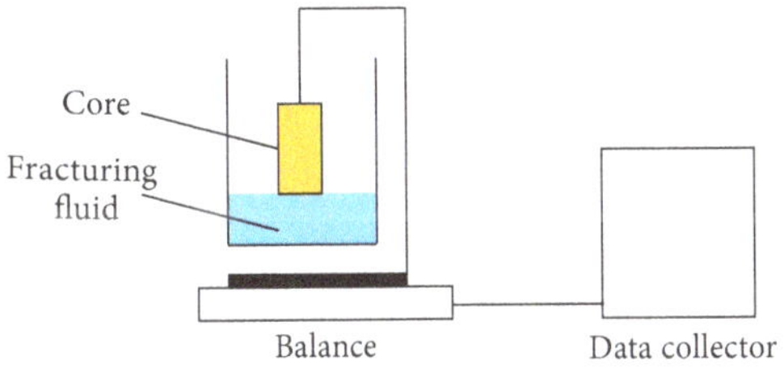

FIGURE 8: Schematic of spontaneous imbibition experiment (cocurrent imbibition).

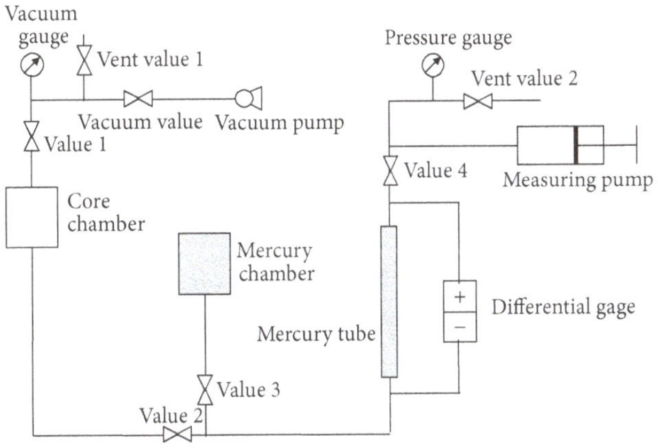

FIGURE 9: Schematic of the mercury intrusion experiment.

±5%. Place a calibration sample on imbibition experiment and note the evaporation rate for liquid. The calibration sample is a glass cylinder, with diameter 25 mm. (b) All samples are dried at temperature 338 K for 12 h before the experiment until the mass remained unchanged. Note the mass of dried samples. (c) Place the samples on imbibition experiment. When the sample contacts liquid at surface, start to record timer and data collector. (d) Correct sample's imbibition curve by calibration sample's data to eliminate the influence of liquid evaporation.

In mercury intrusion experiment, we can derive the pore size distribution of samples. The schematic diagram of mercury intrusion experiment is shown in Figure 9. Because mercury intrusion has irreversible negative effect on samples, the mercury intrusion should be arranged after permeability tests and imbibition tests. The procedure of experiment is as follows.

(a) All samples are dried at temperature 338 K for 12 h before the experiment until the mass remained unchanged.

(b) Place the samples in core chamber, close value 2 and vent value 1, open value 1, vacuum value, and start vacuum pump for 20 minutes.

(c) Open values 3 and 4, close vent value 2, and retreat measuring pump to absorb mercury. Then close value 4 and

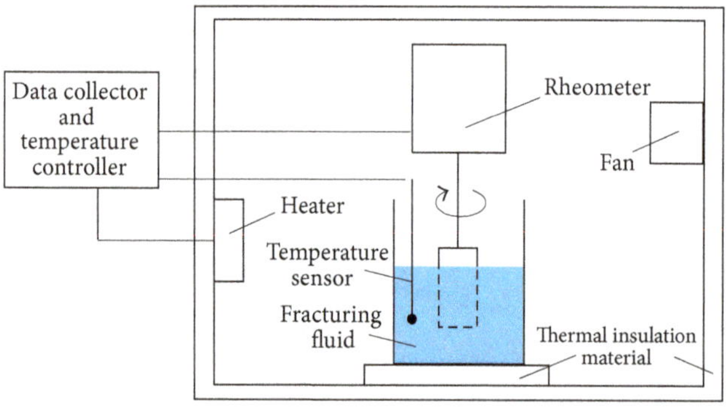

FIGURE 10: Schematic of viscosity instrument in this experiment.

open vent value 2 to displace air in pump. By repeating the above steps, the pump and line are filled with mercury.

(d) Keeping the environment temperature at 298 K, closing value 3 and value 1 and vent value 2, and opening values 2 and 4, the volume of injected mercury in sample increases with the increase of pumping pressure. The relationship between volume of injected mercury and pressure is recorded.

(e) When mercury cannot be injected in core sample, the retreating curve can be recorded by retreating measuring pump and recording the relationship between volume of injected mercury and pressure.

The schematic of viscosity instrument is shown in Figure 10. The measuring range for the rheometer is 1~200 mPa·s, with ±5% accuracy. The two typical fracturing fluid were prepared. This experiment can test the viscosity-temperature curve of liquid.

Conflicts of Interest

The authors declared no potential conflicts of interest with respect to the research, authorship, and/or publication of this article.

Acknowledgments

This study was supported by the National Natural Science Foundation of China (41572141).

References

[1] T. T. Palisch, M. C. Vincent, and R. J. Handren, "Slickwater fracturing: Food for thought," *SPE Production and Operations*, vol. 25, no. 3, pp. 327–344, 2010.

[2] S. A. Holditch, "Tight Gas Sands," *Journal of Petroleum Technology*, vol. 58, no. 06, pp. 86–93, 2006.

[3] J. Cai, B. Yu, M. Zou, and L. Luo, "Fractal characterization of spontaneous co-current imbibition in porous media," *Energy & Fuels*, vol. 24, no. 3, pp. 1860–1867, 2010.

[4] J. C. Cai and B. M. Yu, "A discussion of the effect of tortuosity on the capillary imbibition in porous media," *Transport in Porous Media*, vol. 89, no. 2, pp. 251–263, 2011.

[5] Y. Li, J. Yao, and Y. Yang, "The effect of the absorbed fluid layer on flow in a capillary tube," *Petroleum Science and Technology*, vol. 32, no. 2, pp. 194–201, 2014.

[6] P. Fakcharoenphol, B. Kurtoglu, H. Kazemi, S. Charoenwongsa, and Y.-S. Wu, "The effect of osmotic pressure on improve oil recovery from fractured shale formations," in *Proceedings of the SPE Unconventional Resources Conference 2014*, April 2014.

[7] C. Li, Y. Shen, H. Ge, S. Su, and Z. Yang, "Analysis of spontaneous imbibition in fractal tree-like network system," *Fractals. Complex Geometry, Patterns, and Scaling in Nature and Society*, vol. 24, no. 3, 1650035, 12 pages, 2016.

[8] C. Li, Y. Shen, H. Ge et al., "Analysis of capillary rise in asymmetric branch-like capillary," *Fractals. Complex Geometry, Patterns, and Scaling in Nature and Society*, vol. 24, no. 2, 1650024, 10 pages, 2016.

[9] Y. Shen, C. Li, H. Ge, X. Yang, and X. Zeng, "Spontaneous imbibition in asymmetric branch-like throat structures in unconventional reservoirs," *Journal of Natural Gas Science and Engineering*, vol. 44, pp. 328–337, 2017.

[10] Y. Shen, H. Ge, C. Li et al., "Water imbibition of shale and its potential influence on shale gas recovery—a comparative study of marine and continental shale formations," *Journal of Natural Gas Science and Engineering*, vol. 35, pp. 1121–1128, 2016.

[11] Y. Feng and K. E. Gray, "A parametric study for wellbore strengthening," *Journal of Natural Gas Science and Engineering*, vol. 30, pp. 350–363, 2016.

[12] Y. Feng and K. E. Gray, "A fracture-mechanics-based model for wellbore strengthening applications," *Journal of Natural Gas Science and Engineering*, vol. 29, pp. 392–400, 2016.

[13] Y. Feng, C. Arlanoglu, E. Podnos, E. Becker, and K. E. Gray, "Finite-element studies of hoop-stress enhancement for wellbore strengthening," *SPE Drilling & Completion*, vol. 30, no. 1, pp. 38–51, 2015.

[14] Y. Feng, J. F. Jones, and K. E. Gray, "A Review on fracture-initiation and -propagation pressures for lost circulation and wellbore strengthening," *SPE Drilling & Completion*, vol. 31, no. 2, pp. 134–144, 2016.

[15] P. Chen, D. Gao, Z. Wang, and W. Huang, "Study on multisegment friction factors inversion in extended-reach well based on an enhanced PSO model," *Journal of Natural Gas Science and Engineering*, vol. 27, pp. 1780–1787, 2015.

[16] P. Chen, D. Gao, Z. Wang, and W. Huang, "Study on aggressively working casing string in extended-reach well," *Journal of Petroleum Science Engineering*, vol. 157, 2017.

[17] J. Cai, E. Perfect, C.-L. Cheng, and X. Hu, "Generalized modeling of spontaneous imbibition based on hagen-poiseuille flow in tortuous capillaries with variably shaped apertures," *Langmuir*, vol. 30, no. 18, pp. 5142–5151, 2014.

[18] Y. Shen, H. Ge, S. Su, D. Liu, Z. Yang, and J. Liu, "Imbibition characteristic of shale gas formation and water-block removal capability," *Scientia Sinica Physica, Mechanica & Astronomica*, 2017.

[19] Y. Shen, H. Ge, M. Meng, Z. Jiang, and X. Yang, "Effect of water imbibition on shale permeability and its influence on gas production," *Energy & Fuels*, vol. 31, no. 5, pp. 4973–4980, 2017.

[20] H. Dehghanpour, H. A. Zubair, A. Chhabra, and A. Ullah, "Liquid intake of organic shales," *ENERGY & FUELS*, vol. 26, no. 9, pp. 5750–5758, 2012.

[21] B. Roychaudhuri, J. Xu, T. T. Tsotsis, and K. Jessen, "Forced and Spontaneous Imbibition Experiments for Quantifying Surfactant Efficiency in Tight Shales," in *Proceedings of the SPE Conference*, 2014.

[22] F. Javadpour, "Nanopores and apparent permeability of gas flow in mudrocks (shales and siltstone)," *Journal of Canadian Petroleum Technology*, vol. 48, no. 8, pp. 16–21, 2009.

[23] C. H. Sondergeld, K. E. Newsham, J. T. Comisky, M. C. Rice, and C. S. Rai, "Petrophysical considerations in evaluating and producing shale gas resources," in *Proceedings of the SPE Unconventional Gas Conference*, Society of Petroleum Engineers, Pittsburgh, Pa, USA, 2010.

[24] F. Javadpour, M. McClure, and M. E. Naraghi, "Slip-corrected liquid permeability and its effect on hydraulic fracturing and fluid loss in shale," *Fuel*, vol. 160, pp. 549–559, 2015.

[25] A. Afsharpoor and F. Javadpour, "Liquid slip flow in a network of shale noncircular nanopores," *Fuel*, vol. 180, pp. 580–590, 2016.

[26] Z. Zhou, X. Cui, and W. Zhang, "The relationship between formation temperature and permeability in a heavy oil reservoir," *Petroleum Science and Technology*, vol. 34, pp. 31–36, 2016.

[27] B. M. Yu and P. Cheng, "A fractal permeability model for bi-dispersed porous media," *International Journal of Heat and Mass Transfer*, vol. 45, no. 14, pp. 2983–2993, 2002.

[28] J.-C. Cai, B.-M. Yu, M.-F. Mei, and L. Luo, "Capillary rise in a single tortuous capillary," *Chinese Physics Letters*, vol. 27, no. 5, Article ID 054701, 2010.

[29] S. Diamond and W. L. Dolch, "Generalized log-normal distribution of pore sizes in hydrated cement paste," *Journal of Colloid and Interface Science*, vol. 38, no. 1, pp. 234–244, 1972.

[30] S. I. Hwang and S. E. Powers, "Using particle-size distribution models to estimate soil hydraulic properties," *Soil Science Society of America Journal*, vol. 67, no. 4, pp. 1103–1112, 2003.

[31] J. Restolho, A. P. Serro, J. L. Mata, and B. Saramago, "Viscosity and surface tension of 1-ethanol-3-methylimidazolium tetrafluoroborate and 1-methyl-3-octylimidazolium tetrafluoroborate over a wide temperature range," *Journal of Chemical & Engineering Data*, vol. 54, no. 3, pp. 950–955, 2013.

[32] M. Zheng, J. Tian, and Á. Mulero, "New correlations between viscosity and surface tension for saturated normal fluids," *Fluid Phase Equilibria*, vol. 360, pp. 298–304, 2013.

[33] R. B. Haj-Kacem, N. Ouerfelli, J. Herráez, M. Guettari, H. Hamda, and M. Dallel, "Contribution to modeling the viscosity Arrhenius-type equation for some solvents by statistical correlations analysis," *Fluid Phase Equilibria*, vol. 383, pp. 11–20, 2014.

[34] D. S. Viswanath, T. K. Ghosh, D. H. L. Prasad, N. V. K. Dutt, and K. Y. Rani, "Viscosity of liquids," *AIChE Journal*, vol. 2, no. 3, pp. 290–295, 2007.

[35] D. Das, A. Messaâdi, N. Dhouibi, N. Ouerfelli, and A. H. Hamzaoui, "Viscosity Arrhenius activation energy and derived partial molar properties in N,N-Dimethylacetamide + water binary mixtures at temperatures from 298.15 to 318.15 K," *Physics and Chemistry of Liquids*, vol. 51, no. 5, pp. 677–685, 2013.

[36] H. Salhi, M. Dallel, Z. Trabelsi, N. O. Alzamel, M. A. Alkhaldi, and N. Ouerfelli, "Viscosity arrhenius activation energy and derived partial molar properties in methanol + n, n-dimethylacetamide binary mixtures the temperatures from 298.15 k to 318.15 k," *Physics Chemistry of Liquids*, vol. 51, pp. 677–685, 2013.

[37] M. Dallel, D. Das, E. S. Bel Hadj Hmida, N. A. Al-Omair, A. A. Al-Arfaj, and N. Ouerfelli, "Derived partial molar properties investigations of viscosity Arrhenius parameters in formamide + N,N-dimethylacetamide systems at different temperatures," *Physics and Chemistry of Liquids*, vol. 52, no. 3, pp. 442–451, 2014.

7

Outer Synchronization of a Modified Quorum-Sensing Network via Adaptive Control

Jianbao Zhang, [1,2,3] **Wenyin Zhang**, [1,2] **Denghua Zhang**, [4]
Chengdong Yang, [1,2,3] **Kongwei Zhu**, [1,2] **and Jianlong Qiu** [2,5]

[1] *School of Information Science and Engineering, Linyi University, Linyi 276005, China*
[2] *Key Laboratory of Complex Systems and Intelligent Computing in Universities of Shandong (Linyi University), Linyi 276005, China*
[3] *Department of Mathematics, Southeast University, Nanjing 210096, China*
[4] *College of Science and Technology, North China Electric Power University, Baoding 071000, China*
[5] *Department of Electrical and Computer Engineering, Faculty of Engineering, King Abdulaziz University, Jeddah 21589, Saudi Arabia*

Correspondence should be addressed to Jianbao Zhang; jianbaozhang@163.com

Academic Editor: Hiroaki Mukaidani

Motivated by the quorum-sensing mechanism of bacteria, this paper modifies the network model by adding unknown parameters and noise disturbances and investigates the problem of outer synchronization via adaptive control. In case there exist three unknown parameters, updating laws are presented to identify the unknown parameters with help of Lyapunov stability theory, and the negative effects of noise disturbances are also compensated by designing adaptive controllers. In addition, we simplify the obtained conditions and carry out two succinct and utilitarian corollaries. Finally, numerical simulations are provided to show the validity of the obtained results.

1. Introduction

During the past decades, it has been discovered that bacteria, such as *Escherichia coli*, could communicate with each other through producing and monitoring one kind of signaling molecules [1, 2]. The signaling molecules could diffuse into different bacteria or the environment, and the bacteria could coordinate their gene expression and activities in response to the concentration of the signaling molecules. Then, the bacteria are coupled with each other by the intercellular signaling molecules [3] and display various social behaviors such as behaving synchronously [4]. Obviously, the related researches have wide application prospects in biopharmaceutical industry and human health. Now, the mechanism of bacterial communication is widely known as quorum sensing, and more and more researchers began to study collective behavior caused by quorum sensing [5, 6]. In this paper, we modify one of the previous network models coupled through quorum sensing and discuss a typical kind of collective behavior based on several recent methods developed in the fields of complex networks and nonlinear dynamics.

Recent years have witnessed the great development in the study of complex networks and its collective dynamics [7, 8]. Synchronization is one of the most typical and most extensively studied kinds of collective dynamics, which implies the stability of zero solution of the synchronization error systems. Therefore, from the point of research methods, there are two main effective theoretical methods, i.e., the famous master stability function method [9, 10] and Lyapunov function method [11–13]. The former can be employed to discuss local stability of the synchronous state, and the latter can be used to explore global stability of the synchronous state. Up to now, dozens of different types of synchronization states have been proposed such as complete synchronization [14], cluster synchronization [15, 16], lag synchronization [17, 18], projection synchronization [19, 20], and outer synchronization [21–23]. Thereinto, outer synchronization has attracted many researchers' interest, which describes the synchronization between two or more networks. For instances, in a model of predator-prey interactions in ecological communities, all the predators form a network system and all the preys form another, and the two networks influence one another's evolution to keep the two species in check [24]. Recently, outer synchronization of the fractional order node dynamics was considered in [21], and outer synchronization under aperiodically adaptive intermittent control was considered in [22]. In many cases, networks can not realize a certain expected synchronization relying on just coupling interaction between different nodes [23]. Therefore, many different kinds of output control methods have been introduced, such as pinning control [25], sliding mode control [26, 27], adaptive control [28, 29], and state feedback control [30]. Thereinto, adaptive control could be used to design controllers for systems with uncertain parameters. Due to great demands from wide applications, many researches have been carried out to investigate synchronization induced by adaptive controllers [31]. It is worth pointing out that there are few researches focused on outer synchronization of networks coupled through quorum sensing.

Motivated by the above discussions, this paper investigates outer synchronization induced by adaptive controllers in quorum-sensing network. At first, we present a modified model of previous quorum-sensing network by adding noise disturbances in case there exist three unknown parameter vectors and the network topology is also unknown. Then, effective adaptive controllers are designed to realize outer synchronization, parameter estimations are designed to identify the unknown parameter vectors, and topology estimations are designed to identify unknown network topology. Based on Lyapunov function method and matrix theory, this paper proves that adaptive outer synchronization is achieved in the quorum-sensing network. To the best of our knowledge, there are few researches focused on this subject by a similar method. In our opinion, there is a certain degree of values both in theory and in practice.

The rest of this paper is organized as follows. In Section 2, the synthetic gene network model coupled through quorum sensing is introduced. In Section 3, several criteria are derived for outer synchronization including the construction of adaptive controllers and parameter estimations. In Section 4, some numerical examples are provided to illustrate the effectiveness of the obtained results. Finally, conclusions are given to summarize the contributions of the paper in Section 5.

2. Problem Formulation

The synthetic gene network in *Escherichia coli* was first proposed by Garcia-Ojalvo J et al. [3]. Consider the network consisting of N cells coupled through quorum sensing. Each cell consists of two basic parts illustrated in Figure 1. The first part is composed of three genes a, b, c that express their respective proteins A, B, C, which inhibit the transcription of the three genes b, c, a, in a cyclic way. The second part of each cell is another gene regulated by protein A, which produces a protein and synthesizes a small molecule known as an autoinducer S_i. The autoinducer S_i can diffuse freely through the cell membrane, which activates the transcription of the genes in first part. For more detailed description, the reader is referred to previous articles [32, 33].

Now, we introduce the quorum-sensing network model. The dynamics of each node consists of the concentrations of three genes and their respective proteins, assume that the ith cell is described by the following equations:

$$\begin{aligned}
\dot{a}_i(t) &= -d_1 a_i(t) + \beta_6 \left[\mu_6 + C_i^m(t)\right]^{-1}, \\
\dot{b}_i(t) &= -d_2 b_i(t) + \beta_4 \left[\mu_4 + A_i^m(t)\right]^{-1}, \\
\dot{c}_i(t) &= -d_3 c_i(t) + \beta_5 \left[\mu_5 + B_i^m(t)\right]^{-1} \\
&\quad + \beta_8 S_i(t) \left[\mu_7 + S_i(t)\right]^{-1}, \\
\dot{A}_i(t) &= -d_4 A_i(t) + \beta_1 a_i(t), \\
\dot{B}_i(t) &= -d_5 B_i(t) + \beta_2 b_i(t), \\
\dot{C}_i(t) &= -d_6 C_i(t) + \beta_3 c_i(t),
\end{aligned} \tag{1}$$

where $a_i(t), b_i(t)$, and $c_i(t)$ are the concentrations of $mRNA$ transcribed from genes a, b, c in the ith cell, respectively; $A_i(t), B_i(t)$, and $C_i(t)$ are the concentrations of the corresponding proteins, respectively; $S_i(t)$ and $S_e(t)$ are the concentrations of the autoinducer AI inside the ith cell and in the environment. The concentration dynamics of the autoinducer are governed by

$$\begin{aligned}
\dot{S}_i(t) &= -d_7 S_i(t) + \beta_7 A_i(t) - \eta\left(S_i(t) - S_e(t)\right), \\
\dot{S}_e(t) &= -d_e S_e(t) + \frac{\eta_e}{N}\sum_{j=1}^{N}\left(S_j(t) - S_e(t)\right),
\end{aligned} \tag{2}$$

where $i = 1, 2, \ldots, N$. In the multicell system (1)-(2), the parameters $d_1, d_2, \ldots, d_7$ and d_e are the dimensionless degradation rates of the chemical molecules; $\beta_1, \beta_2, \beta_3$ are the translation rates of the proteins from the $mRNAs$; $\beta_4, \beta_5, \beta_6$ are the dimensionless transcription rates in the absence of repressor; β_7 is the synthesis rate of AI; $m = 4$ is the Hill coefficient; β_8 is the maximal contribution to the gene c transcription in the presence of saturating amounts of AI; η and η_e measure the diffusion rate of AI inward and outward the cell membrane. With the help of the quasi-steady state approximation $\dot{S}_e(t) = 0$, one gets that the extracellular AI concentration can be approximated as

$$S_e(t) = \frac{q}{N}\sum_{j=1}^{N} S_j(t), \quad q = \frac{\eta_e}{d + \eta_e}, \tag{3}$$

which reduces (2) to the following form:

$$\dot{S}_i(t) = -\left(d_7 + \eta\right) S_i(t) + \beta_7 A_i(t) + \frac{\eta q}{N}\sum_{j=1}^{N} S_j(t). \tag{4}$$

Equations (1)-(4) describe the concentration state of the ith cell in the synthetic gene network model coupled through quorum sensing.

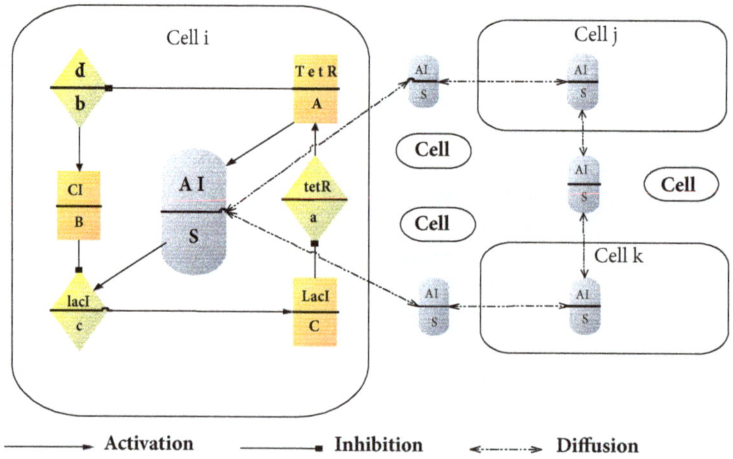

FIGURE 1: Scheme of the repressilator network coupled through signaling molecules, termed quorum sensing. The synchronization scheme of quorum sensing is based on the diffusion of autoinducers (AI) to and from the cells.

Motivated by the quorum-sensing network (1)-(4), we build up a network with three unknown parameter vectors and unknown network topology. The state dynamics are described by the following equations:

$$\begin{aligned} \dot{x}_i(t) &= f_1(x_i(t))\alpha_1 + f_2(x_i(t))\alpha_2 + \beta_8 h(S_i(t)), \\ \dot{S}_i(t) &= -\alpha_3 S_i(t) + \beta_7 A_i(t) + \sum_{j=1}^{N} c_{ij} S_j(t), \end{aligned} \tag{5}$$

where $x_i(t) = (a_i(t), b_i(t), c_i(t), A_i(t), B_i(t), C_i(t))^{\top}$, $\alpha_1 = (d_1, d_2, d_3, d_4, d_5, d_6)^{\top}$, $\alpha_2 = (\beta_6, \beta_4, \beta_5, \beta_1, \beta_2, \beta_3)^{\top}$, and $\alpha_3 = d_7 + \eta$, and the diagonal matrix functions

$$\begin{aligned} f_1(x_i(t)) &= -\operatorname{diag}(a_i(t), b_i(t), c_i(t), A_i(t), B_i(t), \\ &\quad C_i(t)), \\ f_2(x_i(t)) &= \operatorname{diag}\left([\mu_6 + C_i^m(t)]^{-1}, [\mu_4 + A_i^m(t)]^{-1},\right. \\ &\quad \left.[\mu_5 + B_i^m(t)]^{-1}, a_i(t), b_i(t), c_i(t)\right), \\ h(S_i(t)) &= \left(0, 0, S_i(t)[\mu_7 + S_i(t)]^{-1}, 0, 0, 0\right)^{\top}, \end{aligned} \tag{6}$$

where $i = 1, 2, \ldots, N$. To meet the demands of broad applications, the matrix $C = (c_{ij})_{N\times N}$ is a coupling matrix denoting the network topology. The matrix element c_{ij} is defined as follows: if there is a connection from node i to node $j (i \neq j)$, then define the coupling strength as $c_{ij} \neq 0$; otherwise, $c_{ij} = 0$. Let us assume that there are three unknown parameter vectors existing in the node dynamics, α_1, α_2, and α_3, and the network topology matrix $C = (c_{ij})_{N\times N}$ is also unknown.

In order to identify the unknown network topology and parameter vectors, we carry out another network model described by the following equations:

$$\begin{aligned} \dot{\overline{x}}_i(t) &= f_1(\overline{x}_i(t))\overline{\alpha}_1(t) + f_2(\overline{x}_i(t))\overline{\alpha}_2(t) \\ &\quad + \beta_8 h\left(\overline{S}_i(t)\right) + \Delta_1(t) + u_{1i}(t), \\ \dot{\overline{S}}_i(t) &= -\overline{\alpha}_3(t)\overline{S}_i(t) + \beta_7 \overline{A}_i(t) + \sum_{j=1}^{N} \overline{c}_{ij}\overline{S}_j + \Delta_2(t) \\ &\quad + u_{2i}(t), \end{aligned} \tag{7}$$

where $\overline{x}_i(t) = (\overline{a}_i(t), \overline{b}_i(t), \overline{c}_i(t), \overline{A}_i(t), \overline{B}_i(t), \overline{C}_i(t))^{\top}$, $\overline{\alpha}_1(t)$, $\overline{\alpha}_2(t)$, and $\overline{\alpha}_3(t)$ are the estimations of the unknown parameter vectors α_1, α_2, and α_3 in the network (5), $\Delta_1(t), \Delta_2(t)$ are the disturbances, and $u_{1i}(t), u_{2i}(t)$ are the controllers left to be designed later, $i = 1, 2, \ldots, N$.

3. Adaptive Control Schemes for Outer Synchronization

In this section, several criteria are derived for outer synchronization induced by adaptive control schemes. At first, we need to introduce the following two assumptions.

Assumption 1. For any $x = (x^{(1)}, x^{(2)}, \ldots, x^{(6)})^{\top} \in R^6$ and $S \in R$, denote

$$\begin{aligned} F(X, \alpha_1, \alpha_2, \alpha_3) &= \left[(f_1(x)\alpha_1 + f_2(x)\alpha_2 + \beta_8 h(S))^{\top},\right. \\ &\quad \left.-\alpha_3 S + \beta_7 x^{(4)}\right]^{\top} \in R^7, \end{aligned} \tag{8}$$

where $X = (x^{\top}, S)^{\top} \in R^7$. There exists a positive constant L such that the vector function $F(X, \alpha_1, \alpha_2, \alpha_3)$ satisfies that

$$\begin{aligned} &(Y - X)^{\top}[F(Y, \alpha_1, \alpha_2, \alpha_3) - F(X, \alpha_1, \alpha_2, \alpha_3)] \\ &\quad \leq L(Y - X)^{\top}(Y - X) \end{aligned} \tag{9}$$

for any $X, Y \in R^7$.

Assumption 2. The disturbances $\Delta_1(t)$ and $\Delta_2(t)$ are bounded; i.e., there exist two positive constants ρ_1, ρ_2 such that

$$\begin{aligned} \|\Delta_1(t)\| &\le \rho_1, \\ \|\Delta_2(t)\| &\le \rho_2. \end{aligned} \tag{10}$$

Now, we design the state feedback controllers of the following form:

$$\begin{aligned} u_{1i}(t) &= -\delta_{1i}(t)e_{1i}(t) - \gamma_{1i}(t)\operatorname{sign}[e_{1i}(t)], \\ \dot{\delta}_{1i}(t) &= k_{1i}e_{1i}^{\top}(t)e_{1i}(t), \quad k_{1i} > 0, \\ \dot{\gamma}_{1i}(t) &= \xi_{1i}e_{1i}^{\top}(t)\operatorname{sign}[e_{1i}(t)], \quad \xi_{1i} > 0, \\ u_{2i}(t) &= -\delta_{2i}(t)e_{2i}(t) - \gamma_{2i}(t)\operatorname{sign}[e_{2i}(t)] \\ &\quad + \sum_{j=1}^{N} p_{ij}(t)\overline{S}_j(t), \\ \dot{\delta}_{2i}(t) &= k_{2i}e_{2i}^2(t), \quad k_{2i} > 0, \\ \dot{\gamma}_{2i}(t) &= \xi_{2i}\operatorname{sign}[e_{2i}(t)]e_{2i}(t), \quad \xi_{2i} > 0, \\ \dot{p}_{ij}(t) &= -\overline{S}_j(t)e_{2i}(t), \end{aligned} \tag{11}$$

where $e_{1i}(t) = \overline{x}_i(t) - x_i(t)$, $e_{2i}(t) = \overline{S}_i(t) - S_i(t)$, and $i = 1, 2, \ldots, N$. Then, one can prove the following theorem based on Lyapunov function method and matrix theory.

Theorem 3. *Suppose that Assumptions 1 and 2 hold, and the parameter estimations $\alpha_1(t), \alpha_2(t), \alpha_3(t)$ are designed as follows:*

$$\begin{aligned} \dot{\overline{\alpha}}_1(t) &= -\sum_{j=1}^{N} f_1(\overline{x}_j(t))e_{1j}(t), \\ \dot{\overline{\alpha}}_2(t) &= -\sum_{j=1}^{N} f_2(\overline{x}_j(t))e_{1j}(t), \\ \dot{\overline{\alpha}}_3(t) &= \sum_{j=1}^{N} \overline{S}_j(t)e_{2j}(t), \end{aligned} \tag{12}$$

and then the synthetic gene network (5)-(7) with controllers (11) and estimations (12) can achieve outer synchronization.

Proof. Denote $X_i(t) = (x_i^{\top}(t), S_i(t))^{\top}$, $\overline{X}_i(t) = (\overline{x}_i^{\top}(t), \overline{S}_i(t))^{\top}$, $\Delta(t) = (\Delta_1^{\top}(t), \Delta_2(t))^{\top}$, and $U_i(t) = (u_{1i}^{\top}(t), u_{2i}(t))^{\top}$, and $\Gamma \in R^{7\times 7}$ is the inner matrix implying the nodes are coupling through the 7th component, and then the synthetic gene network (5)-(7) can be rewritten as follows:

$$\begin{aligned} \dot{X}_i(t) &= F(X_i(t), \alpha_1, \alpha_2, \alpha_3) + \sum_{j=1}^{N} c_{ij}\Gamma X_j(t), \\ \dot{\overline{X}}_i(t) &= F(\overline{X}_i(t), \overline{\alpha}_1(t), \overline{\alpha}_2(t), \overline{\alpha}_3(t)) \\ &\quad + \sum_{j=1}^{N} \overline{c}_{ij}\Gamma\overline{X}_j(t) + \Delta(t) + U_i(t), \end{aligned} \tag{13}$$

Let $E_i(t) = \overline{X}_i(t) - X_i(t)$, and the following error system can be obtained:

$$\begin{aligned} \dot{E}_i(t) &= F(\overline{X}_i(t), \overline{\alpha}_1(t), \overline{\alpha}_2(t), \overline{\alpha}_3(t)) \\ &\quad - F(X_i(t), \alpha_1, \alpha_2, \alpha_3) \\ &\quad + \sum_{j=1}^{N} [\overline{c}_{ij}\Gamma\overline{X}_j - c_{ij}\Gamma X_j] + \Delta(t) + U_i(t). \end{aligned} \tag{14}$$

Using Assumption 1, one has

$$\begin{aligned} &E_i^{\top}(t)[F(\overline{X}_i(t), \overline{\alpha}_1(t), \overline{\alpha}_2(t), \overline{\alpha}_3(t)) \\ &\quad - F(X_i(t), \alpha_1, \alpha_2, \alpha_3)] \le E_i^{\top}(t) \\ &\quad \cdot [F(\overline{X}_i(t), \overline{\alpha}_1(t), \overline{\alpha}_2(t), \overline{\alpha}_3(t)) \\ &\quad - F(\overline{X}_i(t), \alpha_1, \alpha_2, \alpha_3) + LE_i(t)] = E_i^{\top}(t) \\ &\quad \cdot [((f_1(\overline{x}_i(t))\tilde{\alpha}_1(t) + f_2(\overline{x}_i(t))\tilde{\alpha}_2(t))^{\top}, \\ &\quad - \tilde{\alpha}_3(t)\overline{S}_i(t))^{\top} + LE_i(t)] = \sum_{p=1}^{2} e_{1i}^{\top}(t) f_p(\overline{x}_i(t)) \\ &\quad \cdot \tilde{\alpha}_p(t) - \tilde{\alpha}_3(t)\overline{S}_i(t)e_{2i}(t) + L\sum_{p=1}^{2} e_{pi}^{\top}(t)e_{pi}(t). \end{aligned} \tag{15}$$

Consider the following Lyapunov function:

$$\begin{aligned} V(t) &= \frac{1}{2}\Bigg\{\sum_{i=1}^{N} E_i^{\top}(t)E_i(t) + \sum_{p=1}^{3} \tilde{\alpha}_p^{\top}(t)\tilde{\alpha}_p(t) \\ &\quad + \sum_{p=1}^{2}\sum_{i=1}^{N} \frac{1}{k_{pi}}[\delta_{pi}(t) - \delta_p^*]^2 \\ &\quad + \sum_{p=1}^{2}\sum_{i=1}^{N} \frac{1}{\xi_{pi}}[\gamma_{pi}(t) - \gamma_p^*]^2 \\ &\quad + \sum_{j=1}^{N}\sum_{i=1}^{N} [p_{ij}(t) + \overline{c}_{ij} - c_{ij}]^2\Bigg\}, \end{aligned} \tag{16}$$

where $\tilde{\alpha}_p(t) = \overline{\alpha}_p(t) - \alpha_p$, $p = 1, 2, 3$, where δ_p^*, γ_p^*, $p = 1, 2$, are positive constants chosen arbitrarily. With the help of controllers (11) and estimations (12), the derivative of $V(t)$ along the trajectories of (5)-(7) can be calculated as follows:

$$\dot{V}(t) = \sum_{i=1}^{N} E_i^{\top}(t)\dot{E}_i(t) - \sum_{p=1}^{2}\tilde{\alpha}_p^{\top}(t)\sum_{j=1}^{N} f_p(\overline{x}_j(t))e_{1j}(t)$$
$$+ \tilde{\alpha}_3^{\top}(t)\sum_{j=1}^{N}\overline{S}_j(t)e_{2j}(t)$$
$$+ \sum_{p=1}^{2}\sum_{i=1}^{N}\left[\delta_{pi}(t) - \delta_p^*\right]e_{pi}^{\top}(t)e_{pi}(t) \quad (17)$$
$$+ \sum_{p=1}^{2}\sum_{i=1}^{N}\left[\gamma_{pi}(t) - \gamma_p^*\right]e_{pi}^{\top}(t)\,\mathrm{sign}\left[e_{pi}(t)\right]$$
$$- \sum_{j=1}^{N}\sum_{i=1}^{N}\left[p_{ij}(t) + \overline{c}_{ij} - c_{ij}\right]\overline{S}_j(t)e_{2i}(t).$$

Noticing (14) and inequality (15), one has

$$\dot{V}(t) \le L\sum_{p=1}^{2}\sum_{i=1}^{N} e_{pi}^{\top}(t)e_{pi}(t)$$
$$+ \sum_{i=1}^{N}\sum_{j=1}^{N} e_{2i}(t)\left[\overline{c}_{ij}\overline{S}_j(t) - c_{ij}S_j(t)\right]$$
$$+ \sum_{i=1}^{N} E_i^{\top}(t)\left[\Delta(t) + U_i(t)\right]$$
$$+ \sum_{p=1}^{2}\sum_{i=1}^{N}\left[\delta_{pi}(t) - \delta_p^*\right]e_{pi}^{\top}(t)e_{pi}(t)$$
$$+ \sum_{p=1}^{2}\sum_{i=1}^{N}\left[\gamma_{pi}(t) - \gamma_p^*\right]e_{pi}^{\top}(t)\,\mathrm{sign}\left[e_{pi}(t)\right] \quad (18)$$
$$- \sum_{j=1}^{N}\sum_{i=1}^{N}\left[p_{ij}(t) + \overline{c}_{ij} - c_{ij}\right]e_{2i}(t)\overline{S}_j(t)$$
$$\le L\sum_{p=1}^{2}\sum_{i=1}^{N} e_{pi}^{\top}(t)e_{pi}(t) + \sum_{i=1}^{N}\sum_{j=1}^{N} c_{ij}e_{2i}(t)e_{2j}(t)$$
$$+ \sum_{i=1}^{N} E_i^{\top}(t)\Delta(t) - \sum_{p=1}^{2}\sum_{i=1}^{N}\delta_p^* e_{pi}^{\top}(t)e_{pi}(t)$$
$$- \sum_{p=1}^{2}\sum_{i=1}^{N}\gamma_p^* e_{pi}^{\top}(t)\,\mathrm{sign}\left[e_{pi}(t)\right].$$

Denoting $e_2(t) = (e_{21}(t), e_{22}(t), \ldots, e_{2N}(t))^{\top} \in R^N$, one gets

$$\dot{V}(t)$$
$$\le (L - \delta_1^*)\sum_{i=1}^{N} e_{1i}^{\top}(t)e_{1i}(t)$$
$$+ e_2^{\top}(t)\left[(L - \delta_2^*)I_N + C\right]e_2(t)$$
$$+ \sum_{p=1}^{2}\sum_{i=1}^{N}\left[e_{pi}^{\top}(t)\Delta_p(t) - \gamma_p^* e_{pi}^{\top}(t)\,\mathrm{sign}\left[e_{pi}(t)\right]\right]$$
$$\le (L - \delta_1^*)\sum_{i=1}^{N} e_{1i}^{\top}(t)e_{1i}(t)$$
$$+ e_2^{\top}(t)\left[(L - \delta_2^*)I_N + C\right]e_2(t)$$
$$+ \sum_{p=1}^{2}\sum_{i=1}^{N}\left(\rho_p - \gamma_p^*\right)\left\|e_{pi}(t)\right\|_1, \quad (19)$$

where $\|e_{pi}(t)\|_1 = e_{pi}^{\top}(t)\,\mathrm{sign}[e_{pi}(t)]$. Notice that the constants $\delta_1^*, \delta_2^*, \gamma_1^*, \gamma_2^*$ are chosen arbitrarily, and we can choose them sufficiently large such that $L - \delta_1^* < 0$, $\rho_p - \gamma_p^* < 0$, $p = 1, 2$, and the matrix $(L - \delta_2^*)I_N + C$ is negative semidefinite. According to this, we have

$$\dot{V}(t) < 0. \quad (20)$$

Thus, based on Lyapunov stability methods, the error dynamical system (14) is globally asymptotically stable. Therefore, the synthetic gene network (5)-(7) with controllers (11) and estimations (12) achieves outer synchronization.

The proof is completed. □

If one or two of the unknown parameter vectors in the network (5) are given constants, Theorem 3 still holds after modifying the conditions slightly. For instance, supposing that the parameter vectors α_1 and α_2 are given, we modify network (7) as follows:

$$\dot{\overline{x}}_i(t) = f_1(\overline{x}_i(t))\alpha_1 + f_2(\overline{x}_i(t))\alpha_2 + \beta_8 h(\overline{S}_i(t)) + \Delta_1(t) + u_{1i}(t),$$
$$\dot{\overline{S}}_i(t) = -\overline{\alpha}_3(t)\overline{S}_i(t) + \beta_7\overline{A}_i(t) + \sum_{j=1}^{N}\overline{c}_{ij}\overline{S}_j + \Delta_2(t) + u_{2i}(t). \quad (21)$$

Then the following corollary holds.

Corollary 4. *Suppose that Assumptions 1 and 2 hold. If the parameter estimation $\alpha_3(t)$ is designed as follows:*

$$\dot{\overline{\alpha}}_3(t) = \sum_{j=1}^{N}\overline{S}_j(t)e_{2j}(t), \quad (22)$$

then the synthetic gene network (5)-(21) with controllers (11) and estimations (22) can achieve outer synchronization.

If we do not consider the disturbances $\Delta_1(t)$ and $\Delta_2(t)$ in network (7), Theorem 3 still holds after modifying controllers (11) slightly. Then, we obtain the following corollary.

Corollary 5. *Suppose that Assumptions 1 and 2 hold, and the disturbances in network (7) satisfy that $\Delta_1(t) = 0$ and $\Delta_2(t) = 0$. If the controllers $u_{1i}(t)$, $u_{2i}(t)$ are designed as follows:*

$$
\begin{aligned}
u_{1i}(t) &= -\delta_{1i}(t)\, e_{1i}(t), \\
\dot{\delta}_{1i}(t) &= k_{1i} e_{1i}^{\top}(t)\, e_{1i}(t), \quad k_{1i} > 0, \\
u_{2i}(t) &= -\delta_{2i}(t)\, e_{2i}(t) + \sum_{j=1}^{N} p_{ij}(t)\, \overline{S}_j(t), \\
\dot{\delta}_{2i}(t) &= k_{2i} e_{2i}^{2}(t), \quad k_{2i} > 0, \\
\dot{p}_{ij}(t) &= -\overline{S}_j(t)\, e_{2i}(t),
\end{aligned}
\tag{23}
$$

where $i = 1, 2, \ldots, N$, then the synthetic gene network (5)-(7) with controllers (23) and estimations (12) can achieve outer synchronization.

The proof of Corollaries 4 and 5 is similar to that of Theorem 3; therefore, it is omitted here.

4. Numerical Simulations

In this section, we carry out some numerical simulations on the following synthetic gene network consisting of 6 cells:

$$
\begin{aligned}
\dot{x}_i(t) &= f_1(x_i(t))\,\alpha_1 + f_2(x_i(t))\,\alpha_2 + \beta_8 h(S_i(t)), \\
\dot{\overline{x}}_i(t) &= f_1(\overline{x}_i(t))\,\overline{\alpha}_1(t) + f_2(\overline{x}_i(t))\,\overline{\alpha}_2(t) \\
&\quad + \beta_8 h(\overline{S}_i(t)) + \Delta_1(t) + u_{1i}(t), \\
\dot{S}_i(t) &= -\alpha_3 S_i(t) + \beta_7 A_i(t) + \sum_{j=1}^{6} c_{ij} S_j(t), \\
\dot{\overline{S}}_i(t) &= -\overline{\alpha}_3(t)\,\overline{S}_i(t) + \beta_7 \overline{A}_i(t) + \sum_{j=1}^{6} \overline{c}_{ij} \overline{S}_j + \Delta_2(t) \\
&\quad + u_{2i}(t),
\end{aligned}
\tag{24}
$$

where the functions $f_1(x_i(t)), f_2(x_i(t)), h(S_i(t))$ are defined in network (5), the parameters are given as $(\mu_4, \mu_5, \mu_6, \mu_7) = (0.2, 0.2, 0.2, 0.2), m = 4, \beta_7 = 0.018, \beta_8 = 1, \Delta_1(t), \Delta_2(t)$ are the disturbances, and the coupling matrices are given as

$$
C = \begin{pmatrix}
-5 & 1 & 2 & 0 & 1 & 1 \\
1 & -4 & 0 & 1 & 2 & 0 \\
2 & 0 & -6 & 2 & 0 & 2 \\
0 & 1 & 2 & -4 & 1 & 0 \\
1 & 2 & 0 & 1 & -4 & 0 \\
1 & 0 & 2 & 0 & 0 & -3
\end{pmatrix},
\quad
\overline{C} = \begin{pmatrix}
-6 & 2 & 0 & 1 & 1 & 2 \\
2 & -6 & 1 & 2 & 1 & 0 \\
0 & 1 & -5 & 1 & 2 & 1 \\
1 & 2 & 1 & -5 & 0 & 1 \\
1 & 1 & 2 & 0 & -5 & 1 \\
2 & 0 & 1 & 1 & 1 & -5
\end{pmatrix}.
\tag{25}
$$

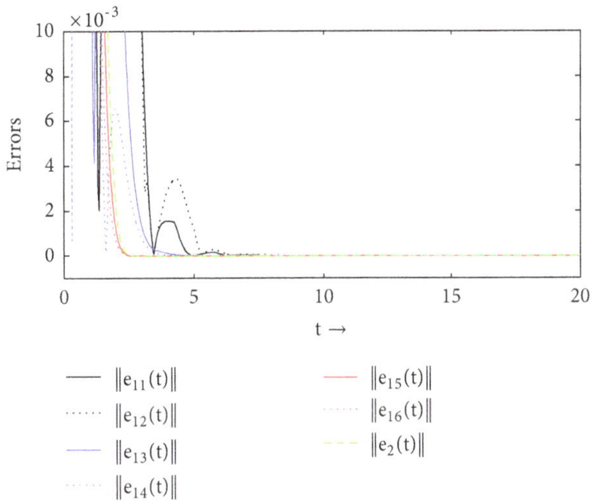

FIGURE 2: Time evolutions of synchronization errors $\|e_{1i}(t)\|$ and $\|e_2(t)\|$, $i = 1, 2, \ldots, 6$.

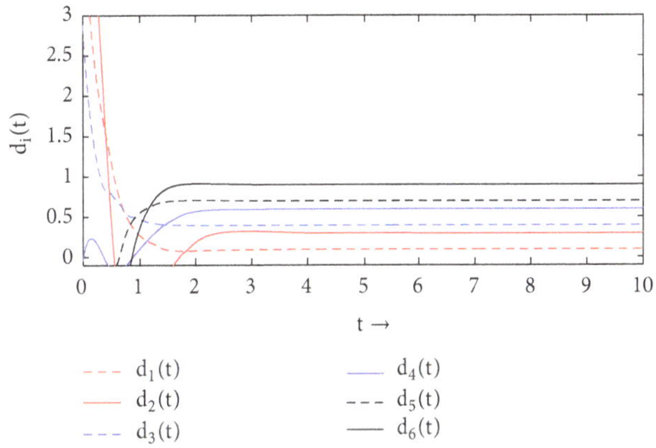

FIGURE 3: Time evolutions of the parameter estimations $\overline{\alpha}_1(t) = (d_1(t), d_2(t), \ldots, d_6(t))^{\top}$ in (12).

Consider the actual meaning of the network, the true values of the unknown parameters are

$$
\begin{aligned}
\alpha_1 &= (0.1, 0.3, 0.4, 0.6, 0.7, 0.9)^{\top}, \\
\alpha_2 &= (2, 1.9, 1.5, 0.2, 0.6, 0.9)^{\top}, \\
\alpha_3 &= 0.42,
\end{aligned}
\tag{26}
$$

estimations (12) are adopted for $\overline{\alpha}_1(t)$, $\overline{\alpha}_2(t)$, $\overline{\alpha}_3(t)$, and controllers (11) are adopted for $u_{1i}(t), u_{2i}(t)$.

By setting the initial values of network (24) randomly in $[0, 1]$ and with the feedback gain taken as $k_{1i} = k_{2i} = 1$, we plot Figure 2 to show the time evolutions of outer synchronization errors $\|e_{1i}(t)\| = \|\overline{x}_i(t) - x_i(t)\|$ and $\|e_{2i}(t)\| = \|\overline{S}_i(t) - S_i(t)\|$, $i = 1, 2, \ldots, 6$. It can be seen that the two errors both go to zero quickly after a short transient period, and network (24) reaches outer synchronization. Figure 3 depicts the time evolutions of the parameter estimations $\overline{\alpha}_1(t)$, which displays the perfect identification performance. Figure 4 shows the time evolutions of the parameter estimations

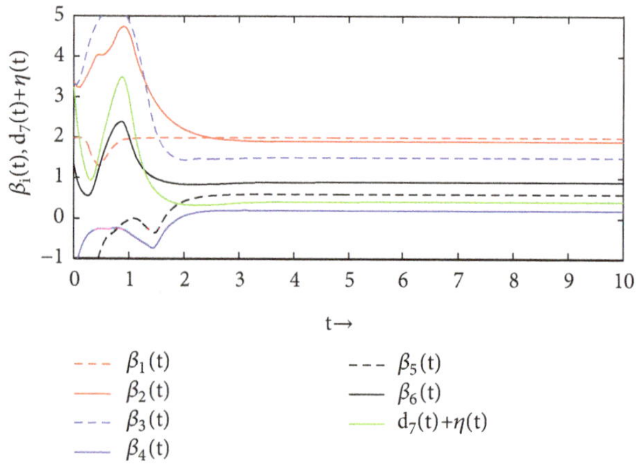

FIGURE 4: Time evolutions of the parameter estimations $\overline{\alpha}_2(t) = (\beta_1(t), \beta_2(t), \ldots, \beta_6(t))^\top$ and $\overline{\alpha}_3(t) = d_7(t) + \eta(t)$ in (12).

$\overline{\alpha}_2(t), \overline{\alpha}_3(t)$, and further illustrates the effectiveness of the parameter estimations (12). From the three figures, it is clearly observed that the unknown parameters of the network is estimated successfully by the control schemes of Theorem 3.

5. Conclusions

This paper builds a model of quorum-sensing network with disturbances, unknown parameter vector, and network topology and investigates the problem of outer synchronization between two quorum-sensing networks. In case that some systems' parameters are unknown in actual applications, adaptive parameter updating laws are designed to estimate the true values of those unknown parameters. Similarly, updating laws are also presented for the unknown elements of the network coupling matrix. Finally, some adaptive controllers are adopted to realize outer synchronization between two quorum-sensing networks. The validity of the proposed control schemes and updating laws is demonstrated by several numerical simulations.

Conflicts of Interest

The authors declare that there are no conflicts of interest regarding the publication of this article.

Acknowledgments

Project is supported by National Natural Science Foundation of China (nos. 11447005, 61771230, and 61663006), Shandong Provincial Key Research and Development Program of China (no. 2017CXGC0701), Shandong Provincial Natural Science Foundation (no. ZR2016FM40, ZR2016JL021), Jiangsu Planned Projects for Postdoctoral Research Funds (no. 1701017A), and the Fundamental Research Funds for the Central Universities (no. 9161718004).

References

[1] J. D. Dockery and J. P. Keener, "A mathematical model for quorum sensing in *Pseudomonas aeruginosa*," *Bulletin of Mathematical Biology*, vol. 63, no. 1, pp. 95–116, 2001.

[2] M. E. Taga and B. L. Bassler, "Chemical communication among bacteria," *Proceedings of the National Acadamy of Sciences of the United States of America*, vol. 100, no. 24, pp. 14549–14554, 2003.

[3] J. Garcia-Ojalvo, M. B. Elowitz, and S. H. Strogatz, "Modeling a synthetic multicellular clock: repressilators coupled by quorum sensing," *Proceedings of the National Academy of Sciences of the United States of America*, vol. 101, no. 30, pp. 10955–10960, 2004.

[4] R. Wang and L. Chen, "Synchronizing genetic oscillators by signaling molecules," *Journal of Biological Rhythms*, vol. 20, no. 3, pp. 257–269, 2005.

[5] E. V. Nikolaev and E. D. Sontag, "Quorum-sensing synchronization of synthetic toggle switches: a design based on monotone dynamical systems theory," *PLoS Computational Biology*, vol. 12, no. 4, Article ID e1004881, 2016.

[6] B. L. Bassler, "Quorum sensing and its control," *The FASEB Journal*, vol. 30, no. 1, 2016.

[7] W. Sun, S. Wang, G. Wang, and Y. Wu, "Lag synchronization via pinning control between two coupled networks," *Nonlinear Dynamics*, vol. 79, no. 4, pp. 2659–2666, 2015.

[8] P. Zhou, S. Cai, J. Shen, and Z. Liu, "Adaptive exponential cluster synchronization in colored community networks via aperiodically intermittent pinning control," *Nonlinear Dynamics*, vol. 92, no. 3, pp. 905–921, 2018.

[9] L. M. Pecora and T. L. Carroll, "Master stability functions for synchronized coupled systems," *Physical Review Letters*, vol. 80, no. 10, pp. 2109–2112, 1998.

[10] A. Brechtel, P. Gramlich, D. Ritterskamp, B. Drossel, and T. Gross, "Master stability functions reveal diffusion-driven pattern formation in networks," *Physical Review E: Statistical, Nonlinear, and Soft Matter Physics*, vol. 97, no. 3, Article ID 032307, 2018.

[11] Z.-Y. Sun, M.-M. Yun, and T. Li, "A new approach to fast global finite-time stabilization of high-order nonlinear system," *Automatica*, vol. 81, pp. 455–463, 2017.

[12] X.-J. Xie, Z.-J. Li, and K. Zhang, "Semi-global output feedback control for nonlinear systems with uncertain time-delay and output function," *International Journal of Robust and Nonlinear Control*, vol. 27, no. 15, pp. 2549–2566, 2017.

[13] Y. Li, Y. Sun, and F. Meng, "New criteria for exponential stability of switched time-varying systems with delays and nonlinear disturbances," *Nonlinear Analysis: Hybrid Systems*, vol. 26, pp. 284–291, 2017.

[14] L. M. Pecora and T. L. Carroll, "Synchronization in chaotic systems," *Physical Review Letters*, vol. 64, no. 8, pp. 821–824, 1990.

[15] A. Hu, J. Cao, M. Hu, and L. Guo, "Cluster synchronization in directed networks of non-identical systems with noises via random pinning control," *Physica A: Statistical Mechanics and its Applications*, vol. 395, pp. 537–548, 2014.

[16] J. Zhang, Z. Ma, and G. Chen, "Robustness of cluster synchronous patterns in small-world networks with inter-cluster co-competition balance," *Chaos: An Interdisciplinary Journal of Nonlinear Science*, vol. 24, no. 2, Article ID 023111, 2014.

[17] H. Liu, W. Sun, and G. Al-mahbashi, "Parameter identification based on lag synchronization via hybrid feedback control in uncertain drive-response dynamical networks," *Advances in Difference Equations*, vol. 2017, no. 1, article 122, 2017.

[18] Z. Wang, L. Huang, and X. Yang, "Adaptive modified function projective lag synchronization for two different chaotic systems with stochastic unknown parameters," *Mediterranean Journal of Mathematics*, vol. 13, no. 3, pp. 1391–1405, 2016.

[19] Abdesselem Boulkroune, Sarah Hamel, Farouk Zouari, Abdelkrim Boukabou, and Asier Ibeas, "Output-feedback controller based projective lag-synchronization of uncertain chaotic systems in the presence of input nonlinearities," *Mathematical Problems in Engineering*, vol. 2017, Article ID 8045803, 12 pages, 2017.

[20] J. Guan, "Function projective synchronization of a class of chaotic systems with uncertain parameters," *Mathematical Problems in Engineering*, vol. 2012, Article ID 431752, 5 pages, 2012.

[21] M. M. Asheghan and J. n. Míguez, "Robust global synchronization of two complex dynamical networks," *Chaos: An Interdisciplinary Journal of Nonlinear Science*, vol. 23, no. 2, Article ID 023108, 2013.

[22] S. Cai, X. Lei, and Z. Liu, "Outer synchronization between two hybrid-coupled delayed dynamical networks via aperiodically adaptive intermittent pinning control," *Complexity*, vol. 21, no. S2, pp. 593–605, 2016.

[23] L. Wang, J. Zhang, and W. Sun, "Adaptive outer synchronization and topology identification between two complex dynamical networks with time-varying delay and disturbance," *IMA Journal of Mathematical Control and Information*, 2018.

[24] X. Wu, W. X. Zheng, and J. Zhou, "Generalized outer synchronization between complex dynamical networks," *Chaos: An Interdisciplinary Journal of Nonlinear Science*, vol. 19, no. 1, Article ID 013109, 9 pages, 2009.

[25] C. Yu, J. Qin, and H. Gao, "Cluster synchronization in directed networks of partial-state coupled linear systems under pinning control," *Automatica*, vol. 50, no. 9, pp. 2341–2349, 2014.

[26] X. Chen, J. H. Park, J. Cao, and J. Qiu, "Adaptive synchronization of multiple uncertain coupled chaotic systems via sliding mode control," *Neurocomputing*, vol. 273, pp. 9–21, 2018.

[27] Junbiao Guan and Kaihua Wang, "Sliding mode control and modified generalized projective synchronization of a new fractional-order chaotic system," *Mathematical Problems in Engineering*, vol. 2015, Article ID 941654, 9 pages, 2015.

[28] W. Sun and L. Peng, "Observer-based robust adaptive control for uncertain stochastic Hamiltonian systems with state and input delays," *Lithuanian Association of Nonlinear Analysts. Nonlinear Analysis: Modelling and Control*, vol. 19, no. 4, pp. 626–645, 2014.

[29] Z.-Y. Sun, C.-H. Zhang, and Z. Wang, "Adaptive disturbance attenuation for generalized high-order uncertain nonlinear systems," *Automatica*, vol. 80, pp. 102–109, 2017.

[30] X. Xie and M. Jiang, "Output feedback stabilization of stochastic feedforward nonlinear time-delay systems with unknown output function," *International Journal of Robust and Nonlinear Control*, vol. 28, no. 1, pp. 266–280, 2018.

[31] Yuling Li, Yixin Yin, and Sen Zhang, "Adaptive Control of Delayed Teleoperation Systems with Parameter Convergence," *Mathematical Problems in Engineering*, vol. 2018, Article ID 1046419, 7 pages, 2018.

[32] M. B. Elowitz and S. Leibier, "A synthetic oscillatory network of transcriptional regulators," *Nature*, vol. 403, no. 6767, pp. 335–338, 2000.

[33] J. Zamora-Munt, C. Masoller, J. Garcia-Ojalvo, and R. Roy, "Crowd synchrony and quorum sensing in delay-coupled lasers," *Physical Review Letters*, vol. 105, no. 26, Article ID 264101, 2010.

8

MAGDM Method with Pythagorean 2-Tuple Linguistic Information and Applications in the HSE Performance Assessment of Laboratory

Chunyu Zhao[1], **Xiyue Tang**[1] **and Lijuan Yuan**[2]

[1]*School of Business, Sichuan Normal University, Chengdu 610101, China*
[2]*School of Mathematical Sciences, Sichuan Normal University, Chengdu 610101, China*

Correspondence should be addressed to Chunyu Zhao; zhaochunyu.sicnu@qq.com

Academic Editor: Anna M. Gil-Lafuente

There is a significant gap between the safety management in the Chinese colleges and many renowned colleges in other countries. The subject of this study is how to assess the performance of health, safety, and environment (HSE) in Chinese college laboratories. The assessment system is established by three parts. First of all, HSE performance assessment indicators for laboratories in Chinese colleges are identified based on the previous studies. Then set valued iteration is used to calculate the weights of the various indicators. Following that, multiple attribute group decision-making (MAGDM) method with Pythagorean 2-tuple linguistic operators is used to assess the laboratory HSE performance in colleges. Finally, taking a college in Sichuan Province as an example, the proposed method is used to assess the laboratory HSE performance. The assessment result shows that the proposed method used in this study is practical and feasible.

1. Introduction

Laboratories are important locations in colleges where experimental teaching, scientific research, and social services are conducted. They cover a wide range of disciplines, involving a large number of students, pathogenic microorganisms, inflammable and explosive chemicals, highly toxic substances, wastes, machining, electrical and electronic materials, instrument, and equipment which pose certain potential safety risks and hazards [1, 2]. Although laboratory safety management and risk precautions are highlighted in colleges [3], accidents have occurred frequently in laboratories in recent years causing casualties and property losses, which threaten the life and property safety of the students, teachers, and the teaching and scientific research environment [1]. There is a significant gap between the safety management in laboratories of the Chinese colleges and many renowned colleges in other countries, where HSE management system is implemented and the gap is reflected in various aspects of the HSE management [3]. HSE performance assessment is an important part of the safety management, which can fully assess the safety management effects and identify any inadequacy so that rectification can be made to ensure to implement dynamic and effective safety management and improve the safety performance continuously [4].

So far significant studies have been conducted and a sound assessment system and method have been formed on the HSE performance assessment, which are widely used in the safety management, risk assessment, environmental protection, occupational health, etc. in the petrochemical, foodstuff, textile, iron and steel, mining, nuclear power industry, etc. [4, 5]. The studies on laboratory HSE in colleges also focus on the establishment of HSE management system [6] and safety culture building [7]. Little concern is given to the study of HSE performance assessment of college laboratories. Previously mainly analytic hierarchy process [8, 9] is used to calculate the weights in terms of assessment methods, and quantitative comparative strategic management model [10], fuzzy comprehensive evaluation [11], dynamic fuzzy assessment [5], fuzzy cognitive map and relative risk analysis [12], etc. are used for the assessment.

We have the following concerns on the HSE assessment for college laboratories in China despite the increasingly proven assessment systems and methods.

(1) There is lack of HSE performance assessment indicators for the college laboratories. No dedicated HSE management institution or committee is created in the Chinese colleges. No specific law, regulation, and operation standard is available. And health, safety, and environmental protection are managed separately by different departments. Collaborative management and law enforcement are inadequate and systematic management is not formed which results in the absence of HSE performance assessment indicators for the college laboratories.

(2) Although the indicator weights are identified in the previous HSE system [11], the weights of the HSE indicator system of laboratories in the Chinese colleges should be redetermined due to the different assessment objects and purposes, the social value of assessment factors, management purpose of the decision-makers, the perception of the evaluators, etc. The element of the judgment matrix is the ratio between two assessment indicators for the previous analytic hierarchy process [8]. The ratios are given by the experts, which make it hard for the judgment matrix to meet the consistence test due to the influence of the experts' perception and preference [13]. In particular, it is harder to pass the consistence test when a number of assessment indicators are present [14].

(3) In addition, HSE performance assessment is conducted prior to actual work. Normally, the experts would conduct the subjective measurement based on the HSE management assessment system [15]. Uncertainties, inadequacies, the experts' indecision, absence of clear boundary and denotation, etc. may have an impact on the accuracy and reliability of the assessment results [5]. Fuzzy comprehensive evaluation method has overcome the disadvantages and has been widely used [16]. The fuzzy sets theory [17] can contain more expert decision-making information, which makes the assessment results more reliable. Only membership is considered in the classical fuzzy multiattribute decision-making [18]. Nonmembership is not considered and the indecision of the decision-makers is not fully reflected. Intuitionistic fuzzy set (IFS) [19] consider membership degree, nonmembership degree, and hesitation to the decision-making preference at the same time. But they are limited to cases where the sum of membership and nonmembership is less than 1 [20], while cases where the sum of membership and nonmembership exceed 1 and the sum of the squares is less than 1 cannot be assessed.

Therefore, first of all, an indicator system is established in this paper for HSE performance assessment of laboratories in Chinese colleges to solve the problems above. Second, the experts tend to provide an interval value for some indicators in terms of the indicator weight calculation comparing with a point value for the individual indicator based on the conventional thinking mode. Set valued iteration [21] has provided the solution, so it is used in this study to determine the indicator weights. Third, Pythagorean 2-tuple linguistic operator [20] can simultaneously meet the conditions of the preferences of membership, nonmembership, and indecision fuzziness; the sum of membership and nonmembership may exceed 1 and the sum of the squares is less than 1, which has extended the range of the previous fuzzy assessment. Since more than one expert is involved in the decision [22], multiple attribute group decision-making (MAGDM) method [23] with Pythagorean 2-tuple linguistic operators is used for the study of the HSE performance assessment of college laboratories.

2. Laboratory HSE Performance Assessment Indicators

An appropriate HSE performance assessment indicator system for college laboratories is the basis of HSE performance assessment, first, the assessment indicators and indicator system should be established properly, objectively, and systematically to ensure the accuracy of the assessment results. In this paper, experts and management responsible for the laboratory safety in colleges are invited and Delphi method is used to reach a consensus based on the actual HSE conditions in colleges considering the previous studies on the laboratory HSE management system [1, 24], HSE performance assessment indicators [5, 11], etc. as well as the new framework for HSE performance measurement and monitoring [25]. A performance assessment system is established including 29 measurement indicators, i.e., organization, objectives and commitment, risk assessment, hazard control, continuous improvement, training, information communication, internal review, rectification, and improvement. This indicator system reflects the HSE management features which highlight precautions, continuous improvement, optimization and motivation, all staff involved, and process control [5, 11, 24]. The specific indicators are shown in Table 1.

3. Pythagorean 2-Tuple Linguistic Operators

In this section, we shall give some definitions of Pythagorean 2-tuple linguistic information, which is the foundation for establishing corresponding group decision-making methods.

To compute with words without loss of information, the 2-tuple linguistic model based on the concept of symbolic translation was proposed in [26, 27]. The model uses a 2-tuple (s_k, α) to represent linguistic information, where $s_k \in S$, α denotes the value of symbolic translation, and $\alpha \in [-0.5, 0.5)$. The specific definition of 2-tuple linguistic model is given as follows.

Definition 1 (see [26]). Let $S = \{s_0, s_1, \cdots, s_g\}$ be a linguistic term set and $\beta \in [0, g]$ be a value representing the result of a symbolic aggregation operation; then the 2-tuple that expresses the equivalent information to β is obtained with the following function:

$$\begin{aligned} \Delta : [0, g] &\longrightarrow S \times [-0.5, 0.5) \\ \Delta(\beta) &= (s_i, \alpha) \end{aligned} \tag{1}$$

with

$$\begin{aligned} s_i, \quad & i = round(\beta) \\ \alpha = \beta - i, \quad & \alpha \in [-0.5, 0.5) \end{aligned} \tag{2}$$

TABLE 1: HSE performance indicators for college laboratories.

u_i	Specific assessment indicators
u_1	College leaders are committed to the HSE management objectives and have clarified the HSE missions.
u_2	A dedicated HSE management institution or committee is created.
u_3	College has specified HSE standards and requirements complying with the relevant laws and regulations and suitable for the laboratory.
u_4	HSE management objectives and indicators can reflect the college's commitment to the laboratory continuous HSE improvement.
u_5	Safety responsibility systems of the college, institutes, departments and laboratories are in place and the responsibilities and obligations of the parties involved are identified.
u_6	Hazard identification, risk assessment and control and preventive and control measures are in place.
u_7	Conventional fund, ear-marked fund or self-financed fund and resources are available for HSE management each year to ensure that the safety measures can be implemented properly.
u_8	HSE management system manuals and procedures are in place and well maintained.
u_9	Management and control process records in details are available.
u_{10}	Relevant trainings and publicity on occupational health, safety and environmental protection management are conducted on a regular basis at the college.
u_{11}	The teachers and students respond to, participate in and follow the HSE management requirements.
u_{12}	The relevant laboratory personnel have undergone trainings and have the HSE management awareness and capabilities.
u_{13}	Teachers and students undergo HSE trainings and tests before they carry out teaching, learning and research activities in the laboratory.
u_{14}	HSE information, data, bulletins etc. are communicated sufficiently among different levels of the management system.
u_{15}	Conditions which may have an impact on the HSE of the community and surrounding residents are communicated and control measures are notified on a timely basis.
u_{16}	Safety equipment and facilities are intact, normal and effective.
u_{17}	PPE and first aid facilities are sound and effective.
u_{18}	Laboratory is clean and tidy. Signs and labels are legible. Articles are classified and placed properly.
u_{19}	Appropriate equipment and facility operation specifications and experiment operation procedures are in place. The management of special equipment complies with the national regulations.
u_{20}	Equipment and raw material suppliers and waste disposal contractors etc. follow the HSE procedures properly.
u_{21}	Procurement, transportation, packaging, storage, use and disposal of hazardous substances etc. stringently follow the applicable national laws, regulations and procedures.
u_{22}	Appropriate safety level qualification, experimental activity qualification and experiment personnel qualification are available for laboratories involving hazardous substance experiments.
u_{23}	HSE performance measurement procedures suitable for the college laboratory are in place and are monitored.
u_{24}	Emergency preparedness and response plans and procedures should be in place to prevent or mitigate any HSE hazards due to accidents or emergencies.
u_{25}	Emergency response plan can be initiated immediately in case of incident or accident.
u_{26}	Plans and procedures should be in place to prevent and correct any failure of the hazard control, emergency preparedness and response procedures.
u_{27}	Have the ability to conduct the incident or accident investigation and root cause analysis and reporting.
u_{28}	HSE management system is reviewed by the college leaders and relevant requirements are rectified on a regular basis if they are inappropriate.
u_{29}	Management measures are modified based on the changes of the laboratory and external circumstances.

where round$(\cdot)$ is the usual round operation, s_i has the closest index label to β, and α is the value of symbolic translation.

Definition 2 (see [26]). Let $S = \{s_0, s_1, \cdots, s_g\}$ be a linguistic term set and (s_i, α) be a 2-tuple; there is a function Δ^{-1}, which can transform a 2-tuple into its equivalent numerical value $\beta \in [0, g]$. The transformation function can be defined as

$$\Delta^{-1} : S \times [-0.5, 0.5) \longrightarrow [0, g]$$
$$\Delta^{-1}(s_i, \alpha) = i + \alpha = \beta. \quad (3)$$

It easily follows from Definitions 1 and 2 that a linguistic term can be considered as a linguistic 2-tuple by adding a value 0 to it as symbolic translation; i.e., $\Delta(s_i) = (s_i, 0)$.

Definition 3 (see [28]). In a finite universe of discourse $X = \{x_1, x_2, \cdots, x_n\}$, a Pythagorean fuzzy set (PFS)P with the structure

$$P = \{\langle x, (\mu_P(x), \nu_P(x))\rangle \mid x \in X\} \tag{4}$$

where $\mu_P : X \longrightarrow [0,1]$ denotes the membership degree and $\nu_P : X \longrightarrow [0,1]$ denotes the nonmembership degree of the element $x \in X$ to the set P, respectively, with the condition that $0 \le (\mu_P(x))^2 + (\nu_P(x))^2 \le 1$.

Definition 4 (see [20]). Assuming nonvoid set X and a linguistic set S exist, $S = \{s_1, s_2, \ldots, s_t\}$, then $\widetilde{P} = \{(s_{\theta(x)}, \rho), (\mu_{\widetilde{P}(x)}, \nu_{\widetilde{P}(x)}), x \in X\}$ is the Pythagorean 2-tuple linguistic operator, where, $s_{\theta(x)} \in S$, $\rho \in [-0.5, 0.5)$, $\mu_{\widetilde{p}}(x), \nu_{\widetilde{p}}(x) \in [0,1]$, and $0 \le (\mu_{\widetilde{p}}(x))^2 + (\nu_{\widetilde{p}}(x))^2 \le 1, \forall x \in X$.

For convenience, $\widetilde{p} = \langle (s_p, \rho), (u_p, \nu_p)\rangle$ is called Pythagorean 2-tuple linguistic number.

Definition 5 (see [20]). Let $\widetilde{p} = \langle (s_p, \rho), (u_p, \nu_p)\rangle$ be Pythagorean 2-tuple linguistic number; a score function S of Pythagorean 2-tuple linguistic number can be represented as follows:

$$s(\widetilde{p}) = \Delta\left(\Delta^{-1}\left(s_{\theta(p)}, \rho\right)\frac{1 + u_p^2 + \nu_p^2}{2}\right), \tag{5}$$

$$\text{where } \Delta^{-1}(s(\widetilde{p})) \in [1, t].$$

Definition 6 (see [20]). If $\widetilde{p}_j = \langle (s_{p_j}, \rho_j), (u_{p_j}, \nu_{p_j})\rangle$, $j = 1, \cdots, m$, then

$$P2TLWA(\widetilde{p}_1, \widetilde{p}_2, \cdots, \widetilde{p}_m) = \bigoplus_{i=1}^{m} (\omega_i \widetilde{p}_i)$$

$$= \left\langle \Delta\left(\sum_{i=1}^{m} \omega_i \Delta^{-1}\left(s_{pi}, \rho_{pi}\right)\right), \left(\sqrt{1 - \prod_{i=1}^{m}\left(1 - (\mu_{pi})^2\right)^{\omega_i}}, \prod_{i=1}^{m}(\nu_{pi})^{\omega_i}\right)\right\rangle \tag{6}$$

4. MAGDM Method with Pythagorean 2-Tuple Linguistic Information

In this section, multiattribute group decision-making (MAGDM) method based on Pythagorean 2-tuple linguistic information is established.

If there is a set of indicators $X = \{x_1, x_2, \cdots, x_m\}$, the corresponding weight is $(w_1, w_2, \cdots, w_m)^T$, where $w_j \in [0,1]$, and $\sum_{j=1}^{m} w_j = 1$. And there are L evaluation experts, whose weight is $(\varpi_1, \varpi_2, \cdots, \varpi_L)^T$, where $\varpi_i \in [0,1]$, and $\sum_{i=1}^{L} \varpi_i = 1$.

4.1. Calculation of Objective Attribute Weight. The system above indicates that all the indicators are in the same hierarchy. So set valued iteration [21] characterized by easy operation and calculation can be used to determine the weights of the indicators. It is a "function-driven" weight calculation method with preference. It corresponds to the weights based on the relevant impotence of the indicators, which can eliminate the personal subjective factors of the evaluators.

Assuming the indicator set $X = \{x_1, x_2, \cdots, x_m\}$ and expert L, the indicator weights are calculated as per the steps of the set valued iteration.

4.1.1. Selection of Indicator Subset. Selecting a positive integer $k (1 \le k \le m)$, each expert selects the indicators strictly as per the following steps. Take expert p as an example.

Step 1. The expert selects the top k indicators he believes from set X to obtain the indicator set $X_{1,p} = \{x_{1,p,1}, x_{1,p,2}, \cdots, x_{1,p,k}\}$.

Step 2. The expert selects the top $2k$ indicators he believes from set X to obtain the indicator set $X_{2,p} = \{x_{2,p,1}, x_{2,p,2}, \cdots, x_{2,p,2k}\}$.

$\cdots \cdots$

Step s. The expert selects the top sk indicators he believes from set X to obtain the indicator set $X_{s,p} = \{x_{s,p,1}, x_{s,p,2}, \cdots, x_{s,p,sk}\}$.

If natural numbers s and r$(1 \le r \le k)$ exist, and $s \cdot k + r = m$, then the indicator selection of expert p is completed and sk indicator sets are obtained.

4.1.2. Calculation of (Indicative) Function. Function $u_{ik}(x_j)$ is used to calculate the times of the various indicators in all the indicator sets selected by the experts:

$$u_{ip}(x_j) = \begin{cases} 1, & x_j \in X_{i,p} \\ 0, & x_j \notin X_{i,p} \end{cases} \tag{7}$$

$$(i = 1, 2, \cdots, s;\ p = 1, 2, \cdots, L)$$

Assume

$$g(x_j) = \sum_{p=1}^{L}\sum_{i=1}^{s} u_{ip}(x_j), \quad j = 1, 2, \cdots, m \tag{8}$$

4.1.3. Calculation of the Weight Coefficient. $g(x_j)$ is normalized to obtain the weight coefficient of indicator x_j:

$$w_j = \frac{g(x_j)}{\sum_{p=1}^{m} g(x_p)}, \quad j = 1, 2, \cdots, m \tag{9}$$

If an indicator is never selected by any expert, it is insignificant and its weight coefficient is adjusted to

$$w_j = \frac{g(x_j) + 1/2m}{\sum_{p=1}^{m}\left(g(x_p) + 1/2m\right)}, \quad j = 1, 2, \cdots, m \tag{10}$$

4.2. MAGDM Method with Pythagorean 2-Tuple Linguistic Weighted Average Operator

Step 1 (establishing Pythagorean 2-tuple linguistic decision matrix). Let evaluation experts score according to Pythagorean 2-tuple linguistic weighted average operator (P2TLWA). First, the grade of evaluation is selected according to the linguistic assessment, and the scope of the deviation is given. Then the degree of membership and nonmembership is scored for the number of 2-tuple linguistic information.

It combines the 2-tuple linguistic operator which allows the use of natural language for scoring and Pythagorean fuzzy number which contains more information. This method is used for scoring and integration, which is closer to the actual conditions. Pythagorean 2-tuple linguistic set consists of 2-tuple linguistic variable and degree of membership and nonmembership. 2-tuple linguistic variable $(s_{\theta(x)}, \rho)$ is used to show the specific score of a certain indicator in a certain scheme. Then Pythagorean membership $\mu_{\tilde{P}}(x)$ is used to express the support extent for the score and nonmembership $v_{\tilde{P}}(x)$ is used to express the opposition extent [20].

If the expert has already determined the score of the number of 2-tuple linguistic information, the membership degree is 1, and the nonmembership degree is 0.

Mark the score of the p expert on the i index as $\tilde{p}_{ip} = \langle (s_{ip}, \rho_{ip}), (\mu_{ip}, v_{ip}) \rangle$ and establish the corresponding Pythagorean 2-tuple linguistic decision matrix $M = (\tilde{p}_{ip})_{m\times L}$..

Step 2 (calculation of each expert's assessment results). Pythagorean 2-tuple linguistic weighted average operator (P2TLWA) [20] is used to calculate each expert's assessment score. The operator definition is as follows:

$$t_p = P2TLWA\left(u_{p1}, u_{p2}, \cdots, u_{pm}\right) = \bigoplus_{i=1}^{m}\left(\omega_i u_{pi}\right)$$
$$= \left\langle \Delta\left(\sum_{i=1}^{m}\omega_i \Delta^{-1}\left(s_{pi}, \rho_{pi}\right)\right), \left(\sqrt{1-\prod_{i=1}^{m}\left(1-\left(\mu_{pi}\right)^2\right)^{\omega_i}}, \prod_{i=1}^{m}\left(v_{pi}\right)^{\omega_i}\right)\right\rangle \tag{11}$$
$$p = 1, 2, \cdots, L$$

where u_{pi} is expert p' score for indicator i. ω_i is the weight of indicator i, $\omega_i \in [0,1]$, $\sum_{i=1}^{m}\omega_i = 1$, where

$$\Delta(\beta) = \begin{cases} s_i, & i = round(\beta) \\ \rho = \beta - i, & \rho \in [-0.5, 0.5) \end{cases} \tag{12}$$
$$\Delta^{-1}(s_i, \rho) = i + \rho$$

Step 3 (calculation of the score). For a Pythagorean 2-tuple linguistic number, according to the results of Step 2, its score is

$$S(t) = \Delta\left(\Delta^{-1}\left(\left(s_{\theta(x)}, \rho\right)\frac{1 + \left(\mu_t\right)^2 - \left(v_t\right)^2}{2}\right)\right) \tag{13}$$

Finally the score set $H = \{S(1), S(2), \cdots, S(L)\}$ is obtained.

Step 4 (calculation of the assessment result). According to the results of Step 3, the assessment result is Q.

$$Q = W * H = \left(\varpi_1, \varpi_2, \cdots, \varpi_L\right)\begin{pmatrix} S(1) \\ S(2) \\ \vdots \\ S(L) \end{pmatrix} = \Delta\left(\sum_{t=1}^{L}\varpi_t \Delta^{-1}\left(S(t)\right)\right) \tag{14}$$

The final results are assessed based on the calculation result Q and assessment rating.

5. Case Study

Taking the HSE performance management of the laboratory in a certain college in Sichuan Province of China as an example, the method above is used and five experts are invited to select and assess the performance indicators shown in Table 1. One expert is the safety person-in-charge of the college experiment equipment section. One expert is a professor specializing in safety management. The other three experts are the directors of chemical, biological, and engineering laboratories. The weight calculation and assessment are as follows.

5.1. Weight Calculation. First, the experts are asked to select the most significant 5 indicators from the assessment indicators and then the second most significant 5 indicators from the remaining indicators, etc., until the last 4 indicators are left. The five experts select the indicator sets as per the rules, which are shown in Table 2.

Set valued iteration is used to calculate the indicator scores based on Table 2 using (9) and (10). The results are shown in Table 3.

The calculation results of the various indicator weights are w = (0.0557, 0.0424, 0.0239, 0.0159, 0.0371, 0.0531, 0.0637, 0.0451, 0.0212, 0.0398, 0.0265, 0.0531, 0.0371, 0.0186, 0.0265, 0.0531, 0.0371, 0.0451, 0.0477, 0.0292, 0.0424, 0.0398, 0.0053, 0.0106, 0.0133).

5.2. MAGDM Method with P2TLWA Operator

Step 1 (establishing Pythagorean 2-tuple linguistic decision matrix). The linguistic assessment is rated uniformly. In this assessment activity, the linguistic assessment is rated as $S = \{S_1 = \text{very poor}, S_2 = \text{poor}, S_3 = \text{odinary}, S_4 = \text{good}, S_5 = \text{very good}\}$. The linguistic sets are rated in many different ways, which are suitable for different assessment systems, areas, specifications, etc. An appropriate assessment rating helps reflect the actual conditions.

TABLE 2: Indicator subset selected by the experts.

expert	Subset					
	k	$2k$	$3k$	$4k$	$5k$	r
1	$u_1, u_6, u_7, u_{21}, u_{26}$	$u_{13}, u_{16}, u_{17}, u_{22}, u_{25}$	$u_{10}, u_{12}, u_{18}, u_{19}, u_{24}$	$u_2, u_5, u_8, u_9, u_{20}$	$u_4, u_{11}, u_{14}, u_{15}, u_{28}$	$u_3, u_{23}, u_{27}, u_{29}$
2	$u_1, u_7, u_{12}, u_{16}, u_{19}$	$u_6, u_{21}, u_{22}, u_{25}, u_{26}$	$u_2, u_8, u_{17}, u_{23}, u_{24}$	$u_{10}, u_{11}, u_{13}, u_{15}, u_{18}$	$u_4, u_5, u_{20}, u_{28}, u_{29}$	u_3, u_9, u_{14}, u_{27}
3	$u_6, u_7, u_8, u_{10}, u_{12}$	$u_{21}, u_{22}, u_{25}, u_{26}, u_{29}$	$u_{15}, u_{16}, u_{17}, u_{19}, u_{20}$	$u_{13}, u_{18}, u_{23}, u_{24}, u_{27}$	$u_1, u_2, u_5, u_{14}, u_{28}$	u_3, u_4, u_9, u_{11}
4	u_1, u_2, u_3, u_5, u_7	$u_4, u_6, u_{10}, u_{12}, u_{19}$	$u_8, u_9, u_{13}, u_{18}, u_{21}$	$u_{11}, u_{20}, u_{22}, u_{23}, u_{24}$	$u_{15}, u_{16}, u_{25}, u_{26}, u_{28}$	$u_{14}, u_{17}, u_{27}, u_{29}$
5	$u_1, u_2, u_5, u_{11}, u_{14}$	$u_3, u_7, u_8, u_{22}, u_{23}$	$u_9, u_{12}, u_{13}, u_{15}, u_{25}$	$u_6, u_{16}, u_{17}, u_{18}, u_{19}$	$u_{10}, u_{20}, u_{21}, u_{24}, u_{26}$	$u_4, u_{27}, u_{28}, u_{29}$

TABLE 3: Set valued iteration calculation results.

j	u_1	u_2	u_3	u_4	u_5	u_6	u_7	u_8	u_9	u_{10}	u_{11}	u_{12}	u_{13}	u_{14}	u_{15}	u_{16}
$g(x_j)$	21	16	9	6	14	20	24	17	8	15	10	20	14	7	10	15
j	u_{17}	u_{18}	u_{19}	u_{20}	u_{21}	u_{22}	u_{23}	u_{24}	u_{25}	u_{26}	u_{27}	u_{28}	u_{29}			
$g(x_j)$	12	14	17	9	17	18	11	11	16	15	2	4	5			

The laboratory HSE performance is scored in the form of Pythagorean 2-tuple linguistic number by the five experts based on the 29 indicators. The weights and assessment scores are consolidated and shown in Table 4.

Step 2. The expert scores are integrated based on Table 4 to obtain each expert's Pythagorean 2-tuple linguistic weighted averages which are shown in Table 5.

Step 3. Calculate the scores of each expert's assessment which are shown in Table 6.

Step 4. The weights of the five experts are assumed as $\omega = (0.15, 0.25, 0.15, 0.25, 0.2)$, respectively, due to their different knowledge background and experience. According to (14), we can get the result as follows.

$$Q = 0.15 * 2.546488 + 0.25 * 2.287582 + 0.15 * 1.40895 + 0.25 * 3.622417 + 0.2 * 3.5436 = 2.7795 \quad (15)$$

The final result $(s_3, -0.2205)$ is obtained according to the proposed method. The final assessment result is lower than the ordinary level.

The assessment result is communicated with the relevant safety persons in charge of the college laboratory; ordinary level was evaluated by the college operation evaluation committee. Most previous evaluations mainly focused on the responsibility and obligation of the organization, the construction of experimental conditions, safety training, operational records, risk assessment, without involving the organizational system, management plans and procedures, information communication, accident investigation and analysis, internal review, and so on. The evaluation method was determined according to the experts' opinion.

Using the proposed method, not only the general level evaluation is given, but also the specific grade of the level is further reflected, so that the administrators can know more about the HSE management and take corresponding management decisions. Therefore, the assessment method of this study is more informative and exact.

6. Discussion

The establishment and implementation of HSE management system in college laboratories can protect the environment, improve the safety and health of the laboratories and the experiment conditions, and maintain the legitimate rights and interests of the teachers and students [24]. The implementation of the system will help optimize the laboratory management, enhance the laboratory image, and create a better experimental environment and safety atmosphere. The establishment, implementation, and promotion of HSE management system in the Chinese colleges have been lagging behind. In particular, the Chinese colleges are pursuing the world first-class universities and disciplines in recent years. Most colleges are highlighting the building of laboratories while neglecting the HSE management. The systematic and structured HSE internal management mechanism has not formed. An appropriate HSE management performance assessment system is absent.

The HSE management performance assessment indicator system is established in this study for the college laboratory HSE management. Similar to the previous HSE performance assessment indicators [5, 11], this indicator system includes 29 measurement indicators, i.e., organization objective and commitment, risk assessment, hazard control, continuous improvement, training, information communication, internal review, adjustment, and improvement, which have reflected the purpose and actions for the laboratory accident prevention and HSE improvement of the teachers and students. It reflected the HSE management objective commitment and identified the HSE mission, the safety responsibility system of the college, department, and laboratory and assessment is conducted in terms of the facilities, equipment, personal protection, hazardous substance management, emergency preparedness, environmental protection, etc. The entire assessment indicator system has reflected the HSE management philosophy.

TABLE 4: Performance indicator scoring based on Pythagorean 2-tuple linguistic numbers.

Indicator	Weights	the expert scores 1	2	3	4	5
u_1	0.0557	((s_3,0),(0.9,0.2))	((s_4,0),(0.8,0.3))	((s_1,0),(0.9,0.1))	((s_4,0),(0.8,0.1))	((s_3,0),(0.4,0.6))
u_2	0.0424	((s_2,0),(0.8,0.1))	((s_3,0),(0.7,0.4))	((s_1,0),(0.8,0.2))	((s_4,0),(0.9,0.05))	((s_4,0),(0.7,0.3))
u_3	0.0239	((s_4,0),(0.6,0.5))	((s_3,0),(0.8,0.1))	((s_2,0),(0.8,0.1))	((s_3,0),(0.9,0.05))	((s_3,0),(0.3,0.7))
u_4	0.0159	((s_2,0),(0.8,0.2))	((s_2,0),(0.8,0.1))	((s_1,0),(0.9,0.1))	((s_3,0),(0.7,0.1))	((s_2,0),(0.2,0.8))
u_5	0.0371	((s_4,0),(0.9,0.1))	((s_2,0),(0.9,0.1))	((s_1,0),(0.8,0.2))	((s_5,0),(0.9,0.05))	((s_4,0),(0.8,0.2))
u_6	0.0531	((s_2,0),(0.6,0.5))	((s_3,0),(0.8,0.1))	((s_3,0),(0.8,0.1))	((s_4,0),(0.9,0.05))	((s_2,0),(0.7,0.3))
u_7	0.0637	((s_4,0),(0.8,0.4))	((s_3,0),(0.9,0.1))	((s_1,0),(0.9,0.1))	((s_4,0),(0.9,0.05))	((s_5,0),(0.9,0))
u_8	0.0451	((s_2,0),(0.7,0.3))	((s_1,0),(0.9,0.1))	((s_1,0),(0.2,0.2))	((s_4,0),(0.8,0.1))	((s_2,0),(0.8,0.1))
u_9	0.0212	((s_2,0),(0.8,0.3))	((s_1,0),(0.9,0.1))	((s_1,0),(0.1,0.1))	((s_3,0),(0.7,0.1))	((s_2,0),(0.8,0.1))
u_{10}	0.0398	((s_4,0),(0.9,0.1))	((s_4,0),(0.9,0.2))	((s_1,0),(0.1,0.1))	((s_4,0),(0.9,0.05))	((s_5,0),(0.9,0.1))
u_{11}	0.0265	((s_4,0),(0.8,0.1))	((s_2,0),(0.9,0.2))	((s_1,0),(0.8,0.2))	((s_4,0),(0.8,0.2))	((s_2,0),(0.4,0.6))
u_{12}	0.0531	((s_4,0),(0.9,0.1))	((s_4,0),(0.9,0.1))	((s_3,0),(0.8,0.1))	((s_5,0),(0.9,0.05))	((s_4,0),(0.7,0.2))
u_{13}	0.0371	((s_4,0),(0.8,0.1))	((s_4,0),(0.9,0.1))	((s_3,0),(0.9,0.1))	((s_4,0),(0.8,0.1))	((s_5,0),(1,0))
u_{14}	0.0186	((s_3,0),(0.8,0.2))	((s_2,0),(0.6,0.5))	((s_1,0),(0.9,0.1))	((s_4,0),(0.8,0.1))	((s_2,0),(0.1,0.9))
u_{15}	0.0265	((s_3,0),(0.5,0.5))	((s_2,0),(0.5,0.5))	((s_2,0),(0.2,0.2))	((s_3,0),(0.7,0.1))	((s_2,0),(0.3,0.5))
u_{16}	0.0398	((s_4,0),(0.9,0.1))	((s_3,0),(0.9,0.2))	((s_3,0),(0.8,0.1))	((s_5,0),(0.9,0.05))	((s_4,0),(0.8,0.1))
u_{17}	0.0318	((s_3,0),(0.9,0.1))	((s_3,0),(0.8,0.3))	((s_2,0),(0.8,0.1))	((s_4,0),(0.7,0.1))	((s_4,0),(0.8,0.1))
u_{18}	0.0371	((s_4,0),(0.9,0.1))	((s_3,0),(0.8,0.2))	((s_3,0),(0.8,0.2))	((s_5,0),(0.9,0.05))	((s_4,0),(0.7,0.2))
u_{19}	0.0451	((s_4,0),(0.9,0.1))	((s_2,0),(0.7,0.3))	((s_3,0),(0.8,0.1))	((s_5,0),(0.9,0.05))	((s_4,0),(0.8,0.1))
u_{20}	0.0239	((s_3,0),(0.5,0.5))	((s_2,0),(0.5,0.5))	((s_1,0),(0.1,0.1))	((s_4,0),(0.7,0.1))	((s_2,0),(0.4,0.5))
u_{21}	0.0451	((s_4,0),(0.6,0.5))	((s_4,0),(0.9,0.1))	((s_3,0),(0.1,0.1))	((s_5,0),(0.9,0.05))	((s_4,0),(0.6,0.3))
u_{22}	0.0477	((s_4,0),(0.9,0.1))	((s_3,0),(0.5,0.5))	((s_1,0),(0.7,0.2))	((s_4,0),(0.7,0.1))	((s_5,0),(0.9,0))
u_{23}	0.0292	((s_1,0),(0.8,0.1))	((s_3,0),(0.9,0.2))	((s_1,0),(0.8,0.1))	((s_4,0),(0.9,0.05))	((s_2,0),(0.8,0.1))
u_{24}	0.0292	((s_1,0),(0.9,0.1))	((s_2,0),(0.6,0.4))	((s_1,0),(0.8,0.2))	((s_4,0),(0.9,0.05))	((s_4,0),(0.8,0.1))
u_{25}	0.0424	((s_2,0),(0.8,0.3))	((s_2,0),(0.8,0.3))	((s_1,0),(0.8,0.1))	((s_4,0),(0.9,0.05))	((s_4,0),(0.8,0.1))
u_{26}	0.0398	((s_2,0),(0.8,0.2))	((s_2,0),(0.8,0.3))	((s_2,0),(0.8,0.1))	((s_4,0),(0.8,0.05))	((s_4,0),(0.8,0.1))
u_{27}	0.0053	((s_2,0),(0.7,0.4))	((s_2,0),(0.9,0.2))	((s_2,0),(0.8,0.1))	((s_4,0),(0.9,0.05))	((s_2,0),(0.7,0.2))
u_{28}	0.0106	((s_2,0),(0.5,0.5))	((s_2,0),(0.7,0.3))	((s_2,0),(0.7,0.2))	((s_4,0),(0.9,0.05))	((s_2,0),(0.6,0.3))
u_{29}	0.0133	((s_2,0),(0.6,0.4))	((s_2,0),(0.9,0.1))	((s_2,0),(0.9,0.1))	((s_4,0),(0.9,0.05))	((s_3,0),(0.6,0.4))

TABLE 5: Experts' Pythagorean 2-tuple linguistic weighted averages.

Expert	1	2	3
P2TLWA	((s3,0.0901),(0.2862,0.1856))	((s3,-0.2360),(0.8322,0.1931))	((s2,-0.2280),(0.7781,0.1233))
Expert	4	5	
P2TLWA	((s4,0.1698),(0.8612,0.0649))	((s4,-0.4564),(1,0))	

TABLE 6: The scores of each expert's assessment.

Expert	1	2	3	4	5
Score	2.546488	2.287582	1.40895	3.622417	3.5436

Set valued iteration [21] is used to determine the weights, which has avoided the shortcoming that it is hard to have direct accurate assignment for the various indicator weights. The calculation of the indicator weights in this study indicates that the most important five indicators are conventional fund, ear-marked fund, or self-financed fund and resources are available for HSE management each year to ensure that the safety measures can be implemented properly (u_7); college leaders are committed to the HSE management objectives and have clarified the HSE missions (u_1); Hazard identification, risk assessment and control, and preventive and control measures are in place (u_6); the relevant laboratory personnel have undergone trainings and have the HSE management awareness and capabilities (u_{12}). Appropriate safety level qualification, experimental activity qualification, and experiment personnel qualification are available for laboratories involving hazardous substance experiments (u_{22}), as shown in Table 4. It also indicates that the HSE management fund is inadequate for the Chinese college laboratories and the college leaders are not serious enough about it, which complies with the Chinese actual conditions. At the same time, it also shows that the hazard identification, assessment, and control, laboratory personnel HSE awareness and management capability, hazardous substance management, etc. in the HSE management are important to the HSE of the teachers and students, which is consistent with the HSE management philosophy.

Pythagorean 2-tuple linguistic set assessment method [20] is used considering the uncertainty of expert assessment. Both membership (support) and nonmembership (opposition) are considered in the assessment of the indicators. And cases where the sum is greater than 1 but the sum of the squares is less than 1 are considered. This method properly reflected the hesitation when people conduct the assessment and considered the different preference of the decision-makers. In theory, it can obtain more informative and exact results for the HSE performance assessment of college laboratories.

But, on the other hand, there are still some limitations with the proposed method. (1) In the proposed method, it contains 29 assessment indicators that are indicative and reflect the comprehensive assessment of the HSE performance. However, there may be some deficiencies affected by the knowledge and the HSE concept of the experts. (2) And the weights of assessment indicators should be recalculated during assessment for laboratories of different disciplines and majors due to the different highlights of HSE management. (3) In addition, each indicator's assessment rating, membership, and nonmembership should be available when completing the forms and they should meet the appropriate value range requirements. The assessment experts may feel that they are complicated.

7. Conclusion

The college laboratory HSE performance assessment system is established based on the previous studies considering the current state of the HSE management of the college laboratories.

The assessment system contains three important parts including (1) creation of the indicator system, (2) determination of the indicator weight, and (3) assessment of the HSE performance. In terms of the assessment indicators, it is tailored for the college laboratory HSE management. Set valued iteration [21] is used to determine the weights. And MAGDM method with Pythagorean 2-tuple linguistic operators [20] is used for assessment.

The proposed assessment system in this study is used to carry out an empirical study on the HSE performance management taking the laboratory in a certain college as an example. The results have indicated that the assessment system proposed in this study is informative and exact. In our future work, we shall continue to do laboratory HSE performance assessment in college with other fuzzy decision-making methods according to its limitations [29–34].

Conflicts of Interest

The authors declare that they have no conflicts of interest.

Acknowledgments

The authors are grateful for the support provided by programs with the Humanities and Social Sciences Foundation of Ministry of Education of the People's Republic of China (17XJA630003) and Experimental Technology and Management Project of Sichuan Normal University (SYJS2017025).

References

[1] E. F. Jorgensen, "Development and psychometric evaluation of the Research Laboratory Safe Behavior Survey (RLSBS)," *Journal of Chemical Health and Safety*, vol. 24, no. 5, pp. 38–43, 2017.

[2] E. Czornyj, D. Newcomer, I. Schroeder, N. L. Wayne, and C. A. Merlic, "Proceedings of the 2016 Workshop Safety By Design - Improving safety in research laboratories," *Journal of Chemical Health and Safety*, vol. 25, no. 4, pp. 36–49, 2018.

[3] D. Zohar, "Thirty years of safety climate research: Reflections and future directions," *Accident Analysis & Prevention*, vol. 42, no. 5, pp. 1517–1522, 2010.

[4] A. Shamaii, M. Omidvari, and F. H. Lotfi, "Health, safety and environmental unit performance assessment model under uncertainty (case study: steel industry)," *Environmental Modeling & Assessment*, vol. 189, no. 1, 2017.

[5] L. Yan, L. Zhang, W. Liang, W. Li, and M. Du, "Key factors identification and dynamic fuzzy assessment of health, safety and environment performance in petroleum enterprises," *Safety Science*, vol. 94, pp. 77–84, 2017.

[6] Y. E. Jihong, C. Ying, Z. Huawen, . Yimin, and X. Yupeng, "Discussion on the Construction of HSE Management System in Oil and Gas Engineering Laboratory," *Industrial safety and environmental protection*, vol. 39, no. 6, pp. 95-96, 2013.

[7] K. Rengarajan, "Laboratory self-inspection program participation as an indication of improved safety culture at Emory University," *Journal of Chemical Health and Safety*, vol. 19, no. 4, pp. 15–19, 2012.

[8] T. L. Saaty and L. G. Vargas, "Uncertainty and rank order in the analytic hierarchy process," *European Journal of Operational Research*, vol. 32, no. 1, pp. 107–117, 1987.

[9] P. Yarahmadi, S. Dashti, and G. R. Sabzghabaei, "Assessment and ranking of contractors from the point of view HSE performance using Multi-criteria decision making method (AHP and TOPSIS) in Imam Khomeini port complex," *Journal of Occupational Hygiene Engineering*, vol. 4, no. 4, pp. 70–80, 2018.

[10] M. Abbaspour, S. Toutounchian, E. Roayaei, and P. Nassiri, "A strategic management model for evaluation of health, safety and environmental performance," *Environmental Modeling & Assessment*, vol. 184, no. 5, pp. 2981–2991, 2012.

[11] W. Li, W. Liang, L. Zhang, and Q. Tang, "Performance assess-

ment system of health, safety and environment based on experts' weights and fuzzy comprehensive evaluation," *Journal of Loss Prevention in the Process Industries*, vol. 35, pp. 95–103, 2015.

[12] J. Kang, J. Zhang, and J. Gao, "Improving performance evaluation of health, safety and environment management system by combining fuzzy cognitive maps and relative degree analysis," *Safety Science*, vol. 87, pp. 92–100, 2016.

[13] E. Szmidt and J. Kacprzyk, "Distances between intuitionistic fuzzy sets," *Fuzzy Sets and Systems*, vol. 114, no. 3, pp. 505–518, 2000.

[14] A. S. Girsang, C.-W. Tsai, and C.-S. Yang, "Rectifying the Inconsistent Fuzzy Preference Matrix in AHP Using a Multi-Objective BicriterionAnt," *Neural Processing Letters*, vol. 44, no. 2, pp. 519–538, 2016.

[15] Z. J. Liao, "A literature review of firm's safety performance: its connotation, measurement and factors affecting it," *China Safety Science Journal*, vol. 25, no. 11, pp. 139–144, 2015.

[16] Y. J. Gu, K. L. Chen, and K. Yang, "Fuzzy comprehensive evaluation method based on analytic hierarchy process for falt risk analysis of power plant equipment," *IEEE*, vol. 3, pp. 443–448, 2008.

[17] L. A. Zadeh, "Fuzzy sets," *Information and Computation*, vol. 8, pp. 338–353, 1965.

[18] R. A. Ribeiro, "Fuzzy multiple attribute decision making: a review and new preference elicitation techniques," *Fuzzy Sets and Systems*, vol. 78, no. 2, pp. 155–181, 1996.

[19] K. T. Atanassov, "Intuitionistic fuzzy sets," *Fuzzy Sets and Systems*, vol. 20, no. 1, pp. 87–96, 1986.

[20] G. Wei, M. Lu, F. E. Alsaadi, T. Hayat, and A. Alsaedi, "Pythagorean 2-tuple linguistic aggregation operators in multiple attribute decision making," *Journal of Intelligent & Fuzzy Systems: Applications in Engineering and Technology*, vol. 33, no. 2, pp. 1129–1142, 2017.

[21] D. Krassowska and M. C. Zdun, "Embeddability of homeomorphisms of the circle in set-valued iteration groups," *Journal of Mathematical Analysis and Applications*, vol. 433, no. 2, pp. 1647–1658, 2016.

[22] T. Yang, Y. Kuo, D. Parker, and K. H. Chen, "A Multiple Attribute Group Decision Making Approach for Solving Problems with the Assessment of Preference Relations," *Mathematical Problems in Engineering*, vol. 2015, Article ID 849897, 10 pages, 2015.

[23] P. Liu, S.-M. Chen, and J. Liu, "Multiple attribute group decision making based on intuitionistic fuzzy interaction partitioned Bonferroni mean operators," *Information Sciences*, vol. 411, pp. 98–121, 2017.

[24] J.-L. Marendaz, J.-C. Suard, and T. Meyer, "A systematic tool for Assessment and Classification of Hazards in Laboratories (ACHiL)," *Safety Science*, vol. 53, pp. 168–176, 2013.

[25] P. Amir-Heidari, R. Maknoon, B. Taheri, and M. Bazyari, "A new framework for HSE performance measurement and monitoring," *Safety Science*, vol. 100, pp. 157–167, 2017.

[26] F. Herrera and L. Martínez, "An approach for combining linguistic and numerical information based on the 2-tuple fuzzy linguistic representation model in decision-making," *International Journal of Uncertainty, Fuzziness and Knowledge-Based Systems*, vol. 8, no. 5, pp. 539–562, 2000.

[27] F. Herrera and L. Martínez, "A 2-tuple fuzzy linguistic representation model for computing with words," *IEEE Transactions on Fuzzy Systems*, vol. 8, no. 6, pp. 746–752, 2000.

[28] R. R. Yager, "Pythagorean membership grades in multicriteria decision making," *IEEE Transactions on Fuzzy Systems*, vol. 22, no. 4, pp. 958–965, 2014.

[29] G. Wei, H. Gao, and Y. Wei, "Some q-rung orthopair fuzzy Heronian mean operators in multiple attribute decision making," *International Journal of Intelligent Systems*, vol. 33, no. 7, pp. 1426–1458, 2018.

[30] Z. Ma and Z. Xu, "Symmetric Pythagorean Fuzzy Weighted Geometric/Averaging Operators and Their Application in Multicriteria Decision-Making Problems," *International Journal of Intelligent Systems*, vol. 31, no. 12, pp. 1198–1219, 2016.

[31] G. W. Wei, M. Lu, X. Y. Tang, and Y. Wei, "Pythagorean Hesitant Fuzzy Hamacher Aggregation Operators and Their Application to Multiple Attribute Decision Making," *International Journal of Intelligent Systems*, vol. 33, no. 6, pp. 1197–1233, 2018.

[32] H. Gao, G. Wei, and Y. Huang, "Dual Hesitant Bipolar Fuzzy Hamacher Prioritized Aggregation Operators in Multiple Attribute Decision Making," *IEEE Access*, vol. 6, pp. 11508–11522, 2017.

[33] G. W. Wei, "Some similarity measures for picture fuzzy sets and their applications," *Iranian Journal of Fuzzy Systems*, vol. 15, no. 1, pp. 77–89, 184, 2018.

[34] G. Deschrijver, C. Cornelis, and E. E. Kerre, "On the representation of intuitionistic fuzzy t-norms and t-conorms," *IEEE Transactions on Fuzzy Systems*, vol. 12, no. 1, pp. 45–61, 2004.

9

Solubility Optimal System for Supercritical Fluid Extraction Based on a New Nonlinear Temperature-Pressure Decoupling Model Constructed with Unequal-Interval Grey Optimal Models and Peng-Robinson Models

Binglin Li and Wen You

School of Electrical and Electronic Engineering, Changchun University of Technology, Changchun 130012, China

Correspondence should be addressed to Wen You; youwen@mail.ccut.edu.cn

Academic Editor: Ivan Giorgio

This paper presents a new solubility optimal system to improve the efficiency of supercritical fluid extraction (SFE). The major contribution is a nonlinear temperature-pressure decoupling model constructed with unequal-interval grey optimal models (UEIGOMs) and Peng-Robinson models (PRMs). The linear parts of temperature and pressure process can be constructed with UEIGOM, respectively. The nonlinear parts of temperature and pressure process can be described by PRMs, respectively. The whole nonlinear model cannot be input-output decoupled resulting from the singularity of decoupling matrix for PRM. This problem on input-output nondecoupling can be transformed to the problem on disturbance decoupling for a class of MIMO nonlinear systems. Therefore, the whole nonlinear coupling model can be disturbance decoupled. Furthermore, solubility optimal method is presented in the paper; it can calculate the optimal pressure according to the given temperature, namely, optimal working points, to maximize solubility for SFE process. The feasibility, effectiveness, and practicality of the proposed nonlinear temperature-pressure decoupling model constructed with UEIGOMs and PRMs are verified by SFE experiments in biphenyl. Experiments using the designed solubility optimal system are carried out to demonstrate the effectiveness in control scheme, simplicity in structure, and flexibility in implementation for the proposed solubility optimal system based on a new nonlinear temperature-pressure coupling model constructed with UEIGOMs and PRMs.

1. Introduction

Extraction of a material using a supercritical fluid is called supercritical fluid extraction (SFE). SFE, which is a contamination-free extraction technology in food science and chemical industry, is of central importance in biomaterial processing. During the separation process, the solvency of SFE can be modified by adjusting temperature, pressure, moisture contents, and so on [1–4]. Temperature and pressure play a crucial role in SFE process. The model of temperature and pressure process, which is nonlinear, is composed of linear and nonlinear parts. The linear part, which can be obtained easily, is SISO model of temperature or pressure. The nonlinear part is the coupling relationship between temperature and pressure. Numerous methods and models have been proposed to describe SFE process. Cubic equation of state (EoS) with simplified inner structure and generalized form is one of the most widely used models to describe the temperature, pressure, and time behaviors for fluid. vdW EoS, which is used in calculation of vapor-liquid equilibrium, was proposed by van der Waals in 1873 [5]. RK EoS was proposed by Redlich and Kwong in 1949 [6]. RK EoS was improved by Soave, and RKS EoS was proposed in 1979 [7]. RP EoS was proposed by Peng and Robinson in 1976 [8]. PR EoS is widely used and contrasted with the other three models. A hybrid model, which is constructed with a radial basis function (RBF) model and RP EoS, was proposed to keep all the physical information in PR model and optimize the binary interaction parameter in the PR model [9, 10]. Combined with operating cost, safety index, and yield rate of extraction calculated by hybrid model, an optimal control system can be designed. However, temperature has a momentous effect on pressure, in temperature and pressure control process, and vice versa. The coupling relationship between temperature and pressure in SFE process is not considered in the proposed optimal control system. The performance of temperature and pressure control has an effect on yield rate and solubility of extraction. Therefore, combining with PR EoS, study on modeling of temperature and pressure coupling and decoupling model has great significance with new theories and methods for improving SFE work efficiency to obtain maximal yield rate and solubility of extraction.

In this work, the linear parts of temperature and pressure models can be described through UEIGOMs with grey technology, respectively. The nonlinear parts of temperature and pressure models are modeled with PR EoS. Discussing the decoupling conditions of input-output of temperature-pressure nonlinear model, the decoupling system is given through state and output transforms. Furthermore, solubility optimal method is presented in the paper; it can calculate the optimal pressure according to the given temperature, namely, optimal working points, to maximize solubility for SFE process. The rest of the paper is organized as follows: In Section 2, temperature-pressure process and solubility modeling are discussed. In Section 3, temperature-pressure decoupling control system and solubility optimal system are designed. Computer simulation results of UEIGOMs of temperature and pressure process, temperature-pressure decoupling control, and solubility optimal system are presented and discussed, respectively, in Section 4. Finally, a conclusion regarding research and future works is made in Section 5.

2. Temperature-Pressure Process and Solubility Modeling

Due to the reaction character of SFE process, the nonlinear strong coupling relationship between temperature and pressure is the main factor affecting the efficiency of SFE. In order to improve the performance of SFE control system effectively, it is necessary to model temperature-pressure process. In this work, the temperature and the pressure processes can be modeled by grey technology, respectively, and their UEIGOMs can be given; the coupling parts can be obtained by PRMs.

2.1. UEIGOM for Temperature-Pressure Process. Grey system theory is based on fewer samples, which are made of some known and unknown information. It can work on extracting the valuable information from the fewer samples. Grey generating technology can provide intermediate information and weaken the randomness of original data; therefore, the model using grey generating technology can give the correct description of the reaction character of SFE process. In grey calculation, accumulated generating operation (AGO) and inverse accumulated generating operation (IAGO) are utilized [11–13]. They can be defined as follows.

Let $\mathbf{x}^{(0)} \in \mathbf{R}^{1\times n}$ be original series and $\mathbf{x}^{(r)} \in \mathbf{R}^{1\times n}$ be r-AGO series, if there is an accumulated generated matrix $\mathbf{A}_1 \in \mathbf{R}^{n\times n}$ that satisfies (1); let $\mathbf{x}^{(r)} \in \mathbf{R}^{1\times n}$ be r-IAGO series, if there is an inverse accumulated generated matrix $\mathbf{A}_2 \in \mathbf{R}^{n\times n}$ that satisfies (2).

$$\mathbf{x}^{(r)} = \mathbf{x}^{(r-1)}\mathbf{A}_1, \quad (1)$$

$$\mathbf{x}^{(r-1)} = \mathbf{x}^{(r)}\mathbf{A}_2, \quad (2)$$

where

$$\begin{gathered}\mathbf{x}^{(0)} = \left[x^{(0)}(1), x^{(0)}(2), \ldots, x^{(0)}(n)\right]; \\ \mathbf{x}^{(r)} = \left[x^{(r)}(1), x^{(r)}(2), \ldots, x^{(r)}(n)\right]; \\ \mathbf{A}_1 = \begin{bmatrix} 1 & 1 & \cdots & 1 \\ 0 & 1 & \cdots & 1 \\ \vdots & \vdots & \ddots & \vdots \\ 0 & 0 & \cdots & 1 \end{bmatrix}; \\ \mathbf{A}_2 = \begin{bmatrix} 1 & -1 & 0 & \cdots & 0 \\ 0 & 1 & -1 & \cdots & 0 \\ 0 & 0 & 1 & \cdots & 0 \\ \vdots & \vdots & \vdots & \ddots & \vdots \\ 0 & 0 & 0 & \cdots & 1 \end{bmatrix}.\end{gathered} \quad (3)$$

2.1.1. Transforming from Unequal-Interval Series to Equal-Interval Series. AGO and IAGO operations can all aim at equal-interval series; however, actual process signals are always unequal-interval series. Therefore, it is necessary to transform to equal-interval series before AGO or IAGO operation.

Let $\mathbf{x}_1^{(0)} = [x_1^{(0)}(t_1), x_1^{(0)}(t_2), \ldots, x_1^{(0)}(t_n)]$, $i = 1, 2, \ldots, n$, be an unequal-interval series; then its average interval can be calculated as follows:

$$\Delta t = \frac{\sum_{i=2}^{n} \Delta t_i}{n-1} = \frac{t_n - t_2}{n-1}, \quad i = 2, 3, \ldots, n. \quad (4)$$

The coefficient between every interval and average interval is given.

$$\mu(t_i) = \frac{t_i - (i-1)\Delta t}{\Delta t}, \quad i = 2, 3, \ldots, n. \quad (5)$$

The total difference value of every interval can be calculated.

$$\Delta x_1^{(0)}(t_i) = \mu(t_i)\left[x_1^{(0)}(t_i) - x_1^{(0)}(t_{i-1})\right], \quad i = 2, 3, \ldots, n. \quad (6)$$

Above all, the equal-interval series $\overline{\mathbf{x}}_2^{(0)} = [\overline{x}_2^{(0)}(1), \overline{x}_2^{(0)}(2), \ldots, \overline{x}_2^{(0)}(i)]$, $i = 1, 2, \ldots, n$, can be obtained, where

$$\begin{gathered}\overline{x}_2^{(0)}(1) = x_1^{(0)}(t_1), \quad i = 1, \\ \overline{x}_2^{(0)}(i) = x_1^{(0)}(t_i) - \Delta x_1^{(0)}(t_i), \quad i = 2, 3, \ldots, n.\end{gathered} \quad (7)$$

2.1.2. Grey Optimal Model. Series $\overline{\mathbf{x}}_2^{(1)}$ is obtained by 1-AGO with series $\overline{\mathbf{x}}_2^{(0)}$; namely, $\overline{\mathbf{x}}_2^{(1)} = \overline{\mathbf{x}}_2^{(0)}\mathbf{A}_1$. Define $\mathbf{Z}^{(1)} = [z^{(1)}(2), z^{(1)}(3), \ldots, z^{(1)}(n)]$ as the black ground value of series $\overline{\mathbf{x}}_2^{(1)}$, where $z^{(1)}(k) = \alpha\overline{x}_2^{(1)}(k-1) + (1-\alpha)\overline{x}_2^{(1)}(k)$, $k = 2, 3, \ldots, n$. Generally, $\alpha = 0.5$; then differential equation can be described as

$$\frac{d\overline{\mathbf{x}}_2^{(1)}}{dt} + a\overline{\mathbf{x}}_2^{(1)} = b, \tag{8}$$

where a is developing coefficient and b is grey input. Utilizing least square method, a and b can be obtained:

$$\hat{\mathbf{a}} = \begin{bmatrix} a \\ b \end{bmatrix} = \left(\mathbf{B}^T\mathbf{B}\right)^{-1}\mathbf{B}^T\mathbf{y}_N, \tag{9}$$

where

$$\mathbf{B} = \begin{bmatrix} -z^{(1)}(2) & 1 \\ -z^{(1)}(3) & 1 \\ \vdots & \vdots \\ -z^{(1)}(n) & 1 \end{bmatrix}; \qquad \mathbf{y}_N = \begin{bmatrix} \overline{x}_2^{(0)}(2) \\ \overline{x}_2^{(0)}(3) \\ \vdots \\ \overline{x}_2^{(0)}(n) \end{bmatrix}. \tag{10}$$

Time response function of GOM can be expressed as follows:

$$\begin{aligned} \hat{\overline{x}}_2^{(1)}(k+1) &= \left(\overline{x}_2^{(0)}(1) - \frac{b}{a} - c\right)e^{-ak} + \frac{b}{a} + c, \\ \hat{\overline{x}}_2^{(0)}(k+1) &= \hat{\overline{x}}_2^{(1)}(k+1) - \hat{\overline{x}}_2^{(1)}(k), \end{aligned} \tag{11}$$

where c is translation value.

2.1.3. Reversion from Equal-Interval Series to Unequal-Interval Series. Considering $k+1 = t_i/\Delta t$, (11) can be transformed for unequal-interval series, namely, UEIGOM:

$$\begin{aligned} \hat{x}_1^{(1)}(t) &= \left(x_1^{(0)}(t_1) - \frac{b}{a} - c\right)e^{-a(t/\Delta t)} + \frac{b}{a} + c, \\ \hat{x}_1^{(0)}(t) &= \hat{x}_1^{(1)}(t) - \hat{x}_1^{(1)}(t - \Delta t), \end{aligned} \tag{12}$$

where $c = ((e^a + 1)/(1 - e^{-2(n-1)a}))\sum_{i=1}^{n-1} q^{(0)}(t_{i+1})e^{-a(t_i/\Delta t)}$; $q^{(0)}(t_i) = x_1^{(0)}(t_i) - \hat{x}_1^{(0)}(t_i)$ is residual error. Therefore, the temperature and the pressure processes can be described as first-order inertia system by UEIGOM, respectively.

2.2. PRM Modeling. Peng-Robinson model is the most commonly exploited model for treating solubility in SFE. The nonlinear strong coupling relationship between temperature and pressure is reflected in the Peng-Robinson equation of state. Therefore, the coupling parts of temperature-pressure process can be embodied by PRM in this work. PRM is given as

$$p = \frac{RT}{V - d_1} - \frac{\Omega_{11}(T)}{V^2 + 2d_1V - d_1^2}, \tag{13}$$

where R is gas constant, T is absolute temperature, V is the molar volume of pure solvent, $\Omega_{11}(T)$ is the parameter describing attractive interactions between molecules, T_c is critical temperature, p_c is critical pressure, ω is acentric factor, T_b is normal boiling point, and d_1 is the parameter describing volume exclusion and repulsive interactions. The parameters in PR model are listed in Table 1. Subscript 1 represents the solvent and subscript 2 represents the solute. Considering that the molar volume of pure solvent is invariable, (13) can be simplified as follows:

$$p = \eta_1 T + \eta_2 T^{1/2} + \eta_3, \tag{14}$$

where

$$\begin{aligned} \eta_1 &= \frac{p_c R\left(V^2 + 2d_1V - d_1^2\right) - 0.457235R^2T_c(V - d_1)f^2(\omega_1)}{p_c\left(V^2 + 2d_1V - d_1^2\right)(V - d_1)}; \\ \eta_2 &= \frac{0.457235R^2T_c^{3/2}\left[2f^2(\omega_1) + 2f(\omega_1)\right]}{p_c\left(V^2 + 2d_1V - d_1^2\right)}; \\ \eta_3 &= -\frac{0.457235R^2T_c^2\left[1 + f(\omega_1)\right]^2}{p_c\left(V^2 + 2d_1V - d_1^2\right)}. \end{aligned} \tag{15}$$

Furthermore, the inverse function of (14) can be obtained as follows:

$$T = g^{-1}(T). \tag{16}$$

Combining UEIGOM with (14) and (16), let $x_1 = T$ and $x_2 = p$ be state variables; the temperature-pressure process model for CO_2 can be described as follows:

$$\begin{aligned} \dot{x}_1 &= a_1x_1 + b_1u_1, \\ \dot{x}_2 &= a_2x_2 + b_2u_2, \\ y_1 &= x_1 + p_1(x_2), \\ y_2 &= x_2 + p_2(x_1), \end{aligned} \tag{17}$$

where $\mathbf{x} \in \mathbf{R}^2$, $\mathbf{u} \in \mathbf{R}^2$, $\mathbf{y} \in \mathbf{R}^2$, $\mathbf{p}(\mathbf{x}) = \left[(\zeta_1\sqrt{\zeta_2x_2 + \zeta_3} + \zeta_4)^2 \;\; \eta_1x_1 + \eta_2{x_1}^{1/2} + \eta_3\right]^T$, and $a_1a_2b_1b_2\eta_1\eta_2\eta_3\zeta_1\zeta_2\zeta_3\zeta_4 \neq 0$, and they are system constants; $\mathbf{x}_0 = [0 \;\; 0]^T$; $\mathbf{p}_0 = \left[(\zeta_1\sqrt{\zeta_3} + \zeta_4)^2 \;\; \eta_3\right]^T$. This temperature-pressure nonlinear model cannot input-output decoupled result from the singularity of decoupling matrix. Namely, input-output decoupling conditions are unsatisfied [14, 15].

TABLE 1: Parameters in PR model.

Parameters	Values of pure component	Values of mixture component
$\Omega(T)$	$\Omega_{11}(T) = \frac{0.457235R^2T_{c,1}^2}{p_{c,1}}\left[1 + f(\omega_1)\left(1 - T_{r,1}^{1/2}\right)\right]^2$	$\Omega_{12}(T) = \frac{0.457235R^2T_{c,1}T_{c,2}}{2\sqrt{p_{c,1}p_{c,2}}}\left[1 + f(\omega_1)\left(1 - T_{r,1}^{1/2}\right)\right]\left[1 + f(\omega_2)\left(1 - T_{r,2}^{1/2}\right)\right]$
$f(\omega)$	$f(\omega_1) = 0.37464 + 1.54226\omega_1 - 0.26992{\omega_1}^2$	$f(\omega_2) = 0.37464 + 1.54226\omega_2 - 0.26992{\omega_2}^2$
ω	$\omega_1 = \left(\frac{3}{7}\left(\frac{T_{b,1}}{T_{c,1} - T_{b,1}}\right)\log p_{c,1} - 1\right)$	$\omega_2 = \left(\frac{3}{7}\left(\frac{T_{b,2}}{T_{c,2} - T_{b,2}}\right)\log p_{c,2} - 1\right)$
T_r	$T_{r,1} = \frac{T}{T_{c,1}}$	$T_{r,2} = \frac{T}{T_{c,2}}$
d	$d_1 = \frac{0.07780RT_{c,1}}{p_{c,1}}$	$d_2 = \frac{0.07780RT_{c,2}}{p_{c,2}}$

Let $\overline{\mathbf{x}} = \mathbf{x} + \mathbf{p}(\mathbf{x}) - \mathbf{p}_0$ and $\overline{\mathbf{y}} = \mathbf{y} - \mathbf{p}_0$; system (17) can be represented as follows:

$$\begin{aligned}\dot{\overline{x}}_1 &= a_1\overline{x}_1 + b_1u_1 + a_1\left[-p_1(x_2) + p_1(x_{20})\right],\\ \dot{\overline{x}}_2 &= a_2\overline{x}_2 + b_2u_2 + a_2\left[-p_2(x_1) + p_2(x_{10})\right],\\ \overline{y}_1 &= \overline{x}_1,\\ \overline{y}_2 &= \overline{x}_2,\end{aligned} \tag{18}$$

where

$$\begin{aligned}\mathbf{f}(\overline{\mathbf{x}}) &= \left[a_1\overline{x}_1 \;\; a_2\overline{x}_2\right]^T;\\ h_1(\overline{\mathbf{x}}) &= \overline{x}_1;\\ h_2(\overline{\mathbf{x}}) &= \overline{x}_2;\\ \mathbf{g}_1(\overline{\mathbf{x}}) &= \left[b_1 \;\; 0\right]^T;\\ \mathbf{g}_2(\overline{\mathbf{x}}) &= \left[0 \;\; b_2\right]^T;\\ \mathbf{e}_1(\mathbf{x}) &= a_1\left[-p_1(x_2) + p_1(x_{20}) \;\; 0\right]^T;\\ \mathbf{e}_2(\mathbf{x}) &= a_2\left[0 \;\; -p_2(x_1) + p_2(x_{10})\right]^T;\\ \overline{\mathbf{x}}_0 &= \left[\left(\zeta_1\sqrt{\zeta_3} + \zeta_4\right)^2 \;\; \eta_3\right]^T.\end{aligned} \tag{19}$$

Definition 1 (see [14, 15]). System (18) is said to have a vector relative degree $\boldsymbol{\rho} = [\rho_1, \ldots, \rho_m] (u \in \mathbf{R}^m)$ for inputs on the initial state $\overline{\mathbf{x}}_0$ if

(i) $L_{\mathbf{g}_j}L_{\mathbf{f}}^k h_i(\overline{\mathbf{x}}) = 0$ for all $\overline{\mathbf{x}}$ in the field $\overline{\mathbf{x}}_0$, $i, j = 1, \ldots, m$, $k < \rho_i - 1$, where

$$\begin{aligned}L_{\mathbf{f}}^k h_i(\overline{\mathbf{x}}) &:= \frac{\partial\left(L_{\mathbf{f}}^{k-1}h_i(\overline{x})\right)}{\partial\overline{\mathbf{x}}}\mathbf{f}(\overline{\mathbf{x}});\\ L_{\mathbf{g}}L_{\mathbf{f}}h_i(\overline{\mathbf{x}}) &:= \frac{\partial\left(L_{\mathbf{f}}h_i(\overline{x})\right)}{\partial\overline{\mathbf{x}}}\mathbf{g}(\overline{\mathbf{x}}),\end{aligned} \tag{20}$$

(ii) decoupling matrix

$$\mathbf{A}(\overline{x}) = \begin{bmatrix} L_{\mathbf{g}_1}L_{\mathbf{f}}^{r_1-1}h_1(\overline{\mathbf{x}}) & \cdots & L_{\mathbf{g}_m}L_{\mathbf{f}}^{r_1-1}h_1(\overline{\mathbf{x}}) \\ \vdots & \cdots & \vdots \\ L_{\mathbf{g}_1}L_{\mathbf{f}}^{r_m-1}h_m(\overline{\mathbf{x}}) & \cdots & L_{\mathbf{g}_m}L_{\mathbf{f}}^{r_1-1}h_m(\overline{\mathbf{x}}) \end{bmatrix}_{m\times m} \tag{21}$$

is nonsingular on the initial state $\overline{\mathbf{x}}_0$.

According to Definition 1, system (18) has a vector relative degree $\boldsymbol{\rho} = [1, 1]$ for inputs.

Proposition 2 (see [14, 15]). *If system (18) has a vector relative degree, namely, decoupling matrix $\mathbf{A}(\overline{x})$ is nonsingular on the $\overline{\mathbf{x}}_0$, then the inputs and outputs of system can be decoupled near $\overline{\mathbf{x}}_0$ through a static state feedback and vice versa.*

Therefore, system (18) can be decoupled through a static state feedback. Consider a feedback control law:

$$u_i = \alpha_i(\overline{\mathbf{x}}) + \beta_{i1}(\overline{\mathbf{x}})v_1 + \beta_{i2}(\overline{\mathbf{x}})v_2, \quad m = 2, \tag{22}$$

where $\alpha(\overline{\mathbf{x}}) := -\mathbf{A}^{-1}(\overline{\mathbf{x}})\mathbf{b}(\overline{\mathbf{x}})$ is an analysis function vector, $\beta(\overline{\mathbf{x}}) = \mathbf{A}^{-1}(\overline{\mathbf{x}})$ is a nonsingular matrix, $\mathbf{b}(\overline{\mathbf{x}}) = [L_{\mathbf{f}}^{\rho_1}h_1(\overline{\mathbf{x}}) \cdots L_{\mathbf{f}}^{\rho_m}h_m(\overline{\mathbf{x}})]^T$, and $\mathbf{v} \in \mathbf{R}^m$ are new input variables. Substituting (22) into (18), the close-loop system can be obtained:

$$\begin{aligned}\dot{\overline{x}}_1 &= \left[a_1\overline{x}_1 + b_1\alpha_1(\overline{\mathbf{x}})\right] + b_1\left[\beta_{11}(\overline{\mathbf{x}})v_1 + \beta_{12}(\overline{\mathbf{x}})v_2\right]\\ &\quad + a_1\left[-p_1(x_2) + p_1(x_{20})\right],\\ \dot{\overline{x}}_2 &= \left[a_2\overline{x}_2 + b_2\alpha_2(\overline{\mathbf{x}})\right] + b_2\left[\beta_{21}(\overline{\mathbf{x}})v_1 + \beta_{22}(\overline{\mathbf{x}})v_2\right]\\ &\quad + a_2\left[-p_2(x_1) + p_2(x_{10})\right],\\ \overline{y}_1 &= \overline{x}_1,\\ \overline{y}_2 &= \overline{x}_2,\end{aligned} \tag{23}$$

where

$$\begin{aligned}\widetilde{\mathbf{f}}(\overline{\mathbf{x}}) &= \left[a_1\overline{x}_1 + b_1\alpha_1(\overline{\mathbf{x}}) \;\; a_2\overline{x}_2 + b_2\alpha_2(\overline{\mathbf{x}})\right]^T;\\ \widetilde{h}_1(\overline{\mathbf{x}}) &= \overline{x}_1;\\ \widetilde{h}_2(\overline{\mathbf{x}}) &= \overline{x}_2;\\ \widetilde{\mathbf{g}}_1(\overline{\mathbf{x}}) &= \left[b_1\beta_{11}(\overline{\mathbf{x}}) \;\; b_2\beta_{21}(\overline{\mathbf{x}})\right]^T;\\ \widetilde{\mathbf{g}}_2(\overline{\mathbf{x}}) &= \left[b_1\beta_{12}(\overline{\mathbf{x}}) \;\; b_2\beta_{22}(\overline{\mathbf{x}})\right]^T.\end{aligned} \tag{24}$$

Theorem 3 (see [14]). *MIMO nonlinear system (18) has the same vector relative degree as MIMO nonlinear system (23) with a static state feedback for inputs; namely,*

$$\boldsymbol{\rho} = \overline{\boldsymbol{\rho}}. \tag{25}$$

Definition 4 (see [14, 15]). System (18) is said to have a vector relative degree $\boldsymbol{\sigma} = [\sigma_1, \ldots, \sigma_m] (u \in \mathbf{R}^m)$ for disturbance on the initial state $\overline{\mathbf{x}}_0$ if $L_{\mathbf{e}_j}L_{\mathbf{f}}^k h_i(\overline{\mathbf{x}}) = 0$ for all $\overline{\mathbf{x}}$ in the field $\overline{\mathbf{x}}_0$, $i, j = 1, \ldots, m$, $k < \sigma_i - 1$; and $L_{\mathbf{e}_j}L_{\mathbf{f}}^{\sigma_i-1}h_i(\overline{\mathbf{x}}) \neq 0$.

According to Definition 1, system (23) has a vector relative degree $\boldsymbol{\sigma} = [\infty, \infty]$ for disturbance.

Theorem 5. *The nonlinear system (18) with $\boldsymbol{\rho}$ and $\boldsymbol{\sigma}$ is said to be disturbance decoupled on the initial state $\overline{\mathbf{x}}_0$ if and only if*

$$\rho_i < \sigma_i, \quad i = 1, \ldots, m. \tag{26}$$

According to Theorem 5, $\rho_1 < \sigma_1$, $\rho_2 < \sigma_2$, system (18) can be disturbance decoupled; namely,

$$\begin{aligned} z_1^1 &= \overline{x}_1,\\ \dot{z}_1^1 &= v_1,\end{aligned}$$

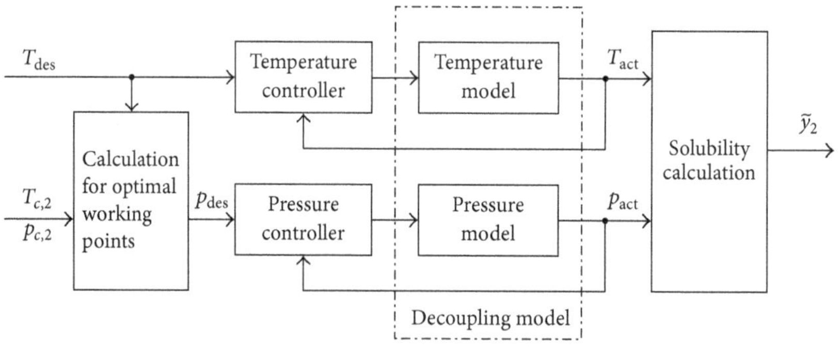

FIGURE 1: Schematic diagram of the proposed solubility optimal system for SFE.

$$\begin{aligned} z_1^2 &= \overline{x}_2, \\ \dot{z}_1^2 &= v_2, \\ \overline{y}_1 &= z_1^1, \\ \overline{y}_2 &= z_1^2. \end{aligned} \tag{27}$$

2.3. Solubility Modeling. The Peng-Robinson equation for the mole fraction, $\tilde{y}_2$, at saturation of a solute of low volatility in SFE can be written as [2]

$$\ln \tilde{y}_2 = \ln \frac{p_v(T)}{p} + \frac{pV_m}{RT} - \ln \phi_2, \tag{28}$$

where $p_v(T)$ is the vapors pressure of the solute, V_m is the volume of the pure solute, and ϕ_2 is the fugacity coefficient, which can be calculated as

$$\begin{aligned} \ln \phi_2 = &-\frac{\Omega_{11}(T)}{2\sqrt{2}RTd_1}\left(\frac{2\Omega_{12}(T)}{\Omega_{11}(T)} - \frac{d_2}{d_1}\right) \\ &\cdot \ln \frac{V + \left(1+\sqrt{2}\right)d_1}{V + \left(1-\sqrt{2}\right)d_1} + \ln \frac{RT}{p(V-d_1)} \\ &+ \frac{d_2}{d_1}\left(\frac{pV}{RT} - 1\right), \end{aligned} \tag{29}$$

where $\Omega_{11}(T)$, $\Omega_{12}(T)$, d_1, and d_2, which are listed in Table 1, can be calculated by combining rules and are given in [2]. $\Omega_{12}(T)$ can be given with a parameter k_{12} in [2], which describes the mixture. It can be calculated by RBF neural network. However, $k_{12} = 0.5$ is set in this work. The solubility optimal control can be carried out through adjusting temperature and pressure parameters, rather than the parameter in PR model.

3. Temperature-Pressure Decoupling Control and Solubility Optimal Control

A solubility optimal system for SFE is presented, and its effectiveness is evaluated through simulation experiments. The overall control scheme is shown in Figure 1.

3.1. Temperature-Pressure Decoupling Control. Temperature and pressure control are playing an increasingly important role in SFE process. In this work, proportional-integral-derivative (PID) controller is chosen as the temperature and pressure controllers. The transfer function of the PID controller is given by

$$G_{\text{PID}}(s) = K_P\left(1 + \frac{1}{T_I s} + T_D s\right), \tag{30}$$

where K_P is proportional gain, T_I is integral time constant, and T_D is derivative time constant.

3.2. Solubility Optimal Calculation. The objective of solubility optimization is to find the optimal working points, namely, the optimal temperature and pressure, to maximize solubility for SFE process. Combining (27) with (28), the yield rate of extraction process can be represented with $\lambda(T, p)$; namely,

$$\tilde{y}_2 = e^{\lambda(T,p)}, \tag{31}$$

where $\lambda(T, p) = \ln(p_v(T)/p + pV_m/RT) - \ln \phi_2$. Hence the optimal calculation is given as

$$\begin{aligned} \frac{\partial \tilde{y}_2}{\partial T} &= \frac{\partial \lambda(T,p)}{\partial T} = 0, \\ \frac{\partial \tilde{y}_2}{\partial p} &= \frac{\partial \lambda(T,p)}{\partial p} = 0, \\ T_l \le T &\le T_h, \quad p_l \le p \le p_h, \end{aligned} \tag{32}$$

where T_l and T_h are the lowest and highest temperature, respectively, and p_l and p_h are the lowest and highest pressure, respectively. Equation (32) can obtain the relationship between temperature and pressure in optimal work points and the optimal molar volume of pure solute. Therefore, the operating parameters can comply with (32).

4. Simulation and Experiment Results and Analysis

Study works are based on SFE optimal system, which is shown in Figure 2. It consists of SFE control equipment and SFE monitoring system. Temperature and pressure control

TABLE 2: Experimental parameters of biphenyl in supercritical CO_2.

Time (min)	0	43	97	147	183	210
Temperature (K)	313.05	316.75	318.45	322.85	327.95	332.85
Pressure (MPa)	8.00	8.40	9.10	9.64	10.20	10.72
Temperature UEIGOM	313.4697	316.6856	320.7714	324.6015	327.3875	329.4927
	$C = 0.2713$, $P = 100\%$					
Pressure UEIGOM	7.8164	8.3336	9.0318	9.7304	10.2666	10.6880
	$C = 0.0958$, $P = 100\%$					

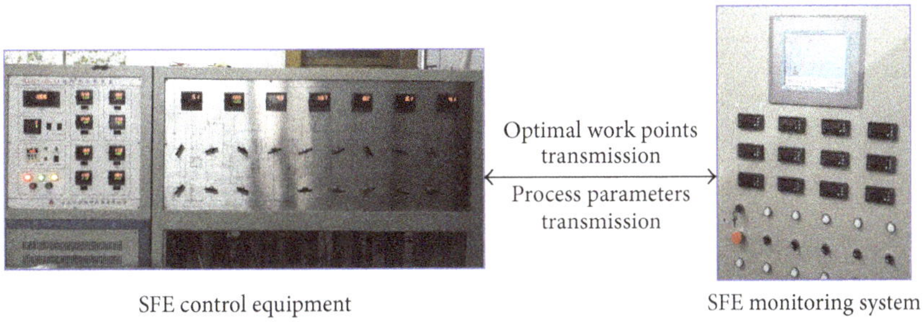

FIGURE 2: Schematic diagram of the proposed solubility optimal system for SFE.

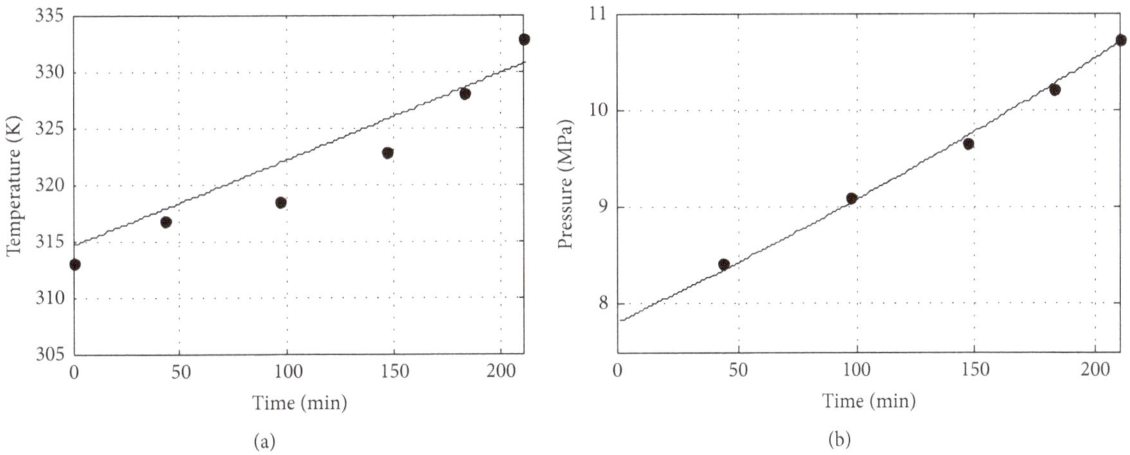

FIGURE 3: UEIGOM of temperature and pressure process. (a) Temperature process. (b) Pressure process.

can be implemented by SFE control equipment, which is improved SFE equipment (HA21-40-11), for SFE process. In SFE monitoring system, the process parameters can be monitored, and the optimal work points of temperature and pressure to maximize solubility can be calculated and sent to SFE control equipment.

4.1. Temperature-Pressure Process Modeling and Decoupling Control. According to SFE experiments in biphenyl, the experimental parameters are listed in Table 2. Define C as proportionality and P as error frequency, which are related to the quality of UEIGOM. C, the smaller the better, generally requires its maximum to not be more than 0.65. P generally requires more than 0.95 and not less than 0.7. It can be seen with data that the temperature and pressure process can be all described with the first order of inertia, completely. The state equations of temperature and pressure process can be obtained by UEIGOM, which can be shown in Figure 3, as follows:

$$\begin{bmatrix} \dot{x}_1 \\ \dot{x}_2 \end{bmatrix} = \begin{bmatrix} -4201.7 & 0 \\ 0 & -666.7 \end{bmatrix} \begin{bmatrix} x_1 \\ x_2 \end{bmatrix} + \begin{bmatrix} 1321800 & 0 \\ 0 & 5210.7 \end{bmatrix} \begin{bmatrix} u_1 \\ u_2 \end{bmatrix}, \qquad (33)$$

$$\begin{bmatrix} y_1 \\ y_2 \end{bmatrix} = \begin{bmatrix} 1 & 0 \\ 0 & 1 \end{bmatrix} \begin{bmatrix} x_1 \\ x_2 \end{bmatrix}.$$

Combining (13) with (16), the temperature and pressure coupling model, namely, input-output coupling model, can be described as follows:

$$\begin{bmatrix} \dot{x}_1 \\ \dot{x}_2 \end{bmatrix}$$

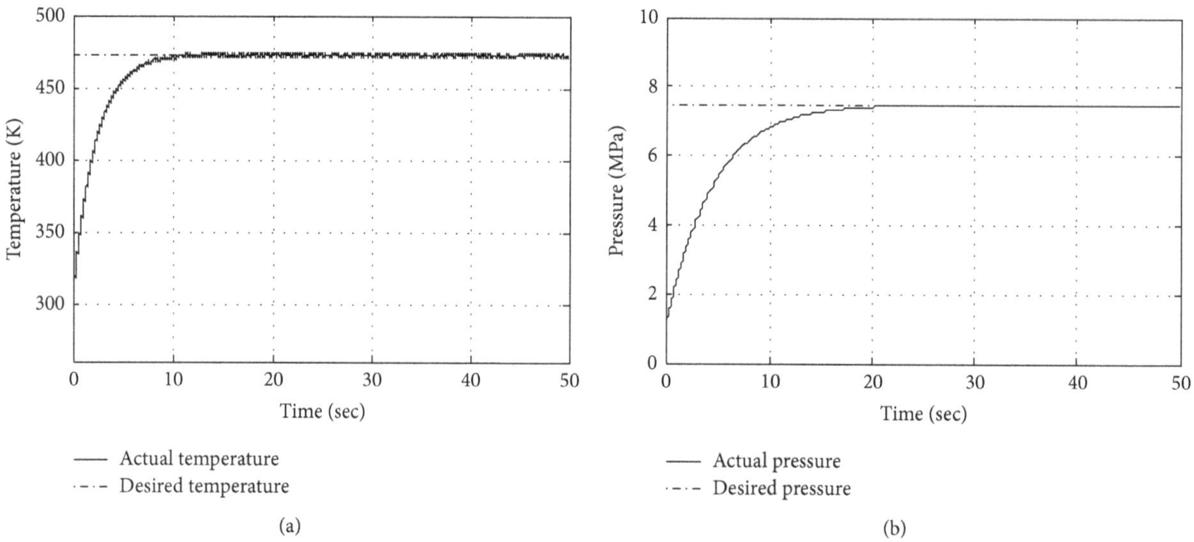

FIGURE 4: Temperature and pressure decoupling close-loop control. (a) Temperature control. (b) Pressure control.

$$= \begin{bmatrix} -4201.7 & 0 \\ 0 & -666.7 \end{bmatrix} \begin{bmatrix} x_1 \\ x_2 \end{bmatrix} + \begin{bmatrix} 1321800 & 0 \\ 0 & 5210.7 \end{bmatrix} \begin{bmatrix} u_1 \\ u_2 \end{bmatrix},$$

$$\begin{bmatrix} y_1 \\ y_2 \end{bmatrix} = \begin{bmatrix} 1 & 0 \\ 0 & 1 \end{bmatrix} \begin{bmatrix} x_1 \\ x_2 \end{bmatrix} + \begin{bmatrix} \left(0.0089158\sqrt{2243.2x_2 + 177211} + 16.422\right)^2 \\ 560.8004x_1 - 18419\sqrt{x_1} + 151160 \end{bmatrix}. \quad (34)$$

According to (27), state equation (34), which can be disturbance decoupled, is decoupled as follows:

$$\begin{bmatrix} \dot{\overline{x}}_1 \\ \dot{\overline{x}}_2 \end{bmatrix} = \begin{bmatrix} -4201.7 & 0 \\ 0 & -666.7 \end{bmatrix} \begin{bmatrix} \overline{x}_1 \\ \overline{x}_2 \end{bmatrix} + \begin{bmatrix} 1321800 & 0 \\ 0 & 5210.7 \end{bmatrix} \begin{bmatrix} u_1 \\ u_2 \end{bmatrix}, \quad (35)$$

$$\begin{bmatrix} \overline{y}_1 \\ \overline{y}_2 \end{bmatrix} = \begin{bmatrix} 1 & 0 \\ 0 & 1 \end{bmatrix} \begin{bmatrix} \overline{x}_1 \\ \overline{x}_2 \end{bmatrix}.$$

Therefore, the temperature and the pressure processes can be controlled independently. In order to verify the feasibility of decoupling models, close-loop control simulations are conducted. The desired temperature is set as 473 K, and the desired pressure is set as 7.5 MPa. The controllers of temperature and pressure process are PID. The simulation results are given in Figure 4. As seen in Figure 4, actual temperature can well track the desired temperature, the same with pressure. The response curve of temperature has slight fluctuation near the desired temperature resulting from the small inertial time constant, but it cannot have an effect on control performance.

TABLE 3: Experimental data of biphenyl in supercritical CO_2.

Temperature (K)	Pressure (MPa)	Solubility (10^{-4})
313.16	6.4727	3.1290
318.25	7.5386	3.9730
323.25	8.5773	4.3240
328.35	9.6285	7.6790
333.15	10.6100	12.0570

4.2. Solubility Optimal. The biphenyl solubility data is listed in Table 3 in pressures 8.0 MPa~10.0 MPa, at temperatures 308 K~338 K. As shown in Table 3, the optimal pressure work points can be changed as the desired temperature changes; at the same time, the solubility in optimal work points can be calculated. The control performance of temperature and pressure is shown in Figures 5–9.

5. Conclusion and Future Works

This paper has presented a new control scheme of solubility optimal system based on a new nonlinear temperature-pressure decoupling model constructed with unequal-interval grey optimal models and Peng-Robinson models. Its performance has been verified through SFE experiments in biphenyl. Conclusions are as follows: Firstly, UEIGOMs of temperature and pressure process are modeled utilizing grey technology, which can extract valuable information from the fewer samples and weaken the randomness of original data. Therefore, UEIGOMs can give the correct description of the reaction character of temperature and

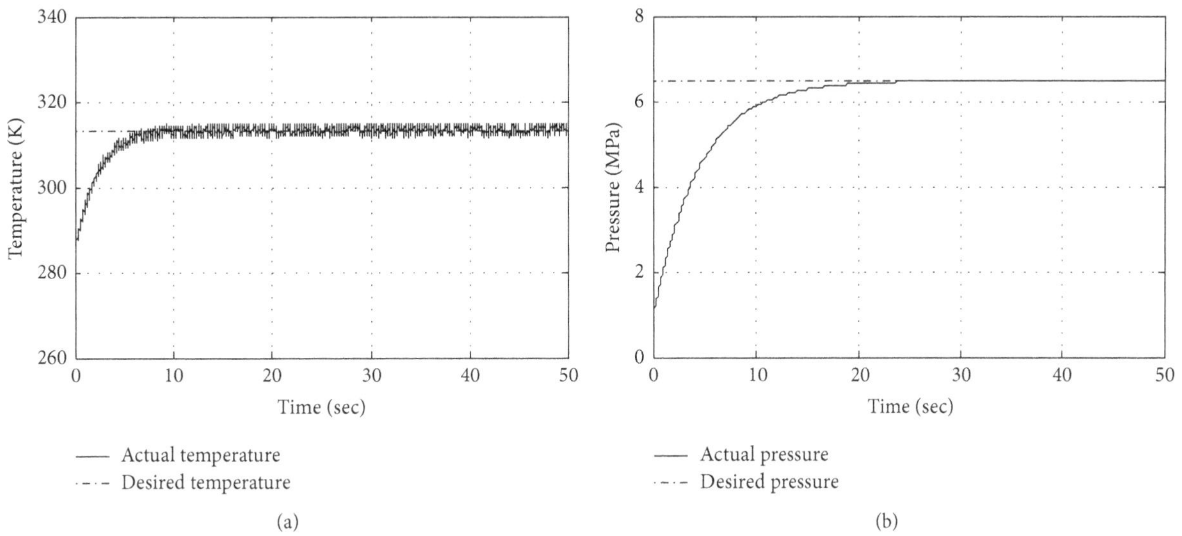

FIGURE 5: Temperature and pressure decoupling close-loop control with $T = 313.16$ K and $p = 6.4727$ MPa. (a) Temperature control. (b) Pressure control.

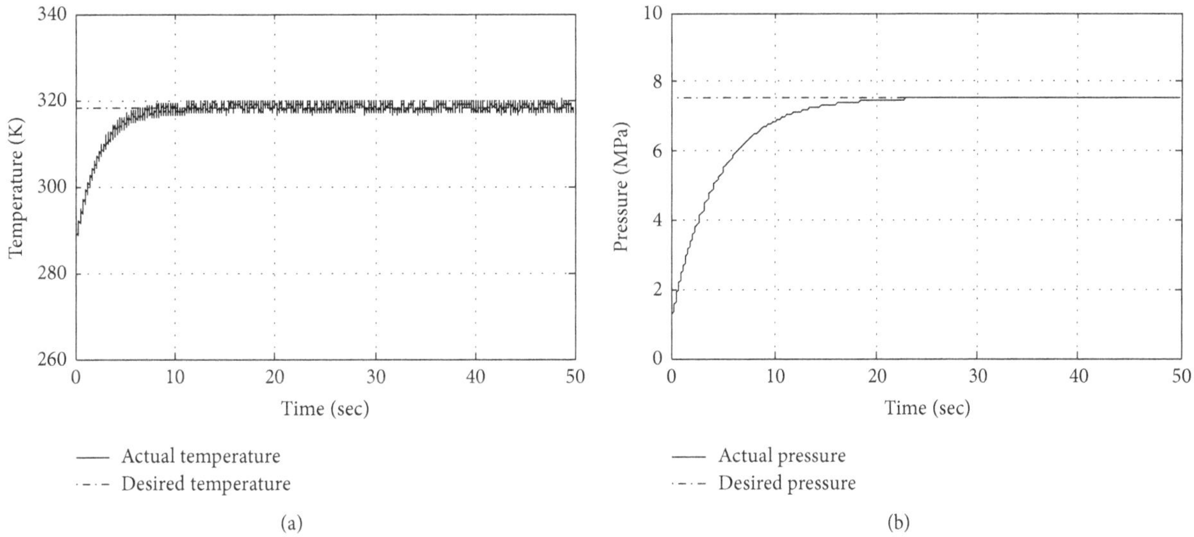

FIGURE 6: Temperature and pressure decoupling close-loop control with $T = 318.25$ K and $p = 7.5386$ MPa. (a) Temperature control. (b) Pressure control.

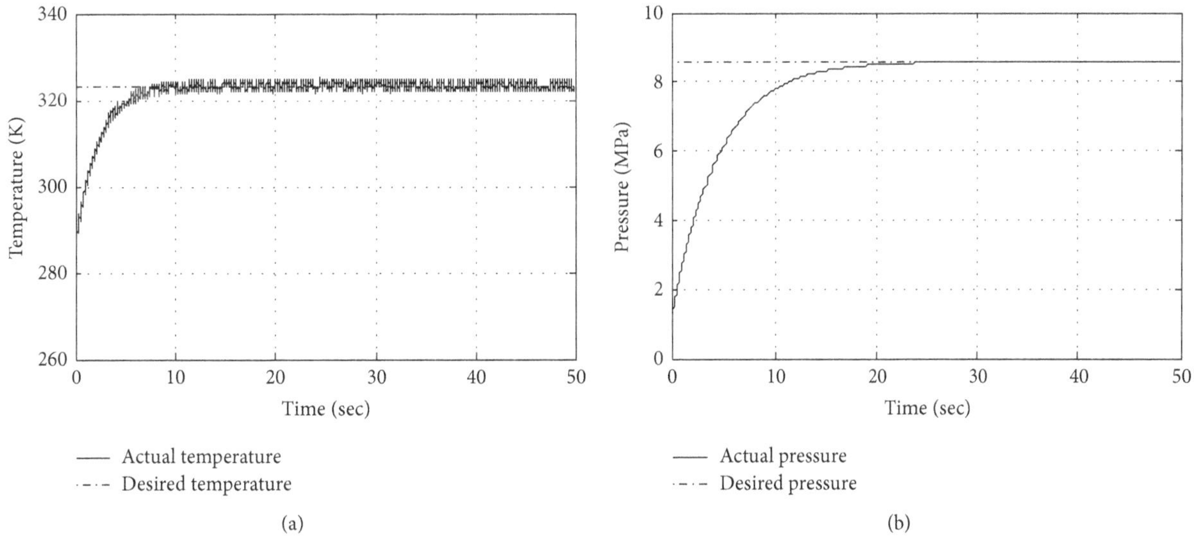

FIGURE 7: Temperature and pressure decoupling close-loop control with $T = 323.25$ K and $p = 8.5773$ MPa. (a) Temperature control. (b) Pressure control.

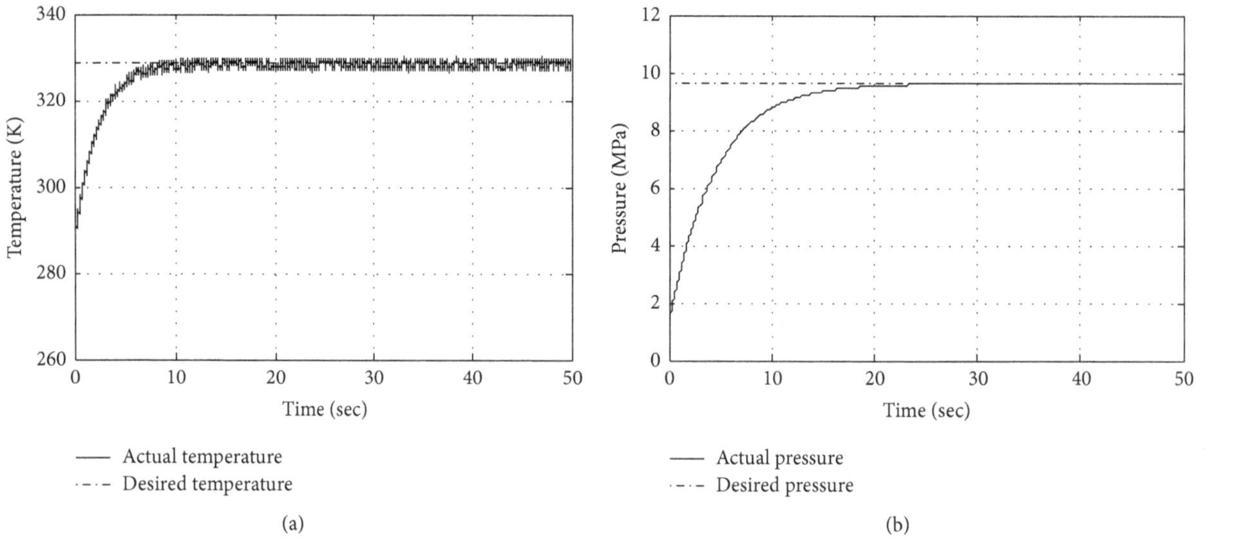

FIGURE 8: Temperature and pressure decoupling close-loop control with $T = 328.35$ K and $p = 9.6285$ MPa. (a) Temperature control. (b) Pressure control.

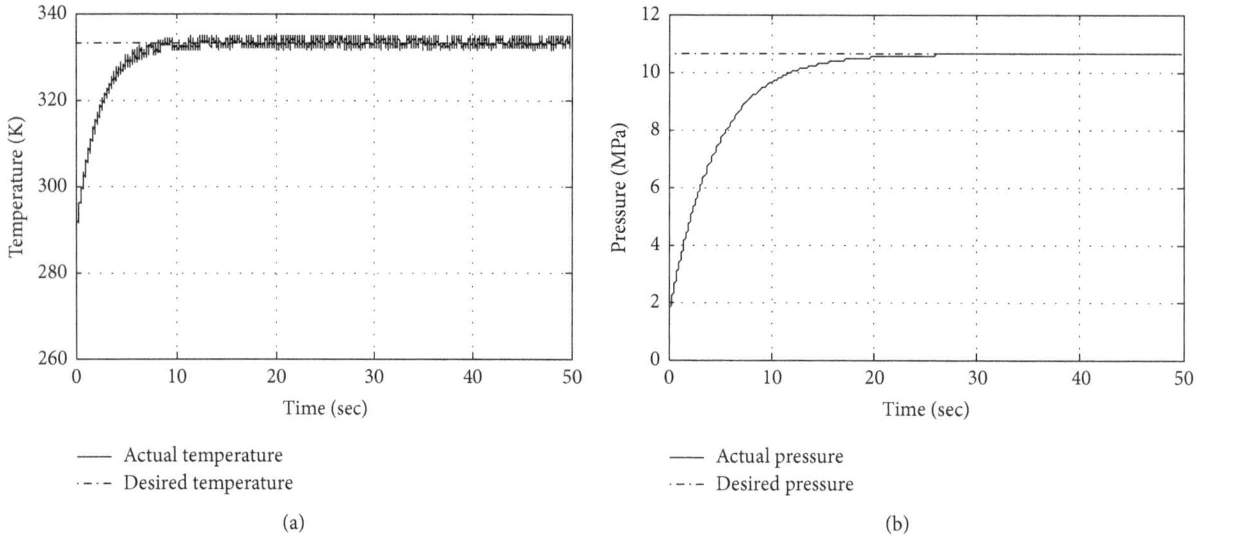

FIGURE 9: Temperature and pressure decoupling close-loop control with $T = 333.15$ K and $p = 10.6100$ MPa. (a) Temperature control. (b) Pressure control.

pressure process in SFE. Secondly, based on the relative degree concept of differential geometry theory on nonlinear system, the nondecoupling problem is studied for a class of MIMO nonlinear systems. The whole nonlinear model cannot satisfy input-output decoupling conditions, namely, singularity of decoupling matrix. By transformation of outputs and state variables, it can be disturbance decoupled. Finally, a solubility optimal method is presented in the work; it can calculate the optimal pressure according to the given temperature, namely, optimal working points, to maximize solubility for SFE process.

In addition, the solubility optimal system needs to be further improved resulting from complex processing technology and the solute species. Hence, processing time, energy consumption, solubility, and so forth are all considered to design a composite performance index for optimizing SFE process. This work needs further research and further improvement.

Conflicts of Interest

The authors declare that they have no conflicts of interest.

Acknowledgments

This work was supported by the Science & Technology Development Project of Jilin Province under Grants 20150204071GX and 20170520062JH and the Science & Technology Research Project of Jilin Province Education Department under Grant 2015117.

References

[1] B. Li, Y. Xu, Y.-X. Jin, Y.-Y. Wu, and Y.-Y. Tu, "Response surface optimization of supercritical fluid extraction of kaempferol glycosides from tea seed cake," *Industrial Crops and Products*, vol. 32, no. 2, pp. 123–128, 2010.

[2] J. Zeng and S. Yang, "Optimal control of supercritical fluid extraction with a hybrid model," in *Proceedings of the 2003 IEEE International Symposium on Intelligent Control*, pp. 346–351, Houston, TX, USA, October 2003.

[3] B. Daneshvand, K. M. Ara, and F. Raofie, "Comparison of supercritical fluid extraction and ultrasound-assisted extraction of fatty acids from quince (Cydonia oblonga Miller) seed using response surface methodology and central composite design," *Journal of Chromatography A*, vol. 1252, pp. 1–7, 2012.

[4] L.-H. Wang, Y.-H. Mei, F. Wang, X.-S. Liu, and Y. Chen, "A novel and efficient method combining SFE and liquid-liquid extraction for separation of coumarins from Angelica dahurica," *Separation and Purification Technology*, vol. 77, no. 3, pp. 397–401, 2011.

[5] G. Schmidt and H. Wenzel, "A modified van der Waals type equation of state," *Chemical Engineering Science*, vol. 35, no. 7, pp. 1503–1512, 1980.

[6] O. Redlich and J. N. Kwong, "On the thermodynamics of solutions; an equation of state," *Chemical Reviews*, vol. 44, no. 1, pp. 233–244, 1949.

[7] G. Soave, "Equilibrium constants from a modified Redlich-Kwong equation of state," *Chemical Engineering Science*, vol. 27, no. 6, pp. 1197–1203, 1972.

[8] D. Y. Peng and D. B. Robinson, "A new two-constant equation of state," *Industrial & Engineering Chemistry Fundamentals*, vol. 15, no. 1, pp. 59–64, 1976.

[9] H. Li and S. X. Yang, "Modeling of supercritical fluid extraction by hybrid Peng-Robinson equation of state and genetic algorithms," in *Proceedings of the 1st International Conference on Communications, Circuits and Systems (ICCCAS '02)*, pp. 1122–1126, July 2002.

[10] S. X. Yang, J. Zeng, C. Guo, and F. C. Sun, "A novel neuro-fuzzy model for supercritical fluid extraction," in *Proceedings of the International Conference Neural Networks and Brain 2005 (ICNN &B '05)*, vol. 3, pp. 1774–1779, 2005.

[11] Y. F. Lian, C. H. Tang, J. Jin, X. D. Li, and Q. Wang, "Predictive control of non equidistance GOM smelting," *Journal of Changchun University of Technology (Natural Science Edition)*, vol. 32, no. 4, pp. 365–369, 2011.

[12] X. P. Xiao, Z. M. Song, and F. Li, *Fundamentals and Applications of Grey Technology*, Science Press, Beijing, China, 2005.

[13] M. Zhou, H. Li, and M. Weijnen, "Grey system: thinking, methods, and models with applications," in *Contemporary Issues in Systems Science and Engineering*, vol. 1, Wiley-IEEE Press, Hoboken, NJ, USA, 2015.

[14] Q. X. Gong, H. G. Zhang, and X. P. Meng, "Disturbance decoupling control with stability for a class of MIMO nonlinear systems," *Control Theory & Applications*, vol. 23, no. 2, pp. 199–203, 2006.

[15] X. H. Xia and W. B. Gao, *Control and Decoupling of Nonlinear System*, Science Press, Beijing, 1997.

10

Optimization of Microwave Vacuum Drying and Pretreatment Methods for *Polygonum cuspidatum*

Wanxiu Xu,[1] **Guanyu Zhu**,[1] **Chunfang Song**,[1,2] **Shaogang Hu**,[1] **and Zhenfeng Li**[1,2]

[1]*Jiangnan University, Wuxi, Jiangsu, China*
[2]*Jiangsu Key Laboratory of Advanced Food Manufacturing Equipment and Technology, Wuxi, Jiangsu, China*

Correspondence should be addressed to Zhenfeng Li; 1010525570@qq.com

Academic Editor: Anna Vila

This study was conducted to optimize the drying process of *Polygonum cuspidatum* slices using an orthogonal experimental design. The combined effects of pretreatment methods, vacuum pressure and temperature of inner material, drying kinetics, color value, and retention of the indicator compounds were investigated. Seven mathematical models on thin-layer drying were used to study and analyze the drying kinetics. Pretreatment method with blanching for 30 s at 100°C increased the intensity of the red color of *P. cuspidatum* slices compared with other pretreatment methods and fresh *P. cuspidatum* slices. *P. cuspidatum* slices dried at 60°C retained more indicator compounds. Furthermore, microwave pretreatment methods, followed by microwave vacuum for 200 mbar at 50°C, resulted in high concentration of indicator compounds, with short drying time and less energy. This optimized condition for microwave vacuum drying and pretreatment methods would be useful for processing *P. cuspidatum*. The Newton, Page, and Wang and Singh models slightly fitted the microwave vacuum drying system. The logarithmic, Henderson and Pabis, two-term, and Midilli et al. models can be used to scale up the microwave vacuum drying system to a commercial scale. The two-term and Midilli et al. models were the best fitting mathematical models for the no-pretreatment case at 600 mbar and 60°C.

1. Introduction

Polygonum cuspidatum is a well-known Chinese herb and is officially listed in the Chinese Pharmacopoeia. This herb has been traditionally used for the treatment of various inflammatory diseases, including hepatitis, tumors, and diarrhea, in Eastern Asian countries, such as China, Korea, and Japan [1]. The phytochemistry of the root of this plant has been well studied. Currently, over 67 compounds from this plant have been isolated and identified [2]. The four kinds of indicator compounds are polydatin, emodin, physcion, and resveratrol [3]. To retain indicator compounds, the roots must be dried. Furthermore, the degree of retention of the indicator compounds was determined by drying conditions.

P. cuspidatum are usually sun-dried. However, this process takes a long time, because it is dependent on the weather. Moreover, the plant can easily become moldy because reaching a low safe moisture content to prevent the growth of molds is difficult, and the retention of indicator compounds is low. Microwave vacuum drying is a promising, rapid, and efficient dehydration method, which yields improved product appearance and quality compared to conventionally dried products [4–12]. In the literature, microwave power and vacuum levels are key factors. As moisture decreases with drying, an increasing number of burnt spots can be found in the last stage of drying process [13]. Li et al. used temperature control to avoid burning [14]. Pretreatment of material by blanching and microwaving can influence product quality. The efficiency of the blanching process is usually based on the inactivation of heat-resistant enzymes, such as peroxidase and polyphenol oxidase. Microwave irradiation has been successfully used in the pretreatment of various types of biomass, including agricultural residues, woody biomass, grass, energy plants, and industrial residuals [15]. Pretreatment methods, vacuum pressure, and temperature of inner materials were shown to be factors influencing the indicator compounds and drying time.

The aims of this paper are as follows:

(1) Design and build a microwave vacuum drying system and test the system.

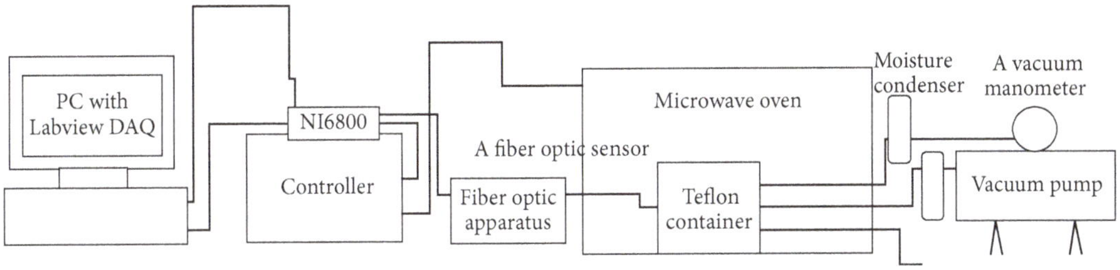

FIGURE 1: Schematic diagram of the microwave vacuum drying system.

(2) Analyze the effect of pretreatment methods, vacuum pressure, and temperature of inner material on indicator compounds and drying kinetics.

(3) Optimize conditions for pretreatment and microwave vacuum drying of *P. cuspidatum*.

(4) Develop mathematical models for the microwave vacuum drying of carrot slices to help in the scaling up of this drying technology.

2. Materials and Methods

2.1. Sample Preparation. Raw and sun-dried *P. cuspidatum* were purchased from farmers (Liu Jianhui, Liu Jiandong) in Jiangxi Province. Raw *P. cuspidatum* was washed, vacuum-packed, and sent to Jiangnan University. The samples were kept in the fridge at 4°C for 3 days. Before they were processed, each group was washed with tap water for 5 min and sliced to 5 mm thickness with a radius of 21.85–24.15 mm. The initial moisture content was determined by drying the samples for 24 h in a hot air oven at 80°C. Samples were dried to a final moisture content of less than 0.11 (dry basis, kg water/kg dry solid). Sample (25 g) was used for each drying experiment. Nine groups were tested and dried. All experiments were performed in triplicate.

2.2. Microwave Vacuum Drying System. A microwave vacuum drying system for drying *P. cuspidatum* was designed by our team and built in our laboratory (Figure 1). The system consisted of a microwave drying unit, a power and temperature control unit, a moisture condenser, a vacuum pump, a vacuum manometer, and a PC-based data acquisition unit. The microwave unit consisted of a reequipped microwave oven. The power and temperature control unit consisted of a fiber optic apparatus, a fiber optic sensor, and the power supply for the microwave oven. The PC-based data acquisition unit was an NI6800 with analog and digital inputs and outputs to record the temperature of inner material and to control the microwave oven power. A 2450 MHz microwave oven (media, MM720KG1-PW) and a hot air generator (Kada, 850) were modified for the experiments as the microwave dryer. The schematic diagram of the system is shown in Figure 1.

Samples were placed in a closed Teflon container, which was a cylinder with 110 mm in height and 120 mm in diameter. A porous basket was also fixed at 30 mm over the bottom of the container. The container was supported with an electronic

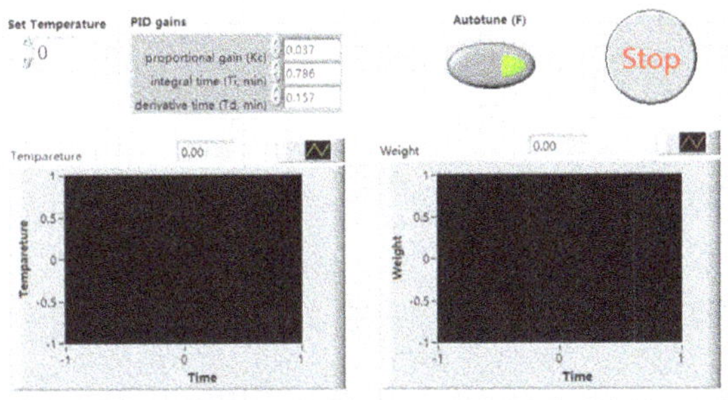

FIGURE 2: LabVIEW program.

balance (Lightever, LBA5200) to measure the weight of the samples.

One fiber optic sensor (Optsensor, ThermAigle-RD HQ-28) was inserted in the center of one of the samples for temperature measurement. The temperature of the center of sample and power control of magnetron was integrated in the DAQ board with a self-developed LabVIEW program, which used Proportion Integral Derivative (PID) control method (Figure 2). The recorded temperature is shown in Figure 3. Temperature was successfully maintained with the appropriate power levels. The maximum deviation in surface temperature was ±3°C, and standard deviations were less than 2°C. All the data were recorded at a time interval of 1 s.

2.3. Orthogonal Experiment Design. Three factors were studied: pretreatment method, temperature of the inner material, and vacuum (Table 1). Each pretreatment method changed the structure of the *P. cuspidatum* in a distinct way. By considering the cost of processing, samples were pretreated in three ways as follows: blanching for 30 s at 100°C, microwaving at 700 W for 10 s, and no pretreatment. The temperature of the inner material had a significant influence on the indicator components. When the temperature of the inner material was higher than 90°C, less indicator compounds were retained and toxic chemicals were produced. Below 40°C, the drying time was too long and the energy consumption was more than that of other drying methods. Therefore, the temperature of the inner material should be controlled in the range of 50°C–80°C. The use of a high vacuum lowered the drying temperature of the inner material. More thermosensitive compounds were retained under a high-temperature vacuum. Three pressures were used: 200, 400, and 600 mbar.

TABLE 1: Orthogonal experiment design.

Expt. number	Pretreatment methods	Vacuum pressure (mbar)	Temperature of inner materials (°C)
(1)	1 (blanching for 30 s at 100°C)	1 (200)	1 (50)
(2)	1	2 (400)	2 (60)
(3)	1	3 (600)	3 (70)
(4)	2 (microwave 700 w for 10 s)	1	2
(5)	2	2	3
(6)	2	3	1
(7)	3 (no pretreatment)	1	3
(8)	3	2	1
(9)	3	3	2

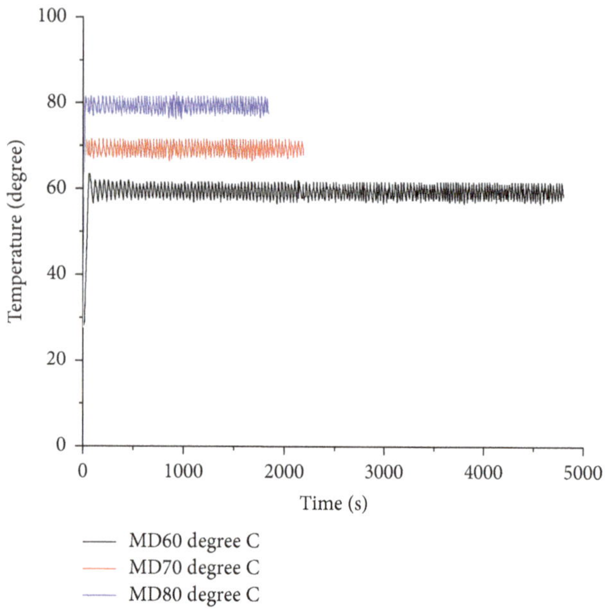

FIGURE 3: Recorded temperature of the microwave vacuum drying system.

2.4. Mathematical Models. The data on the drying of *P. cuspidatum* slices in microwave vacuum dryer were used to study the drying kinetics and to analyze the fit of mathematical models on thin-layer drying, including Newton, Page, logarithmic, Wang and Singh, Henderson and Pabis, two-term, and Midilli et al. models (Table 2). In Table 2, the MR in (i)–(vii) is the moisture ratio, which is defined in

$$\mathrm{MR} = \frac{M_t - M_e}{M_o - M_e}, \quad (1)$$

where M_t, M_o, and M_e are the moisture content (d.b.) at time t, initial moisture content (d.b.), and equilibrium moisture content (d.b.), respectively. k is the drying rate constant (min^{-1}); n, a, b, and c are the drying coefficients (unitless) that have different values depending on the equation and the drying curve; and t is time (min).

When the difference between the moisture content at time t and the equilibrium moisture content is negligible, (1) is reduced to

$$\mathrm{MR} = \frac{M_t}{M_o}. \quad (2)$$

2.5. Color Measurement. Color change is one of the quality criteria for dried products. Color parameters (L^*, a^*, b^*) were measured directly on the surface of fresh, dried, and rehydrated *P. cuspidatum* slices by using a chroma meter with $d/0$ diffuse illumination/0° viewing system (Model CR-300X, Minolta Co., Ltd., Japan). The CIE 1976 (L^*, a^*, b^*) color space was used to estimate the color values of *P. cuspidatum* slice samples. Color values, expressed as L^* (whiteness or brightness/darkness), a^* (redness/greenness), and b^* (yellowness/blueness), were determined for a sample in all drying conditions. The total color difference (ΔE) was calculated based on color values of fresh ($L_f{}^*$, $a_f{}^*$, and $b_f{}^*$) and dried *P. cuspidatum* slices ($L_d{}^*$, $a_d{}^*$, and $b_d{}^*$) based on

$$\Delta E = \sqrt{\left(L_f{}^* - L_d{}^*\right)^2 + \left(a_f{}^* - a_d{}^*\right)^2 + \left(b_f{}^* - b_d{}^*\right)^2}. \quad (3)$$

2.6. Chemicals. Polydatin (3,4′,5-trihydro-hydroxystil-bene-3-b-mono-D-glucoside), emodin (6-ethyl-1,3,8-trihydroxyanthraquinone), physcion(1,8-dihydroxy-3-methoxy-6-methylanthraquinone emodin 3-methyl ether), and resveratrol (3,4′,5-trihydrohydroxystilbene) were provided by the National Institute for the Control of Pharmaceutical and Biological Products (China). For standard solutions, polydatin, emodin, physcion, and resveratrol were dissolved in 50% ethanol and used to prepare standard solutions containing 239.7, 83.68, 10.78, and 62.96 μg/mL, respectively, which were diluted with the same solvent to obtain solutions with different concentrations.

2.7. High Performance Liquid Chromatography (HPLC) Analysis. Samples (200 mg) were ground to powders and were refluxed with 25 ml ethanol (50 : 50, v/v) for 30 minutes. The weight loss was made up with ethanol (50 : 50 v/v). The solution obtained was filtered through a 0.45 μm micropore membrane. Ten μl solution was injected into the HPLC instrument for analysis (Palo Alto, CA, USA). Samples were separated on an Agilent ZORBAX SB-C18 column (250 mm × 4.6 mm, 5 μm particles) (Agilent, USA) together with a C18 guard column. The mobile phase was a gradient prepared from acetonitrile (component A) and 0.1% formic

TABLE 2: Mathematical models of the kinetics of fluidized bed drying.

Model name	Model equation	Reference	Equation number
Newton	$MR = e^{-kt}$	[16]	(i)
Page	$MR = e^{-kt^n}$	[17]	(ii)
Logarithmic	$MR = ae^{-kt} + c$	[18]	(iii)
Wang and Singh	$MR = 1 + at + bt^2$	[19]	(iv)
Henderson and Pabis	$MR = ae^{-kt}$	[20]	(v)
Two-term	$MR = ae^{k_0 t} + be^{k_1 t}$	[21]	(vi)
Midilli et al.	$MR = ae^{-kt^n} + bt$	[22]	(vii)

TABLE 3: Color values of *P. cuspidatum* slices in all drying conditions.

Samples	Color values			
	L^*	a^*	b^*	ΔE
Fresh	62.78	8.25	36.63	-
(1)	51.78	9.34	34.12	11.33
(2)	48.59	9.97	33.90	14.55
(3)	49.3	8.57	34.19	13.70
(4)	51.57	7.05	34.35	11.50
(5)	56.82	7.23	41.05	7.49
(6)	51.94	7.98	36.49	10.84
(7)	55.48	7.04	38.85	7.73
(8)	54.79	7.77	37.48	8.05
(9)	56.25	7.69	41.24	8.01

acid (component B) prepared in water. The elution program was 0–15 min, 15% (A) to 20% (A), and 15–60 min, 20% (A) to 80% (A), and the total acquisition time was 65 min. The mobile phase flow rate was 1.0 mL/min, and the column temperature was set at 25°C [23].

2.8. Correlation Coefficients and Error Analysis. Statistical parameters, such as root mean square error (RMSE) (see (4)), chi-square (χ^2) (see (5)), and correlation coefficient (R^2) (see (6)), were used to estimate the quality of fit of drying models to the observed values.

$$\mathrm{RMSE} = \left[\frac{1}{N}\sum_{i=1}^{N}\left(\mathrm{MR}_{\mathrm{pre},i} - \mathrm{MR}_{\exp,i}\right)^2\right]^{1/2}, \quad (4)$$

where $\mathrm{MR}_{\exp,i}$ is the experimental moisture ratio at time t, $\mathrm{MR}_{\mathrm{pre},i}$ is the predicted moisture ratio at time t, and N is the observation number.

$$\chi^2 = \frac{\sum_{i=1}^{N}\left(\mathrm{MR}_{\exp,i} - \mathrm{MR}_{\mathrm{pre},i}\right)^2}{N - n}, \quad (5)$$

where n is the constant number in drying models.

$$R^2 = 1 - \frac{\sum_{i=1}^{N}\left(\mathrm{MR}_{\mathrm{pre},i} - \mathrm{MR}_{\exp,i}\right)^2}{\sum_{i=1}^{N}\left(\mathrm{MR}_{\exp,i} - \overline{\mathrm{MR}}_{\exp}\right)^2}, \quad (6)$$

where $\overline{\mathrm{MR}}_{\exp}$ is the mean experimentally measured value of MR.

In addition to the parameters mentioned above, the reduced sum square error (SSE) (see (7)) was also used as a criterion to analyze the closeness of fit.

$$\mathrm{SSE} = \frac{1}{N}\sum_{i=1}^{N}\left(\mathrm{MR}_{\mathrm{pre},i} - \mathrm{MR}_{\exp,i}\right)^2. \quad (7)$$

2.9. Statistical Analysis. The statistical analysis of variance of the experiment results was conducted using the SPSS17.0 (SAS Institute Inc., Cary, NC, USA) with a confidence level ($p \leq 0.05$) of 95%. The mathematical models of the drying of carrot slices in microwave vacuum drying were fitted and analyzed by using SPSS.

3. Results and Discussion

3.1. Color Values. The results of color parameters of dried *P. cuspidatum* slices in microwave vacuum drying system are presented in Table 3. The L^* values of all the dried *P. cuspidatum* slices decreased in comparison with the fresh *P. cuspidatum* slices. However, b^* values almost stayed the same in all drying conditions, which shows the yellowness of *P. cuspidatum* slices, and ranged from 33.90 to 41.24. The a^* values varied with different pretreatment methods. Pretreatment method with blanching for 30 s at 100°C made *P. cuspidatum* slices redder than other pretreatment methods and fresh *P. cuspidatum* slices. The total color difference (ΔE) of the dried *P. cuspidatum* slices from all the drying conditions was between 7.49 and 14.55.

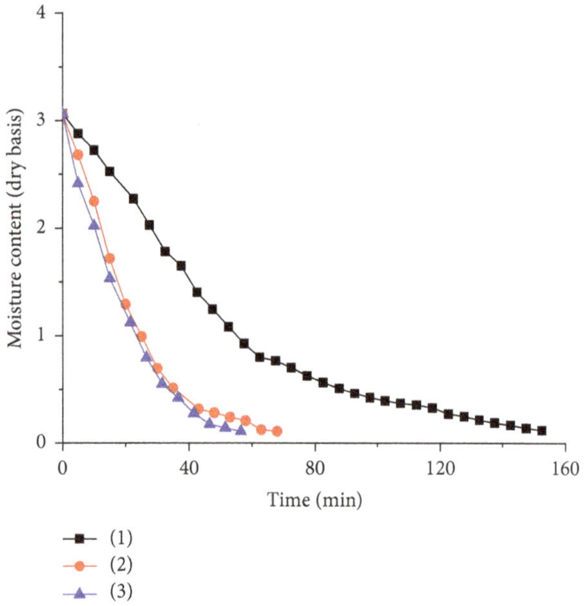

FIGURE 4: Drying curves of *P. cuspidatum* slices at (1) blanching for 30 s at 100°C, vacuum pressure at 200 mbar and 50°C; (2) blanching for 30 s at 100°C, vacuum pressure at 400 mbar and 60°C; and (3) blanching for 30 s at 100°C, vacuum pressure at 600 mbar and 70°C.

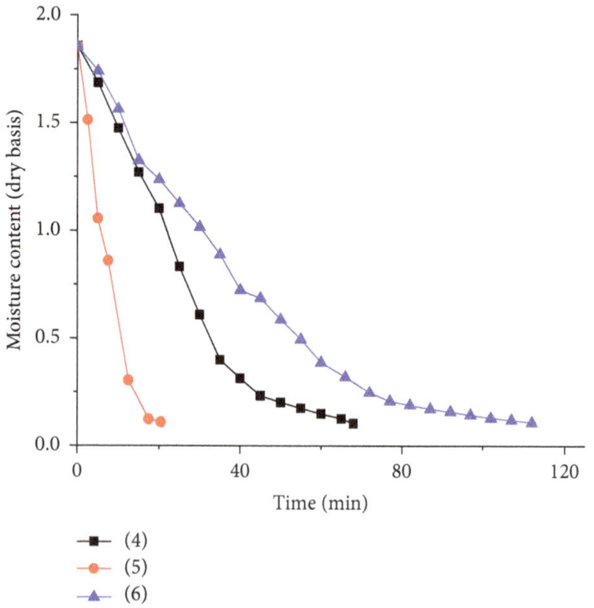

FIGURE 5: Drying curves of *P. cuspidatum* slices at (4) blanching for 30 s at 100°C, vacuum pressure at 200 mbar and 60°C; (5) blanching for 30 s at 100°C, vacuum pressure at 400 mbar and 70°C; and (6) blanching for 30 s at 100°C, vacuum pressure at 600 mbar and 50°C.

3.2. Drying Curves. The pretreatment methods of the nine groups were various; hence, the initial moisture content (dry basis) of each sample was different. The lowest moisture content was found in the pretreated material microwaved at 700 w for 10 s before the drying process (1.86, dry basis), whereas the dried material blanched for 30 s at 100°C had the highest moisture content (3.06, dry basis). Figure 4 shows the change of moisture content for groups (1) to (3). The drying curve for group (2) was similar to that for group (3). The moisture content of group (1) slowly decreased over time. Thus, a low temperature of the inner material and low vacuum pressure required more time for drying to occur. This phenomenon was due to the fact that, at a low temperature, the pressure of water vapor inside the material was low; thus, the water was not easily transferred. Similar observations were reported by Li et al. [14].

Figure 5 shows the drying curves for groups (4) to (6), which were pretreated with microwave at 700 w for 10 s. As the initial moisture was the least, the drying time for group (5) (22 minutes) was the shortest. The use of a high temperature and a high vacuum pressure to the inner material resulted in fast drying. The drying time for group (6) (600 mbar and 50°C) was 1.6 times longer than that for group (4) (600 mbar and 60°C). Thus, the influence of temperature of the inner material on the drying time was more significant than that of vacuum pressure.

Figure 6 shows the drying curves of groups without pretreatments. The initial moisture was 2.33 (dry basis). Vacuum pressure had little influence on drying time; hence, the drying time of group (8) at 50°C was almost 2 times more than that for group (7) at 70°C without considering the vacuum pressure. Furthermore, atmospheric pressure for

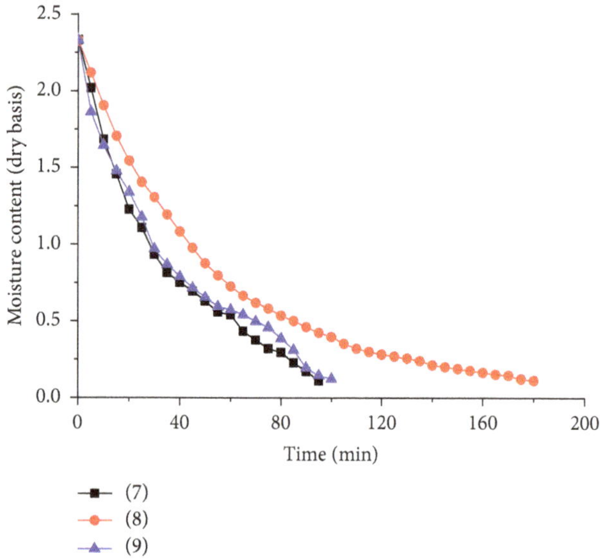

FIGURE 6: Drying curves of *P. cuspidatum* slices at (7) vacuum pressure at 200 mbar and 70°C, (8) vacuum pressure at 400 mbar and 50°C, and (9) vacuum pressure at 600 mbar and 60°C.

group (7) was 200 mbar lower than that for group (8). The temperature of the root had more significant influence on the drying time for microwave vacuum drying.

To compare the influence of pretreatment on the drying time, the groups with the same temperature should be considered. The lengths of drying times of groups with a temperature of 50°C were in the following order: the length of drying time of group (8) was greater than that of groups (1) and (6). In addition, the initial moisture of group (8) with a

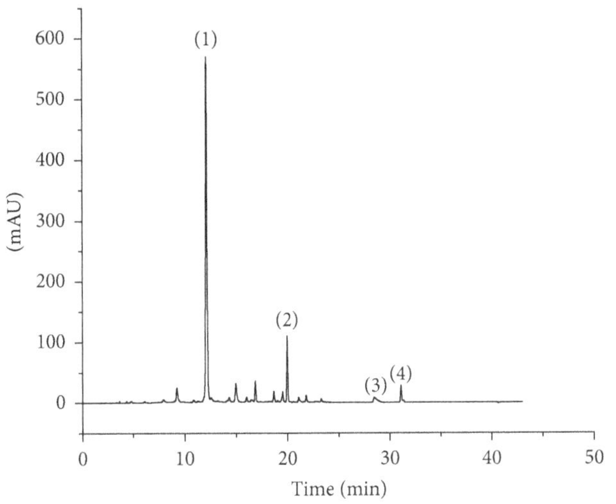

FIGURE 7: HPLC chromatograms of compound retained in sample (1) (peak (1), polydatin; peak (2), resveratrol; peak (3), emodin; and peak (4), physcion).

higher vacuum pressure was less than that of group (1) with lower vacuum pressure. Thus, pretreatment methods with blanching for 30 s at 100°C made dry process faster than no pretreatment. The lengths of drying times of groups with a temperature of 70°C were in the following order: the length of drying time of group (7) was greater than that of groups (3) and (5). In addition, the vacuum pressure of group (3) was lower than that for group (5). Thus, pretreatment with microwave for 10 s made dry process faster than blanching for 30 s.

3.3. Mathematical Models and Statistical Analysis. The drying curves of *P. cuspidatum* slices were fitted into the mathematical models (see (i)–(vii)). The statistical values were calculated to estimate the accuracy of closeness of fit and are shown in Tables 4–6. The best fitting model was determined according to the lowest χ^2, RMSE, and SSE and the highest R^2 values. The R^2 values ranged from 0.2090 to 1.0000 and the SSE values ranged from 0.0065 to 0.0997, respectively (Tables 4–6). The R^2 in Tables 4 and 6 is low. Thus, Newton, Page, and Wang and Singh model were not suitable under all the drying conditions.

The two-term and Midilli et al. models were the best fitting mathematical models for the no-pretreatment case at 600 mbar and 60°C. Moreover, the lowest χ^2 values were 0.0001 and 0.0002. The logarithmic, Henderson and Pabis, two-term, and Midilli et al. models can be used to scale up the microwave vacuum drying system to a commercial scale.

3.4. Effect of Microwave Vacuum Drying and Pretreatment Methods on Concentrations of Indicator Compounds. The concentrations of indicator compounds in sample 1 were analyzed by HPLC (Figure 7). The results of nine groups of experiments are listed in Table 7. The highest concentrations of polydatin were found in samples pretreated by blanching for 30 s and microwave vacuum drying at 400 mbar and 60°C. Emodin and physcion were the highest when samples were pretreated with blanching for 30 s and microwave vacuum drying at 200 mbar and 50°C. Furthermore, resveratrol concentrations were the highest when samples were pretreated with microwave at 700 W for 10 s and dried with microwave vacuum at 600 mbar and 50°C.

3.4.1. Factors Affecting the Retention of Polydatin. The variance of the concentrations of polydatin for each of the factors was analyzed in Table 7. Pretreatment had a significant influence on the retention of polydatin at a confidence level of 0.05, whereas vacuum pressure and temperature of root material had a significant influence on retention at a confidence level of 0.01. The concentrations of polydatin were the highest in groups (1), (2), and (3) (Table 2), which were pretreated with blanching for 30 s. This effect was probably due to the inhibition of the hydrolysis of the polydatin. As the temperature of the root increased, retention decreased, which may be due to the decomposition of polydatin. Moreover, as the temperature of the inner material increased, the moisture content decreased and less polydatin was retained because polydatin is soluble in water. When the vacuum pressure was higher, the temperature of inner material should be lower [24]. Thus, with a low temperature of inner material, less polydatin was retained.

3.4.2. Factors Affecting the Retention of Emodin and Physcion. The effect of the factors on the retention of emodin is in the following order. The effect of temperature of inner material was greater than that in pretreatment methods and use of vacuum pressure (Table 8). Vacuum pressure and pretreatment had no significant influence, whereas the temperature of inner material had a significant influence at a confidence level of 0.1. A high temperature of the root may decrease the retention of emodin. In addition, a high thermal process may affect the stability of bioactive compounds in food, including vitamins and trace elements, as proposed by Hirun et al. [25].

The effect of the different factors on the retention of physcion is discussed in Table 8. Pretreatment and vacuum pressure had no significant influence, whereas the temperature of inner material had a significant influence at a confidence level of 0.1. This finding was the same as that obtained for emodin, thus suggesting the use of low temperature.

3.4.3. Factors Affecting the Retention of Resveratrol. The effect on the factors on the retention of resveratrol is in the following order: the temperature of the root was greater than that in pretreatment and vacuum pressure use (Table 8). The three factors had no significant influence at a confidence level of 0.1. Table 2 shows the least retention of resveratrol in the no-pretreatment case, followed by microwave vacuum drying with a vacuum pressure of 0.02 Mpa and a temperature of 70°C. A high temperature led to chemical changes in the resveratrol. The highest retention was 0.397 mg/g, which was 13 times more than the highest retention (0.029 mg/g) obtained by Zhang et al. [23]. Drying at 50°C caused most of the retention in the groups, but drying time with low temperature of the inner material was long.

TABLE 4: Statistical results of mathematical model for *P. cuspidatum* for groups (1), (2), and (3).

Expt. number	Statistics/coefficient	Models						
		Newton	Page	Logarithmic	Wang and Sing	Henderson and Pabis	Two-term	Midilli et al.
(1)	R^2	0.209	0.319	0.993	0.334	0.992	0.992	0.998
	SSE	0.00653125	0.009969	0.031031	0.010438	0.031	0.031	0.031188
	RMSE	0.080816149	0.099844	0.176157	0.102164	0.176068	0.176068	0.1766
	χ^2	0.648780935	0.577014	0.006232	0.564585	0.006624	0.007097	0.002238
(2)	R^2	0.211	0.302	0.983	0.295	0.982	0.984	0.997
	SSE	0.015071	0.021571	0.070214	0.021071	0.070143	0.070286	0.071214
	RMSE	0.122766	0.146872	0.26498	0.14516	0.264845	0.265115	0.26686
	χ^2	0.821399	0.786975	0.021068	0.795255	0.020394	0.021737	0.004037
(3)	R^2	0.23	0.343	0.997	0.337	0.993	0.998	0.998
	SSE	0.017692	0.026385	0.076692	0.025923	0.076385	0.076769	0.076769
	RMSE	0.133012	0.162433	0.276934	0.161006	0.276378	0.277073	0.277073
	χ^2	0.755287	0.70324	0.003333	0.70912	0.00756	0.00291	0.002067

Table 5: Statistical results of mathematical model for *P. cuspidatum* for groups (4), (5), and (6).

Expt. number	Statistics/coefficient	Models						
		Newton	Page	Logarithmic	Wang and Sing	Henderson and Pabis	Two-term	Midilli et al.
(4)	R^2	0.653	0.771	0.983	0.69	0.979	0.985	0.996
	SSE	0.0653	0.036714	0.04681	0.032857	0.046619	0.046905	0.047429
	RMSE	0.255539	0.19161	0.216355	0.181265	0.215914	0.216575	0.217781
	χ^2	0.120337	0.087957	0.006901	0.118796	0.008169	0.006536	0.001526
(5)	R^2	0.588	0.751	0.989	0.678	0.989	0.99	0.997
	SSE	0.025565	0.0751	0.0989	0.0678	0.0989	0.099	0.0997
	RMSE	0.159891	0.274044	0.314484	0.260384	0.314484	0.314643	0.315753
	χ^2	0.241822	0.129542	0.006555	0.167694	0.005932	0.006765	0.002055
(6)	R^2	0.41	0.727	0.995	0.699	0.991	0.995	0.997
	SSE	0.018636	0.031609	0.043261	0.030391	0.043087	0.043261	0.043348
	RMSE	0.136515	0.177788	0.207992	0.174331	0.207574	0.207992	0.208201
	χ^2	0.3664	0.091569	0.001927	0.111174	0.003075	0.001801	0.001044

TABLE 6: Statistical results of mathematical model for *P. cuspidatum* for groups (7), (8), and (9).

Expt. number	Statistics/coefficient	Models						
		Newton	Page	Logarithmic	Wang and Sing	Henderson and Pabis	Two-term	Midilli et al.
(7)	R^2	0.491	0.513	0.988	0.507	0.983	0.994	0.993
	SSE	0.018885	0.023318	0.044909	0.023045	0.044682	0.045182	0.045136
	RMSE	0.137421	0.152703	0.211918	0.151807	0.211381	0.21256	0.212453
	χ^2	0.344994	0.181661	0.004813	0.183733	0.006275	0.002431	0.002721
(8)	R^2	0.521	0.573	0.989	0.551	0.98	0.994	0.909
	SSE	0.013711	0.022038	0.038038	0.021192	0.037692	0.038231	0.034962
	RMSE	0.117092	0.148454	0.195035	0.145576	0.194145	0.195527	0.18698
	χ^2	0.544095	0.144854	0.003781	0.152392	0.006815	0.002187	0.033678
(9)	R^2	0.657	0.617	0.999	0.573	0.997	1	1
	SSE	0.031286	0.016237	0.026289	0.015079	0.026237	0.026316	0.026316
	RMSE	0.176878	0.127424	0.16214	0.122796	0.161978	0.162221	0.162221
	χ^2	0.214399	0.145253	0.000242	0.161685	0.001093	0.000129	0.000178

TABLE 7: Indicator compounds retained in samples with drying condition.

Number	Polydatin (mg/g)	Emodin (mg/g)	Physcion (mg/g)	Resveratrol (mg/g)
(1)	18.828	18.589	0.128	0.379
(2)	20.656	13.664	0.011	0.228
(3)	17.806	10.186	0.004	0.285
(4)	12.619	14.154	0.027	0.272
(5)	12.037	6.397	0.004	0.068
(6)	13.931	17.397	0.055	0.397
(7)	11.795	5.49	0.026	0.112
(8)	15.013	14.714	0.111	0.377
(9)	12.405	7.513	0.008	0.188

TABLE 8: Variance analysis of the retention of polydatin.

Indicator compounds	R^2	Source of variation	Sum of squares	Degree of freedom	Mean square	F value	Sig.
		Pretreatment	88.853	2	44.427	89.450	0.011
Polydatin	0.991	Vacuum pressure	9.291	2	4.646	9.354	0.097
		Temperature of inner material	15.266	2	7.633	15.369	0.061
		Pretreatment	27.766	2	13.883	2.273	0.306
Emodin	0.939	Vacuum pressure	1.646	2	0.823	0.135	0.881
		Temperature of inner material	157.277	2	78.639	12.873	0.072
		Pretreatment	0.001	2	0.000	0.535	0.652
Physcion	0.944	Vacuum pressure	0.002	2	0.001	1.421	0.413
		Temperature of inner material	0.016	2	0.008	14.791	0.063
		Pretreatment	0.008	2	0.004	0.543	0.648
Resveratrol	0.865	Air pressure	0.006	2	0.003	0.428	0.700
		Temperature of inner material	0.082	2	0.041	5.418	0.156

4. Conclusion

Pretreatment method with blanching for 30 s at 100°C intensified the red color of *P. cuspidatum* slices compared with other pretreatment methods and fresh *P. cuspidatum* slices. The logarithmic, Henderson and Pabis, two-term, and Midilli et al. models can be used to scale up the microwave vacuum drying system to a commercial scale. The temperature of the root had a significant influence on the retention of polydatin, emodin, and physcion at a confidence level of 0.1. The temperature of the root, pretreatment, and vacuum pressure had no significant influence on the retention of resveratrol. In addition, low temperature of inner material retained all the indicator compounds of *P. cuspidatum*. However, low root temperature required a long drying time. The order of drying time with pretreatment was as follows. The drying time of no-pretreatment case was greater than that of blanching for 30 s and microwaving at 700 w for 10 s. Furthermore, the temperature of the root had more significant influence on drying time than vacuum pressure. In conclusion, pretreatment with microwave at 700 w for 10 s, with a low temperature and low vacuum pressure, would retain more indicator compounds and require less drying time.

Conflicts of Interest

The authors declare that there are no conflicts of interest regarding the publication of this paper.

Acknowledgments

The authors acknowledge the Cooperative Innovation Fund of Jiangsu Province (BY2016022-10) and the Colleges and Universities in Jiangsu Province for graduate research and innovation (KYLX_1158).

References

[1] H. Chen, T. Tuck, X. Ji et al., "Quality assessment of Japanese knotweed (Fallopia japonica) grown on Prince Edward Island as a source of resveratrol," *Journal of Agricultural and Food Chemistry*, vol. 61, no. 26, pp. 6383–6392, 2013.

[2] W. Peng, R. Qin, X. Li, and H. Zhou, "Botany, phytochemistry, pharmacology, and potential application of *Polygonum cuspidatum* Sieb.et Zucc.: a review," *Journal of Ethnopharmacology*, vol. 148, no. 3, pp. 729–745, 2013.

[3] P. Ma, K. Luo, Y. Peng et al., "Quality control of polygonum cuspidatum by uplc-pda and related statistical analysis," *Journal of Liquid Chromatography & Related Technologies*, vol. 36, no. 20, pp. 2844–2854, 2013.

[4] A. Figiel, "Drying kinetics and quality of vacuum-microwave dehydrated garlic cloves and slices," *Journal of Food Engineering*, vol. 94, no. 1, pp. 98–104, 2009.

[5] Á. Calín-Sánchez, A. Figiel, A. Wojdyło, M. Szarycz, and Á. A. Carbonell-Barrachina, "Drying of Garlic Slices Using Convective Pre-drying and Vacuum-Microwave Finishing Drying:

Kinetics, Energy Consumption, and Quality Studies," *Food and Bioprocess Technology*, vol. 7, no. 2, pp. 398–408, 2014.

[6] S. Ferenczi, B. Czukor, and Z. Cserhalmi, "Evaluation of microwave vacuum drying combined with hot-air drying and compared with freeze- and hot-air drying by the quality of the dried apple product," *Periodica Polytechnica Chemical Engineering*, vol. 58, no. 2, pp. 111–116, 2014.

[7] A. Figiel and A. Michalska, "Overall quality of fruits and vegetables products affected by the drying processes with the assistance of vacuum-microwaves," *International Journal of Molecular Sciences*, vol. 18, no. 1, article no. 71, 2017.

[8] S. Ambros, S. A. W. Bauer, L. Shylkina, P. Foerst, and U. Kulozik, "Microwave-Vacuum Drying of Lactic Acid Bacteria: Influence of Process Parameters on Survival and Acidification Activity," *Food and Bioprocess Technology*, vol. 9, no. 11, pp. 1901–1911, 2016.

[9] J. de Bruijn, F. Rivas, Y. Rodriguez et al., "Effect of Vacuum Microwave Drying on the Quality and Storage Stability of Strawberries," *Journal of Food Processing and Preservation*, vol. 40, no. 5, pp. 1104–1115, 2016.

[10] M. Zielinska, P. Sadowski, and W. Błaszczak, "Combined hot air convective drying and microwave-vacuum drying of blueberries (Vaccinium corymbosum L.): Drying kinetics and quality characteristics," *Drying Technology*, vol. 34, no. 6, pp. 665–684, 2016.

[11] Y. Tian, S. Wu, Y. Zhao, Q. Zhang, J. Huang, and B. Zheng, "Drying Characteristics and Processing Parameters for Microwave-Vacuum Drying of Kiwifruit (Actinidia deliciosa) Slices," *Journal of Food Processing and Preservation*, vol. 39, no. 6, pp. 2620–2629, 2015.

[12] D. Wray and H. S. Ramaswamy, "Quality Attributes of Microwave Vacuum Finish-Dried Fresh and Microwave-Osmotic Pretreated Cranberries," *Journal of Food Processing and Preservation*, vol. 39, no. 6, pp. 3067–3079, 2015.

[13] N. Therdthai and W. Zhou, "Characterization of microwave vacuum drying and hot air drying of mint leaves (Mentha cordifolia Opiz ex Fresen)," *Journal of Food Engineering*, vol. 91, no. 3, pp. 482–489, 2009.

[14] Z. Li, G. S. V. Raghavan, and V. Orsat, "Temperature and power control in microwave drying," *Journal of Food Engineering*, vol. 97, no. 4, pp. 478–483, 2010.

[15] J. Xu, *Chapter 9 - Microwave Pretreatment*, A. P. N. B. Larroche, Ed., Pretreatment of Biomass, Elsevier, Amsterdam, p. 157-172, 2015.

[16] P. C. Corrêa, F. M. Botelho, G. H. H. Oliveira, A. L. D. Goneli, O. Resende, and S. de Carvalho Campos, "Mathematical modeling of the drying process of corn ears," *Acta Scientiarum—Agronomy*, vol. 33, no. 4, pp. 575–581, 2011.

[17] Q. Shi, Y. Zheng, and Y. Zhao, "Mathematical modeling on thin-layer heat pump drying of yacon (*Smallanthus sonchifolius*) slices," *Energy Conversion and Management*, vol. 71, pp. 208–216, 2013.

[18] M. Özdemir and Y. Onur Devres, "The thin layer drying characteristics of hazelnuts during roasting," *Journal of Food Process Engineering*, vol. 42, no. 4, pp. 225–233, 1999.

[19] C. Chen and P.-C. Wu, "Thin-layer drying model for rough rice with high moisture content," *Journal of Agricultural Engineering Research*, vol. 80, no. 1, pp. 45–52, 2001.

[20] A. Midilli, H. Kucuk, and Z. Yapar, "A new model for single-layer drying," *Drying Technology*, vol. 20, no. 7, pp. 1503–1513, 2002.

[21] S. M. Henderson, "Progress in developing the thin layer drying equation," *Transactions of the ASAE*, vol. 17, no. 6, pp. 1167–1172, 1974.

[22] H. O. Menges and C. Ertekin, "Mathematical modeling of thin layer drying of Golden apples," *Journal of Food Engineering*, vol. 77, no. 1, pp. 119–125, 2006.

[23] W. Zhang, Y. Jia, Q. Huang, Q. Li, and K. Bi, "Simultaneous determination of five major compounds in Polygonum cuspidatum by HPLC," *Chromatographia*, vol. 66, no. 9-10, pp. 685–689, 2007.

[24] N. Hamidi and T. Tsuruta, "Improvement of freezing quality of food by pre-dehydration with microwave-vacuum drying," *Journal of Thermal Science and Technology*, vol. 3, no. 1, pp. 86–93, 2008.

[25] S. Hirun, N. Utama-ang, and P. D. Roach, "Turmeric (Curcuma longa L.) drying: an optimization approach using microwave-vacuum drying," *Journal of Food Science and Technology*, vol. 51, no. 9, pp. 2127–2133, 2014.

11

A Mathematical Method for Determining Optimal Quantity of Backfill Materials Used for Grounding Resistance Reduction

Jovan Trifunovic

Faculty of Electrical Engineering, University of Belgrade, Bulevar Kralja Aleksandra 73, 11000 Belgrade, Serbia

Correspondence should be addressed to Jovan Trifunovic; jovan.trifunovic@etf.rs

Academic Editor: Guido Ala

During installation of grounding system, which represents a significant part of any electrical power system, various backfill materials are used for grounding resistance reduction. The general mathematical method for determining an optimal quantity of backfill materials used for grounding resistance reduction, based on the mathematical tools, 3D FEM modeling, numerical analysis of the obtained results, and the "knee" of the curve concept, as well as on the engineering analysis based on the designer's experience, is developed and offered in this paper. The proposed method has been tested by applying it to a square loop enveloped by a backfill material and buried in a 2-layer soil. The results obtained by the presented method showed a good correlation with the experimentally obtained data from literature. The proposed method can help the designers to avoid the saturation areas in order to maximize efficiency of backfill material usage.

1. Introduction

A proper design of grounding systems is essential to assure the safety of the persons and avoid interruptions of power supply, as well as to protect the electrical and electronic equipment [1]. In order to meet the electrical safety standards, grounding resistance of a grounding system must be lower than the demanded values (in further text denoted as R_{dem}), which can vary from $10\,\Omega$ for lightning protection [2] to below $0.1\,\Omega$ for sites where protective devices must operate very quickly [3]. This is not always easy to obtain, especially in troubled environments (high soil resistivity and/or soil which forms a poor contact with grounding system electrodes) [4]. In such cases grounding system resistance can be decreased by increasing the number of rods or the electrode length or by using appropriate backfill materials. The latter solution, although generally not suitable for large grounding systems, in some cases can be efficient for electrodes covering small areas. Various backfill materials are used in practice to eliminate the contact resistance component of grounding resistance, as well as to reduce the grounding resistance to R_{dem} (usage of bentonite was analyzed in [5, 6]; usage of coconut coir peat, planting-clay soil, and paddy dust was analyzed in [7, 8]; usage of granulated blast furnace slag and fly ash was analyzed in [9] and [10], respectively; usage of waste drilling mud was analyzed in [11]; general analysis of backfill material usage was performed in [12, 13]).

The analysis conducted in [14] indicated that in soils which form a poor contact with the grounding system electrodes the value of grounding resistance could be significantly decreased using a backfill material which is characterized either by ability to provide excellent contact with grounding electrodes or by low resistivity or by proper combination of those 2 features. It means that in such terrains the whole contact resistance component (which is not taken into account by standard engineering methods and formulas, e.g., given in [15, 16]) can be eliminated, and the values of grounding resistance R could be decreased to those which are computed by standard engineering methods and formulas (in further text, for convenience, denoted as the baseline values R_0) using relatively small quantities of a backfill material characterized by ability to achieve perfect contact with both the grounding electrodes and the surrounding soil. In that case the whole contact resistance component would be eliminated even if the backfill material is characterized by the same high resistivity as the surrounding soil [14]. According to findings presented in [17] (based on experimental investigations), the sufficient amount of a backfill material characterized by ability to achieve perfect contact with both the grounding electrodes and the surrounding soil, such as bentonite, which should provide the successful elimination of the whole contact resistance component, is 0.02 m^3 per 1 m of grounding strip (V'_{min}).

However, if the used backfill material is characterized by a lower resistivity than the surrounding soil, the additional decrease of grounding resistance can be achieved. The grounding resistance R of a grounding system decreases with increasing quantity of the used backfill material. Nevertheless, this grounding resistance reduction effect will display the saturation phenomenon when the quantity of the used backfill material increases to a certain level [18]. The best understanding of this saturation phenomenon can be obtained analyzing the effect of the volume V of used backfill material, as well as the effect of its resistivity ρ_{bf}, on the grounding resistance reduction rate δR (%), defined by the following expression:

$$\delta R\ (\%) = \frac{R_0 - R}{R_0} \cdot 100. \tag{1}$$

Optimization methodologies, based on mathematical tools and computer-aided design, are required for minimizing investment costs of power systems, parts of which are power transmission lines [19] and their grounding systems [20]. Hence, such optimization methodologies for determining theoretical maximum efficiency of backfill material usage for grounding rods are offered in [13, 18, 21]. Being based on grounding resistance calculation techniques which are suitable only for solving 2D problems, the application of those optimization methodologies is limited only to grounding systems with geometries characterized by rotational symmetries (e.g., grounding rods surrounded by a cylinder of backfill material), buried in uniform soil, which can be reduced to 2D problems. They cannot be applied to grounding systems characterized by more complex geometries, buried in nonuniform soils, for example, square loop buried in a 2-layer soil (Figure 1) and enveloped by a backfill material (Figure 2).

For such complex geometries, the general mathematical method for determining a backfill material optimal volume, based on 3D FEM modeling, numerical analysis of the obtained results, and the "knee" of the curve concept, as well as on the engineering analysis based on the designer's experience, is developed and offered in this paper. The proposed method is suitable only for cases in which using backfill materials has advantages (in terms of efficiency and cost) in relation to simply increasing the number of rods or the electrode length.

The "knee" of the curve concept was adopted following examples from engineers working in different areas of system design which use the "knee" of the curve (i.e., of the graph of a continuous function which is relevant for the system behavior), representing the border between the saturated and unsaturated area of a curve, in their optimization methodologies. General system design analyses with "knee" concept were given in [22, 23], and usage of "knee" concept in information technologies was explained in [24, 25], in chemical engineering in [26], and in projectile design in [27].

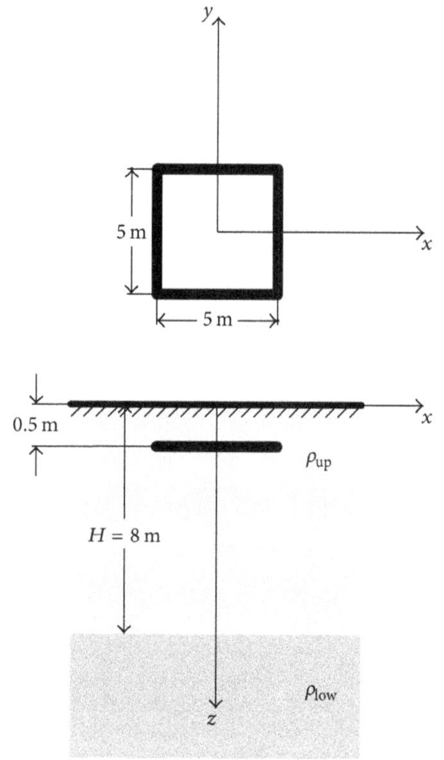

FIGURE 1: The considered grounding loop installed in a 2-layer soil.

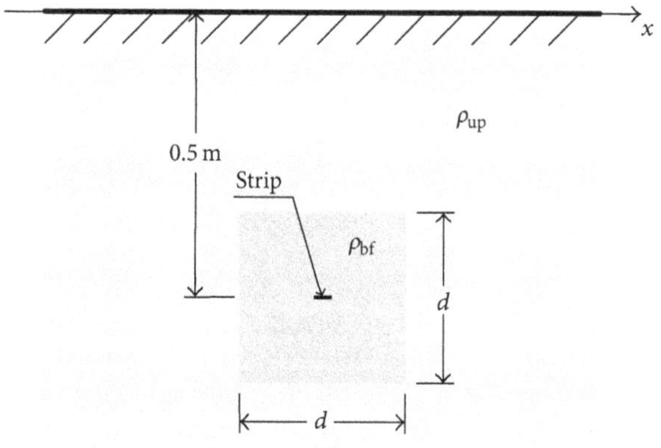

FIGURE 2: Cross-section showing grounding strip of loop surrounded by backfill material.

The experimental setup and the results of measurements of the 2 identical square grounding loops buried in a 2-layer soil (the one backfilled with bentonite suspension and the other conventional), presented in [11], were analyzed using the proposed method. Being based on 3D FEM modeling, the proposed method is suitable for any kind of multilayer soil, as well.

2. The Experimental Setup and the Results of Measurements

As reported in [11], the 2 identical square loops were installed in a former stone bed. The site was described by a 2-layer soil ($\rho_{up} = 170\,\Omega$m, $\rho_{low} = 75\,\Omega$m, and $H = 8$ m (Figure 1)). The upper soil layer was made of stones (karst terrain). The dimensions of the loops (5 m × 5 m) belong to the range of the dimensions of grounding loops which are frequently used as parts of grounding systems of 35 kV transmission line towers and 10 kV/0.4 kV transformer stations. The loops made of rectangular cross-section (30 mm × 4 mm) zinc protected steel strips were installed at a depth of 0.5 m. It was proven in [14] (by 3D FEM modeling of both loops) that the input data and the results of measurements from this experimental setup, reported in [11], were obtained with a reasonable accuracy.

The backfill material of the first loop channel was the excavated material. The grounding resistance of the loop was measured to be $R = 50.2\,\Omega$, and the calculated resistance amounted to $R = 14.6\,\Omega$ [11]. It was suggested in [11] and shown by 3D FEM modeling in [28, 29] that the huge difference between the measured and calculated grounding resistances in this particular case was caused by reduced contact surface between the grounding electrodes and the surrounding soil (i.e., very high contact resistance component), which was not taken into consideration by the applied calculation formula in [11]. Due to the type of soil (stones, karst terrain), it was impossible to achieve good contact between electrodes and soil by compacting the soil above the electrode, which is the common practice for avoiding such very high contact resistance component. It is also possible that when electrodes are subjected to impressed currents by faults or other occurrences (for instance, lightning-related currents), the high values of the associate electric field at the electrode surface naturally would promote a good contact of this surface with the surrounding soil.

The second loop channel was backfilled with 1.2 m^3 of bentonite suspension (0.06 m^3 of bentonite per 1 m of the grounding strip). Resistivity of this backfill material was $\rho_{bf} = \rho_{bentonite} = 2.5\,\Omega$m. The grounding resistance of this loop was measured to be $R = 12.5\,\Omega$ [11]. It is obvious that by the use of bentonite not only is the whole contact resistance component eliminated, but the additional decrease of grounding resistance is achieved.

3. The "Knee" of the Curve Concept

In cases of grounding systems with backfill materials, if the volume of used backfill material corresponds to the values from the saturated area of the relevant curve $\delta R(V)$, it is likely that this volume is oversized, and therefore, material and human or machine efforts may be wasted and investment increased without a justified reason. Therefore, the values of the "knee" point coordinates, V_{knee} and δR_{knee}, practically can be considered as the maximum volume of backfill material, V_{max}, that should be used and the maximum grounding resistance reduction rate, δR_{max}, which realistically can be achieved with the use of the considered backfill material for the considered grounding system at the considered installation site.

On the contrary, if the volume of used backfill material corresponds to the values from the unsaturated area of the relevant curve, it is likely that this volume is undersized, and hence, the opportunity to additionally reduce the grounding resistance of a grounding system with relatively small additional investment, using volume V_{dem} of backfill material ($V_{dem} \leq V_{max}$), sufficient to achieve $R \leq R_{dem}$, could unjustifiably be wasted.

However, it is not always easy to determine the "knee" point of a curve. It should not be read from the graphic because an "optical illusion," produced by using different aspect ratios in the V and δR axes, can occur, misleading an engineer and giving him a false "knee" point data. Therefore, a mathematical approach for finding coordinates of a "knee" point is adopted. It is based on differential calculus and the mathematical definition of curvature for continuous functions, considering the formal definition of a "knee" for continuous functions given in [22], where it is defined as a point of maximum curvature of a curve. The point of the maximum curvature of a curve corresponds to the point of the minimum radius of curvature. The radius of curvature, r, of a graph of a continuous function (curve) at a point is the length of the radius of the circular arc which represents the best approximation of the curve at that point. For any continuous function $y(x)$, if given in Cartesian coordinates and assuming that it is differentiable up to the second order, the radius of curvature at an arbitrary point of its graph can be determined using the following expression [30]:

$$r = \frac{\left(1 + (dy/dx)^2\right)^{3/2}}{\left|d^2y/dx^2\right|}. \qquad (2)$$

The "knee" point of a curve, that is, the point characterized by the minimum radius of curvature, can be determined using the equation

$$\frac{dr}{dx} = 0, \qquad (3)$$

and satisfying the condition

$$\frac{d^2r}{dx^2} > 0. \qquad (4)$$

While the radius of curvature is well defined for continuous functions, it is not well defined for discrete data sets. Note that in the considered case only the discrete data sets, several pairs of values $(V, \delta R)$ for each value of ρ_{bf}, can be obtained by 3D FEM calculations. In the discrete case, the radius of curvature and the "knee" point of a curve can be determined by fitting a suitable continuous function to the available data, followed by the application of (2)–(4) to that function.

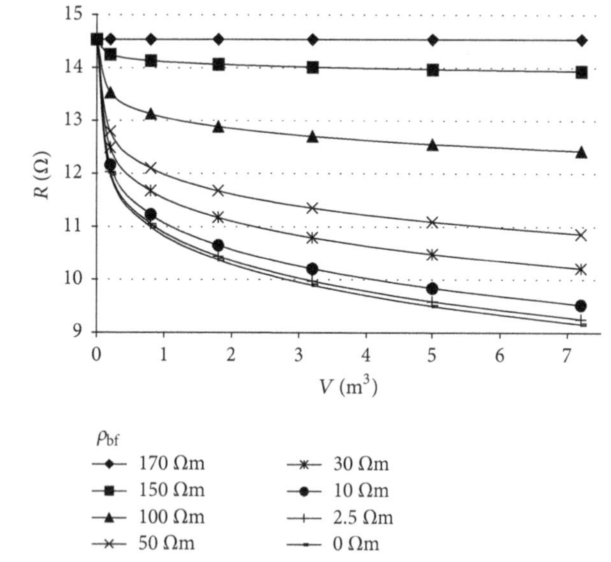

FIGURE 3: The effect of the volume of a backfill material and its resistivity on the grounding resistance of the considered loop.

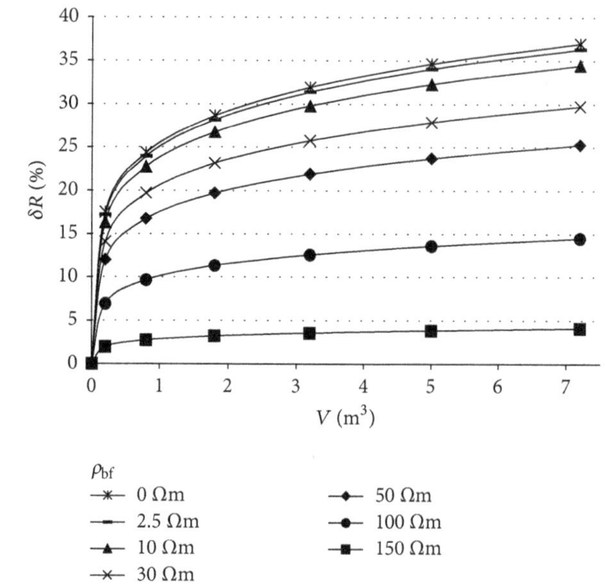

FIGURE 4: The effect of the volume of a backfill material and its resistivity on the grounding resistance reduction rate.

4. Determining the Curve $\delta R(V)$ and the Values of the "Knee" Point Coordinates for the Considered Grounding Loops

The described experimental setup (Figure 1) was 3D modeled applying FEM. The used model is described in detail in [14, 28, 31]. Note that any kind of multilayer soil could also easily be modeled using 3D FEM, in the same manner as it was done in case of 2-layer soil for the described experimental setup. Backfill material was modeled as subdomain which surrounds the grounding loop, cross-section of which is shown in Figure 2.

Values of resistivity of backfill material were varied from $0\,\Omega$m (ideal conducting material) to $170\,\Omega$m (ρ_{up}), as well as the subdomain's dimension d (Figure 2) from 0.1 m to 0.6 m, and the grounding resistance of the considered loop was calculated for each case using FEM. The variation of the grounding resistance of considered loop R as a function of volume V of backfill material, for its various resistivities ρ_{bf}, is presented in Figure 3. The volume V of backfill material is calculated as

$$V = d^2 \cdot P \tag{5}$$

(P is the grounding loop perimeter, 20 m in the considered case).

It is obvious from curves shown in Figure 3 that the values of the grounding resistance lower than the baseline value ($R_0 = 14.53\,\Omega$, calculated for the considered case using 3D FEM) can be obtained with backfill materials characterized by different values of their resistivities, as long as their resistivities are lower than the resistivity of surrounding soil ($\rho_{up} = 170\,\Omega$m).

The pairs of values ($V, \delta R$) for various values of ρ_{bf}, calculated for the considered case using 3D FEM modeling, are presented at the diagram shown in Figure 4 as points marked by different items for each value of ρ_{bf}. As expected, for lower resistivities of the backfill material, the lower grounding resistances of the considered loop were obtained. Nevertheless, the saturation phenomenon in grounding resistance reduction effect obviously appears even if backfill material is an ideal conducting material ($\rho_{bf} = 0\,\Omega$m). According to the diagram shown in Figure 4, it would be very difficult to achieve the value of $\delta R = 40\%$ in the considered case, even if large volume of an ideal conducting material was used as a backfill material. For backfill materials characterized by higher resistivities, δR gets saturated at its smaller values, even for very small used volumes of a backfill material.

Analyzing values presented at the diagram shown in Figure 4, it was perceived that the grounding resistance reduction rate δR (%), as a function of the used volume V (m^3) of backfill material, can be approximated by the following expression:

$$\delta R = a \cdot \ln (b \cdot V + 1), \tag{6}$$

where a and b are positive parameters, different for each value of ρ_{bf}, which describe the shape of the $\delta R(V)$ curve in the considered case. This function was chosen among other candidate functions because it was the most suitable one for the determination of the radius of curvature. It is differentiable up to the second order and its first and second derivatives were easily determined:

$$\begin{aligned} \frac{d\delta R}{dV} &= \frac{a \cdot b}{b \cdot V + 1}, \\ \frac{d^2\delta R}{dV^2} &= -\frac{a \cdot b^2}{(b \cdot V + 1)^2}. \end{aligned} \tag{7}$$

The parameters a and b, which describe the shape of the $\delta R(V)$ curve in the considered case, were determined by fitting the continuous function expressed by (6) through the points presented at the diagram shown in Figure 4, using the method of least squares and the iterative calculation method. The values of the determined a and b parameters are given in Table 1 for several values of ρ_{bf}.

The radius of curvature $r(V)$ at arbitrary point of a curve $\delta R(V)$ can be determined by applying (2) on (6), which gives

$$r = \frac{\left(1 + (d\delta R/dV)^2\right)^{3/2}}{\left|d^2\delta R/dV^2\right|}, \tag{8}$$

and then incorporating (7) into (8), which after rearranging becomes

$$r = \frac{\left((a^2 \cdot b^2) / \left((b \cdot V + 1)^2 + 1\right)\right)^{3/2} \cdot (b \cdot V + 1)^2}{a \cdot b^2}. \tag{9}$$

The first and the second derivative of $r(V)$ function, respectively, are

$$\begin{aligned} \frac{dr}{dV} &= \frac{\left((a^2 \cdot b^2) / (b \cdot V + 1)^2 + 1\right)^{1/2} \cdot \left(2 \cdot b^2 \cdot V^2 - a^2 \cdot b^2 + 4 \cdot b \cdot V + 2\right)}{a \cdot b \cdot (b \cdot V + 1)}, \\ \frac{d^2r}{dV^2} &= \frac{2 \cdot a^4 \cdot b^4 + a^2 \cdot b^4 \cdot V^2 + 2 \cdot a^2 \cdot b^3 \cdot V + a^2 \cdot b^2 + 2 \cdot b^4 \cdot V^4 + 8 \cdot b^3 \cdot V^3 + 12 \cdot b^2 \cdot V^2 + 8 \cdot b \cdot V + 2}{a \cdot \left((a^2 \cdot b^2) / (b \cdot V + 1)^2 + 1\right)^{1/2} \cdot (b \cdot V + 1)^4}. \end{aligned} \tag{10}$$

The "knee" point of $\delta R(V)$ function can be determined by using the equation

$$\frac{dr}{dV} = 0 \tag{11}$$

and satisfying the condition

$$\frac{d^2r}{dV^2} > 0. \tag{12}$$

The solution of (11) which is realistic and which satisfies the condition expressed by (12) is

$$V_{knee} = \frac{\sqrt{2} \cdot a \cdot b - 2}{2 \cdot b}. \tag{13}$$

This solution represents the V coordinate of the "knee" point of $\delta R(V)$ function. The δR coordinate of the "knee" point of $\delta R(V)$ function can be determined introducing the solution expressed by (13) into (6), which after rearranging becomes

$$\delta R_{knee} = a \cdot \ln\left(\frac{\sqrt{2} \cdot a \cdot b}{2}\right). \tag{14}$$

The values of the "knee" point coordinates, V_{knee} and δR_{knee}, are given in Table 1 for several values of ρ_{bf}. They are calculated incorporating the corresponding a and b parameters into (13) and (14).

5. Determining a Backfill Material Optimal Volume for the Considered Grounding Loops by Engineering Analysis

For determining an optimal volume, V_{opt}, of used backfill material, a designer must be able to estimate the influence of use of arbitrary volume, V, of a backfill material on grounding resistance, R, of a grounding system. For the considered grounding loops it can be done by incorporating value of δR (obtained by (6)), along with the baseline value of R_0, into the following expression obtained rearranging (1):

$$R = R_0 \cdot \left(1 - \frac{\delta R}{100}\right). \tag{15}$$

By incorporating (6) into (15), which after rearranging becomes

$$R = R_0 \cdot \ln\left(\frac{e}{(b \cdot V + 1)^{a/100}}\right), \tag{16}$$

direct $R(V)$ dependence is obtained, which can be used for the same purpose (e is the base of the natural logarithm).

For the considered loops, if it is assumed that $R_{dem} = 10\ \Omega$, $\rho_{bf} = \rho_{bentonite} = 2.5\ \Omega$m and that the sufficient amount of bentonite, which should provide the successful elimination of the whole contact resistance component, is $V'_{min} = 0.02\ m^3$ per 1 m of grounding strip (according to findings presented in [17], based on experimental investigations), characteristic backfill material volumes are $V_{min} = P \cdot V'_{min} = 0.4\ m^3$, $V_{max} = V_{knee} = 3.8\ m^3$ (obtained using (13), $V'_{max} = 0.19\ m^3$ of bentonite per 1 m of strip), and $V_{dem} = 2.97\ m^3$ (obtained using (16), $V'_{dem} \approx 0.15\ m^3$ of bentonite per 1 m of strip).

Comparing the experimentally obtained values of the grounding resistance ($R = 50.2\ \Omega$ for the conventional loop and $R = 12.5\ \Omega$ for the loop backfilled with $1.2\ m^3$ of bentonite suspension ($0.06\ m^3$ per 1 m of strip)) with the baseline value of $R_0 = 14.53\ \Omega$ calculated using FEM and the values of $R = 11.55\ \Omega$, $R = 10.71\ \Omega$, and $R = 9.81\ \Omega$, calculated using (16) for loop backfilled with $V_{min} = 0.4\ m^3$, $V = 1.2\ m^3$, and $V_{max} = 3.8\ m^3$ of bentonite suspension, respectively, it can be concluded that the amount of $0.06\ m^3$ of bentonite suspension per 1 m of the grounding strip used in experimental setup [11] was sufficient to eliminate the whole contact resistance component, but if that was the only goal, it could be achieved using 3 times smaller volume ($0.02\ m^3$ per 1 m of strip [17]).

TABLE 1: The values of the determined a and b parameters and the "knee" point coordinates for various resistivities of backfill material, for the considered case.

ρ_{bf} (Ωm)	a	b	V_{knee} (m^3)	δR_{knee} (%)
0	5.48	109.43	3.87	33.17
2.5	5.39	109.43	3.80	32.51
10	5.11	109.69	3.61	30.58
30	4.41	110.96	3.11	25.77
50	3.74	112.39	2.63	21.28
100	2.14	115.67	1.51	11.06
150	0.60	118.42	0.42	2.37

If the goal was to achieve $R_{dem} = 10\,\Omega$, nearly 2.5 times larger volume (0.15 m^3 of bentonite per 1 m of strip) should be used. However, it must be taken into consideration that there is some difference between the experimentally obtained value of grounding resistance of the loop backfilled with $V = 1.2\,\text{m}^3$ bentonite and the one calculated using (16) ($R = 12.5\,\Omega$ and $R = 10.71\,\Omega$, respectively), because the input data and the results of measurements from this experimental setup, reported in [11], were obtained only with a reasonable, not perfect, accuracy. Therefore, during the design process, a designer should definitely foresee usage of somewhat larger volumes of backfill material than those estimated by (16), in order to compensate potential errors caused by inaccuracy of certain input parameters and their seasonal variations, but certainly not larger than V_{max} (= 3.8 m^3 in the considered case), because the increase of volume over this value could not provide significant positive effect but would only needlessly increase the installation costs. All things considered, in the end, it is again on the designer to estimate the optimal volume, V_{opt}, of a backfill material, on the basis of the existing situation on the site, the impact of considered grounding system on the safety of people and equipment, and the available budget, as well as on the basis of personal experience and engineering sense.

In the considered case, if the position of the grounding system required $R_{dem} = 10\,\Omega$, the logical designer's choice would be to adopt, for example, $V_{opt} = 3.3\,\text{m}^3$ ($\approx 1.1 \cdot V_{dem} < V_{max} = 3.8\,\text{m}^3$; additional 10% of backfill material to compensate, to some extent, potential errors caused by inaccuracy of certain input parameters and their seasonal variations). If at the position of the grounding system $R_{dem} = 15\,\Omega$ was acceptable (which is often the case in the power system in Serbia), the logical designer's choice would be to adopt, for example, $V_{opt} = 0.6\,\text{m}^3$ ($= 1.5 \cdot V_{min}$, enough to eliminate the entire contact resistance and additional 50% of backfill material to compensate potential errors). Either way, it can be concluded that the used volume $V = 1.2\,\text{m}^3$ of bentonite was wrong choice (undersized if $R_{dem} = 10\,\Omega$, meaning that the additional work would be required after installation; oversized if $R_{dem} = 15\,\Omega$, meaning that the installation costs were increased without a justified reason).

6. Method for Determining a Backfill Material Optimal Volume

The procedure for obtaining (6), (13), (14), and (16) and the determination of necessary coefficients, as well as engineering analysis of the results, which are presented in Sections 4 and 5 for considered grounding loop, represent a new mathematical method for determining optimal quantity of backfill materials used for grounding resistance reduction, which can be applied to any type of grounding system, with various dimensions, placed in any soil structure, with various quantities and characteristics of backfill materials. To summarize, the method consists of the following 6 steps:

(1) 3D FEM modeling of the considered soil structure, grounding system, and backfill material,

(2) calculation (using 3D FEM) of several pairs of values $(V, \delta R)$ for backfill materials (characterized by ρ_{bf}) which are available for grounding system construction on the desired location (Figures 3 and 4 in the considered case),

(3) finding and fitting a suitable $\delta R(V)$ continuous function to a set of the obtained $(V, \delta R)$ points for each backfill material ((6) in the considered case),

(4) determining the "knee" point (V_{knee}, δR_{knee}) applying mathematical approach based on differential calculus and the mathematical definition of curvature to the obtained continuous function $\delta R(V)$ ((13) and (14) and Table 1 in the considered case),

(5) determining $R(V)$ dependence ((16) in the considered case) and using it to calculate characteristic backfill material volumes (V_{min}, V_{max}, and V_{dem}), as well as values of grounding resistances that correspond to them,

(6) conducting engineering analysis, on the basis of the existing situation on the site, the impact of considered grounding system on the safety of people and equipment, and the available budget, as well as on the basis of designer's personal experience and engineering sense, in order to estimate the optimal volume, V_{opt}, of a backfill material.

The proposed method can help grounding system designers to avoid the saturation phenomenon when using a backfill material for grounding resistance reduction and maximize the efficiency of its use. Although it does not calculate the exact optimum quantity of a backfill material (which is impossible in a practical sense), it represents a new tool (a kind of which was not available neither in standards nor in scientific and professional literature) for conducting technical and techno-economic analysis, the results of which

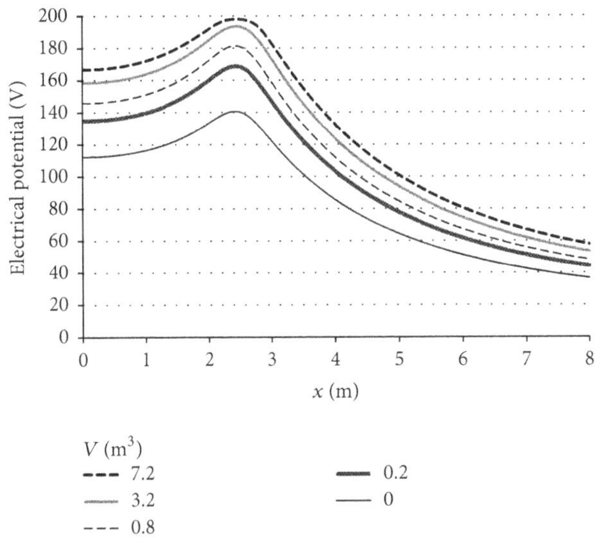

FIGURE 5: Potential distributions along a line situated at the ground surface, presented for various volumes of backfill material (bentonite).

can assist in assessing the optimum quantity of a backfill material that should be used. However, the practical problems could occur related to implementing the optimized results in real conditions. In some cases it could be difficult, or even impossible, to construct, in real conditions of soil, holes with the dimensions determined with the proposed method for the perfect disposal of the optimized volume of backfill material.

The author's present attempts are focused on simplifying the presented method. The room for possible simplification may lie in the fact that the "knee" value of the volume of backfill material is almost perfectly linearly dependent from the resistivity of backfill material, according to the data presented in Table 1. Hence, it is possible that only the "knee" value of the volume of backfill material for $\rho_{\text{bf}} = 0$ ($V_{\text{knee}}(\rho_{\text{bf}} = 0)$) has to be calculated using the proposed method, in order to draw a straight line between the points $(V_{\text{knee}}, \rho_{\text{bf}}) = (V_{\text{knee}}(\rho_{\text{bf}} = 0), 0)$ and $(V_{\text{knee}}, \rho_{\text{bf}}) = (0, \rho_{\text{up}})$ on diagram with V_{knee} and ρ_{bf} axis, and read from that line the "knee" value of the volume for any ρ_{bf}. However, before the mentioned simplification could be applied, the linear dependence of the "knee" value of the volume from the resistivity of backfill material has to be checked for other types of grounding systems and soil structures.

7. Distribution of Electrical Potential at the Ground Surface

The influence of the used volume of backfill material on the distribution of electrical potential at the ground surface above the buried grounding system (during the earth fault) was also investigated. The diagram shown in Figure 5 contains curves representing the potential distributions along the line between the points $(x, y) = (0\text{ m}, 0\text{ m})$ and $(x, y) = (8\text{ m}, 0\text{ m})$ belonging to the ground surface ($z = 0$ m), calculated using 3D FEM for various volumes of bentonite ($\rho_{\text{bf}} = \rho_{\text{bentonite}} = 2.5\ \Omega\text{m}$) and for electrical potential of the grounding system electrodes that equals 200 V.

The diagram shown in Figure 5 illustrates how the volume of used backfill material (bentonite) affects the potential distribution on the ground surface during earth fault. It is obvious that the touch voltage, which represents the potential difference between the ground potential rise (electrical potential of the grounding system electrodes) and the ground surface potential at a point where a person is standing while at the same time having a hand in contact with a grounded structure, is decreasing with the increase of backfill material volume (for the same electrical potential of the grounding system electrodes). The step voltage, representing the difference in ground surface potentials experienced by a person bridging a distance of 1 m, is obviously increasing with the increase of backfill material volume (for the same electrical potential of the grounding system electrodes). However, to precisely determine the touch and step voltages—the parameters that indicate the quality of the grounding system—as well as electrical potential of grounding system electrodes in a specific case, conditions in the corresponding electrical circuit must be taken into consideration and deeper analyses have to be performed, which will also be the subject of future author's work.

8. Conclusions

A new mathematical method for determining an optimal quantity of backfill materials used for grounding resistance reduction, based on 3D FEM modeling, numerical analysis of the obtained results, and the "knee" of the curve concept, as well as on the engineering analysis based on the designer's experience, is developed and offered in this paper. Being based on 3D FEM, to the best of the author's knowledge, it is the first such method offered in literature which is general and can be applied to any geometry of electrodes and soil composition.

The proposed method has been tested by applying it to the experimental setup presented in literature [11], containing square loop enveloped by a backfill material and buried in a 2-layer soil. The results obtained by the presented method provide the designer with a range of the optimal volume for any backfill material which is available. It helps a designer to make decision whether to use the volume of backfill material closer to the lower limit of range and try to eliminate only contact resistance component with a smaller investment or to use the volume closer to the upper limit of range and achieve the additional decrease of the grounding resistance with a somewhat higher investment.

However, the proposed method is suitable only for cases in which using backfill materials has advantages (in terms of efficiency and cost) in relation to simply increasing the number of rods or the electrode length. Also, in some cases it could be difficult, or even impossible, to construct, in real conditions of soil, holes with the dimensions determined with the proposed method for the perfect disposal of the optimized volume of backfill material.

Conflicts of Interest

The author declares that there are no conflicts of interest related to this paper.

Acknowledgments

This research was partially supported by the Ministry of Education, Science and Technological Development of the Republic of Serbia (Project TR 36018).

References

[1] I. Colominas, J. París, D. Fernández, F. Navarrina, and M. Casteleiro, "A numerical simulation tool for multilayer grounding analysis integrated in an open-source CAD interface," *International Journal of Electrical Power & Energy Systems*, vol. 45, no. 1, pp. 353–361, 2013.

[2] IEC 62305-3, *Protection against lightning Part 3: Physical damage to structures and life hazard*, Protection against lightning - Part 3, Physical damage to structures and life hazard, 2010.

[3] G. Eduful, J. E. Cole, and P. Y. Okyere, "Optimum mix of ground electrodes and conductive backfills to achieve a low ground resistance," in *Proceedings of the IEEE 2nd International Conference on Adaptive Science Technology*, pp. 140–145, Accra, Ghana, 2009.

[4] S. C. Lim, C. Gomes, and M. Z. A. Ab Kadir, "Electrical earthing in troubled environment," *International Journal of Electrical Power & Energy Systems*, vol. 47, no. 1, pp. 117–128, 2013.

[5] Z. R. Radakovic, M. V. Jovanovic, V. M. Milosevic, and N. M. Ilic, "Application of earthing backfill materials in desert soil conditions," *IEEE Transactions on Industry Applications*, vol. 51, no. 6, pp. 5288–5297, 2015.

[6] Z. R. Radakovic and M. B. Kostic, "Behaviour of grounding loop with bentonite during a ground fault at an overhead line tower," *IEE Proceedings Generation, Transmission and Distribution*, vol. 148, no. 4, pp. 275–278, 2001.

[7] J. Jasni, L. K. Siow, M. Z. A. Ab Kadir, and W. F. Wan Ahmad, "Natural materials as grounding filler for lightning protection system," in *Proceedings of the 30th International Conference on Lightning Protection (ICLP '10)*, Italy, September 2010.

[8] N. Kumarasinghe, "A low cost lightning protection system and its effectiveness," in *Proceedings of the 20th International Lightning Detection Conference and 2nd International Lightning Meteorology Conference*, Tucson, AZ, USA, 2008.

[9] L.-H. Chen, J.-F. Chen, T.-J. Liang, and W.-I. Wang, "A study of grounding resistance reduction agent using granulated blast furnace slag," *IEEE Transactions on Power Delivery*, vol. 19, no. 3, pp. 973–978, 2004.

[10] S. Chen, L. Chen, C. Cheng, and J. Chen, "An Experimental Study on the Electrical Properties of Fly Ash in the Grounding System," *International Journal of Emerging Electric Power Systems*, vol. 7, no. 2, 2006.

[11] M. B. Kostic, Z. R. Radakovic, N. S. Radovanovic, and M. R. Tomasevic-Canovic, "Improvement of electrical properties of grounding loops by using bentonite and waste drilling mud," *IEE Proceedings Generation, Transmission and Distribution*, vol. 146, no. 1, pp. 1–6, 1999.

[12] H. E. Martínez, E. L. Fuentealba, L. A. Cisternas, H. R. Galleguillos, J. F. Kasaneva, and O. A. De La Fuente, "A new artificial treatment for the reduction of resistance in ground electrode," *IEEE Transactions on Power Delivery*, vol. 19, no. 2, pp. 601–608, 2004.

[13] A. A. Al-Arainy, N. H. Malik, M. I. Qureshi, and Y. Khan, "Grounding pit optimization using low resistivity materials for applications in high resistivity soils," *International Journal of Emerging Electric Power Systems*, vol. 12, no. 1, article 3, 2011.

[14] J. Trifunović, "The algorithm for determination of necessary characteristics of backfill materials used for grounding resistances of grounding loops reduction," *Journal of Electrical Engineering*, vol. 63, no. 6, pp. 373–379, 2012.

[15] *IEEE Guide for Safety in AC Substation Grounding, ANSI/IEEE Std. 80-1986*, 1986.

[16] *IEEE Guide for Safety in AC Substation Grounding, ANSI/IEEE Std. 80-2000*, 2000.

[17] H. Kutter and W. Lange, "Grounding improvement by using bentonite," *Elektrie*, vol. 21, pp. 421–424, 1967.

[18] L.-H. Chen, J.-F. Chen, T.-J. Liang, and W.-I. Wang, "A research on used quantity of ground resistance reduction agent for ground systems," *European Transactions on Electrical Power*, vol. 20, no. 4, pp. 408–421, 2010.

[19] S. K. Teegala and S. K. Singal, "Optimal costing of overhead power transmission lines using genetic algorithms," *International Journal of Electrical Power & Energy Systems*, vol. 83, pp. 298–308, 2016.

[20] H. M. Khodr, "Optimal methodology for the grounding systems design in transmission line using mixed-integer linear programming," *Electric Power Components and Systems*, vol. 38, no. 2, pp. 115–136, 2010.

[21] Y. Khan, F. R. Pazheri, N. H. Malik, A. A. Al-Arainy, and M. I. Qureshi, "Novel approach of estimating grounding pit optimum dimensions in high resistivity soils," *Electric Power Systems Research*, vol. 92, pp. 145–154, 2012.

[22] V. Satopaa, J. Albrecht, D. Irwin, and B. Raghavan, "Finding a "kneedle" in a haystack: Detecting knee points in system behavior," in *Proceedings of the 31st International Conference on Distributed Computing Systems Workshops (ICDCSW '11)*, pp. 166–171, USA, June 2011.

[23] I. Das, "On characterizing the "knee" of the Pareto curve based on normal-boundary intersection," *Journal of Structural Optimization*, vol. 18, no. 2-3, pp. 107–115, 1999.

[24] F. A. Gonzalez-Horta, R. A. Enriquez-Caldera, J. M. Ramirez-Cortes, J. Martínez-Carballido, and E. Buenfil-Alpuche, "Mathematical model for the optimal utilization percentile in M/M/1 systems: a contribution about knees in performance curves," in *Proceedings of the 3rd International Conference on Adaptive and Self-Adaptive Systems and Applications*, Rome, Italy, 2011.

[25] S. Salvador and P. Chan, "Determining the number of clusters/segments in hierarchical clustering/segmentation algorithms," in *Proceedings of the 16th IEEE International Conference on Tools with Artificial Intelligence (ICTAI '04)*, pp. 576–584, USA, November 2004.

[26] Z. Jia and M. G. Ierapetritou, "Generate Pareto optimal solutions of scheduling problems using normal boundary intersection technique," *Computers & Chemical Engineering*, vol. 31, no. 4, pp. 268–280, 2007.

[27] A. F. Hathaway and J. R. Burnett, "Sabot front borerider stiffness vs. dispersion: Finding the knee in the curve," *Shock and Vibration*, vol. 8, no. 3-4, pp. 193–201, 2001.

[28] J. Trifunovic and M. Kostic, "Analysis of influence of imperfect contact between grounding electrodes and surrounding soil on electrical properties of grounding loops," *Electrical Engineering*, vol. 96, no. 3, pp. 255–265, 2014.

[29] J. Trifunovic and M. Kostic, "An algorithm for estimating the grounding resistance of complex grounding systems including contact resistance," *IEEE Transactions on Industry Applications*, vol. 51, no. 6, pp. 5167–5174, 2015.

[30] I. Newton, *The method of fluxions and infinite series: with its application to the geometry of curve-lines*, London, UK, 1736.

[31] J. Trifunovic and M. Kostic, "Quick calculation of the grounding resistance of a typical 110 kV transmission line tower grounding system," *Electric Power Systems Research*, vol. 131, pp. 178–186, 2016.

12

Coupled THM and Matrix Stability Modeling of Hydroshearing Stimulation in a Coupled Fracture-Matrix Hot Volcanic System

Yong Xiao,[1,2] Jianchun Guo,[2] Hehua Wang,[1,3] Lize Lu,[1,3] John McLennan,[4] and Mengting Chen[5]

[1]*China Zhenhua Oil Co., Ltd, Beijing, China*
[2]*State Key Laboratory on Oil and Gas Reservoir Geology and Exploitation, Southwest Petroleum University, Chengdu, China*
[3]*Chengdu Northern Petroleum Exploration and Development Technology Co., Ltd., Chengdu, Sichuan, China*
[4]*Energy & Geoscience Institute, University of Utah, UT, USA*
[5]*Borehole Operation Branch Office of Sinopec Southwest Petroleum Engineering Co., Ltd., Sichuan, China*

Correspondence should be addressed to Jianchun Guo; guojianchun@vip.163.com

Academic Editor: Mohammed Nouari

A coupled thermal-hydraulic-mechanical (THM) model is developed to simulate the combined effect of fracture fluid flow, heat transfer from the matrix to injected fluid, and shearing dilation behaviors in a coupled fracture-matrix hot volcanic reservoir system. Fluid flows in the fracture are calculated based on the cubic law. Heat transfer within the fracture involved is thermal conduction, thermal advection, and thermal dispersion; within the reservoir matrix, thermal conduction is the only mode of heat transfer. In view of the expansion of the fracture network, deformation and thermal-induced stress model are added to the matrix node's in situ stress environment in each time step to analyze the stability of the matrix. A series of results from the coupled THM model, induced stress, and matrix stability indicate that thermal-induced aperture plays a dominant role near the injection well to enhance the conductivity of the fracture. Away from the injection well, the conductivity of the fracture is contributed by shear dilation. The induced stress has the maximum value at the injection point; the deformation-induced stress has large value with smaller affected range; on the contrary, thermal-induced stress has small value with larger affected range. Matrix stability simulation results indicate that the stability of the matrix nodes may be destroyed; this mechanism is helpful to create complex fracture networks.

1. Introduction

With the increase of energy consumption, the exploration and development of oil and gas has gradually shifted from the conventional to unconventional reservoir, such as volcanic reservoirs [1, 2]. The volcanic reservoir has the characteristics of high temperature gradient and rock mechanical strength, and natural fractures are an important indicator of economic viability. The most important consideration for oil and gas or geothermal energy development in volcanic reservoirs is the use of hydraulic stimulation technology to establish a fracture network to effectively conduct fluid between injection and production wells [3–7]. In hydraulic fracturing associated with oil & gas or enhanced geothermal energy extraction, significant differences in surface pumping pressure and the injectivity index are common, even in the same field, as shown in Table 1.

The purpose of fracturing is to improve the conductivity of natural fractures and possibly extend or develop new fractures [4–8]. As Table 1 indicates, hydraulic fracturing can include extreme (high or low) injection pressure and often there are relatively low injection rates with large total injected volumes. Researchers [9, 10] found that two factors contributed to the difficultly of hydraulic stimulation: (1) with the propensity for reactivation of existing faults, and extension or shearing of natural fractures, hydraulic stimulation in volcanic reservoirs is sometimes referred to as hydroshearing; (2) these stimulation behaviors encompassed in thermal-hydraulic-mechanical (THM) or thermal-hydraulic-chemical (THC), such as shear dilation and thermomechanical aperture closure.

The injected water flow pattern of the volcanic reservoir is mainly controlled by the network of preexisting and induced

Table 1: Injectivity index during stimulations in hydrocarbon and geothermal granitic formations.

Different Energy Development	Project	Depth m	Lithology	Max. Inj. Rate m^3/min	Inj. Pressure MPa/m	Total Injection Volume m^3	Injectivity Indices kg/sec·MPa	Reference
Oil and Gas	Liaohe: #1	3022	Granite Gneiss	6.4	0.0218	572	1.62	/
	Liaohe: #2	4772		5.5	0.0145	498	1.33	/
	Liaohe: #3	4188		4.4	0.0205	480	0.85	/
Geothermal	GPK2	3600	Granite	3.6	0.0036	30000	4.62	(Baria, 2009)
	Habanero 1	4421	Granite	1.8	0.0147	20000	1.00	[21]
	Habanero 4	4205	Granite	3.2	0.0114	27000	16.00	[21]
	RRG-9 ST-1	1524	Granite	1.0	0.0036	/	3.03	[22]

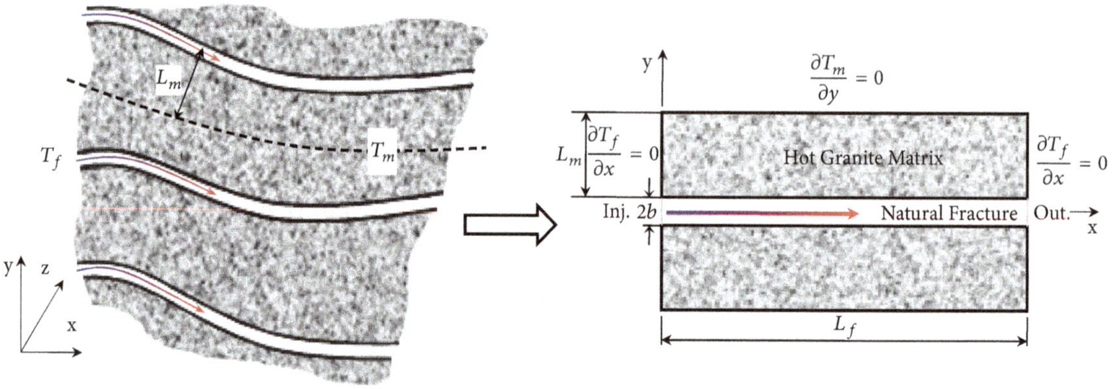

FIGURE 1: Idealized hot matrix-natural fracture system.

natural fractures and is affected by changes in conductivity resulting from heat transfer, shear dilation, and chemical processes. Many detailed studies of the precipitation/dissolution processes and their impact on fracture aperture variation in fractured volcanic reservoirs can be found in the literature [11–13]. The results show, in short-term hydroshearing fracturing, the effect of chemical action on the aperture change is smaller than thermal and mechanical processes, so it can be neglected. Therefore, when employing hydroshearing in the volcanic reservoir, aperture changes due to shear dilation and thermal stress result in the decrease of net pressure and growth of contract area between fracture planes, and this in turn affects fluid flows, heat transfer, and shear behaviors [14, 15]. To explain these behaviors, the authors built a modeling structure that couples thermal, hydraulic, and mechanical effects (THM) and apply this simulator to a hot granite reservoir to examine the importance of this coupling.

Because of the complexities of the numerical modeling, deformation and thermal-induced stress are often neglected and the matrix nodes assume stability and no secondary fractures form [16]. This paper describes a displacement discontinuity method [17] to solve the deformation-induced stress. The fracture is considered as a displacement discontinuity over a finite-length segment, which is treated as the inner boundary condition of the problem [18, 19]. After the given induced stress, the new stress state of the matrix nodes can be updated. The new stress state consists of three parts: initial stress, deformation-induced stress, and thermal-induced stress. The new stress state will affect the stability of the rock matrix and help to create complex fracture networks.

2. Physical System and Governing Equations

The hot and fractured volcanic oil & gas reservoir is always considered nonpermeability, and natural fractures within the matrix are the channels for fluid flow and heat exchange. Based on the simple principle of reducing the extremely complex system to an idealized one, an idealized system for integrating hot matrix and natural fracture is established to study the fracture deformation and their induced stress. The geometry of the conceptually idealized model corresponding to a fracture-matrix coupled system is shown in Figure 1. The upstream boundary along the fracture axis represents the injection well (inlet), and the downstream boundary is the production well (outlet).

2.1. Fracture Fluid Flow Model and Heat Transfer

2.1.1. Fracture Fluid Flow Model. It is assumed that the fracture aperture varies in space and time, and the interface of the fracture does not leak-off so fluid flows along the natural fracture. The fracture is assumed to correspond to a parallel-plate system, and the laminar flow is valid. Without considering the slip effect, the velocity is integrated in the fracture cross section [20]. For this fracture system, the momentum balance indicates that the average flow velocity per unit length is proportional to the pressure gradient. The fracture fluid flow model is

$$u_x(x,t) = \frac{w^2(x,t)}{12\mu}\left(-\frac{\partial P(x,0,t)}{\partial x}\right) \tag{1}$$

where u is the average velocity in the fracture per unit length, w is the average aperture or width of the fracture unit length, μ is the viscosity of the injected fluid which is assumed to be a static value, $P(x,0,t)$ is the fluid pressure within the fracture (y=0), x is the position, and t is the time. Assuming that there is no leak-off and the injected fluid is incompressible, according to the continuity equation, fluid discharge and changes per unit height can be written as

$$\begin{gathered} q(x,t) = w(x,t)\,u_x(x,t) \\ \frac{\partial q(x,t)}{\partial x} = 0 \end{gathered} \tag{2}$$

where $q(x, t)$ is the volume flow. The fracture deformation and heat transfer are based on the balance of energy and mass, respectively. The flow equations will be coupled with the fracture constitutive model and thermal-induced aperture change model.

2.1.2. Heat Transfer Equations. Temperature differences drive the heat transfer between the hot reservoir matrix and injected cold water. The main mechanisms involved are thermal conduction, thermal advection, and thermal dispersion

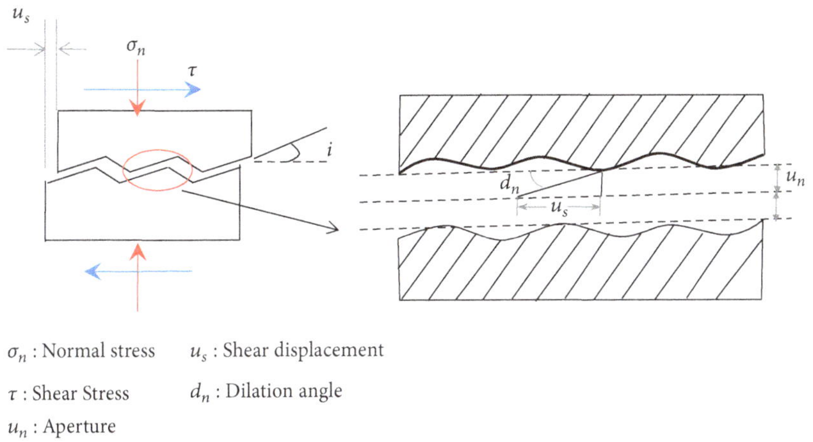

FIGURE 2: Graphs of the fracture aperture caused by shear displacement and dilation.

within the fracture, conduction-limited thermal transport from the reservoir matrix to the fracture. Within the reservoir matrix, thermal conduction is the only mode of heat transfer [20].

The following general assumptions are used in this simulation for calculating the fracture's temperature profile in each time step: (1) specific heat capacities of the matrix and injected water are not functions of temperature (in other words, they are static parameters); (2) there is no fluid leak-off form the interface into rock matrix (i.e., pressure equilibrium exists between the fracture and the reservoir matrix). The governing equations used in this fracture and the reservoir matrix heat transfer studies are given as

$$\rho_f c_f \frac{\partial \vec{T}_f}{\partial t} = -\rho_f c_f \vec{u} \nabla \vec{T}_f + K_{f,eff} \nabla^2 \vec{T}_f + \alpha_f h_{mf} \nabla \vec{T}_m \Big|_{y=b} \tag{3}$$

$$\rho_s c_s \frac{\partial \vec{T}_m}{\partial t} = K_{m,eff} \nabla^2 \vec{T}_m - \alpha_f h_{mf} \nabla \vec{T}_m \Big|_{y=b}$$

where t is the time; T_m and T_f are temperature in the matrix and fracture, separately; ρ_s and ρ_f are rock and water density; c_s and c_f are rock and water heat capacity; $K_{m,eff}$ is the rock thermal conductivity.

2.2. Fracture Constitutive Model and Deformation-Induced Stress

2.2.1. Fracture Constitutive Model. The rough surface of a natural fracture is always undulating. During the hydroshearing stimulation, the effective normal stress of the fracture surface will continue to decrease, and the permanent shear displacement will be generated. The interface of the natural fracture will no longer completely overlap under normal stress and the dilation aperture w_d will increase because of shear displacement, as shown in Figure 2.

In 1985, Barton [23] summarized a nonlinear peak shear strength equation that was based on the joint surface roughness coefficient and the compressive strength (as represented by JRC and JCS):

$$\tau = \sigma_n \tan \left[\phi_r + JRC \log_{10} \left(\frac{JCS}{\sigma_n} \right) \right] \tag{4}$$

where $\emptyset_r$ is the internal friction angle, the angle within the brackets is called the friction angle or residual friction angle. The latter term $JRC \log_{10} (JCS/\sigma_n)$ is called the dilation angle which is closely related to the normal aperture of the fracture surface. The simplest relation between shear strain and stress before peak strength can be described in a linear equation as

$$\tau = k_s u_s \quad 0 \le u_s \le u_s^p \tag{5}$$

$$k_s = \frac{\sigma_n \tan \left[\phi_r + JRC_n \log_{10} (JCS_n/\sigma_n) \right]}{(L_n/500) \left[JRC_n/L_n \right]^{0.33}} \quad 0 \le u_s \le u_s^p \tag{6}$$

where k_s is the shear stiffness, $u_s{}^P$ is shear displacement at peak strength, and L_n is the length of the fracture sample. Dilation is defined as a function of shear displacement and tangent of the dynamic dilation angle; the equation is shown as

$$du_n = \tan(d_n)\, du_s = \tan(d_n) \frac{1}{k_s} d\tau \tag{7}$$

where d_n is the dynamic dilation angle. Barton and Choubey [24] estimated the value of the mobilized dilation using the following empirically derived equation.

$$d_{n-dynamic} = \frac{1}{2JRC_{n-dynamic} \log_{10} (JCS_n/\sigma_n')} \tag{8}$$

2.2.2. Deformation-Induced Stress. The elastic displacement discontinuity method (DDM) is an indirect boundary element method to cope those problems involving pure elastic nonporous media containing discontinuous surfaces or thin fractures [17]. The elastic DDM is based on analytical solution for the infinite two-dimensional homogeneous and

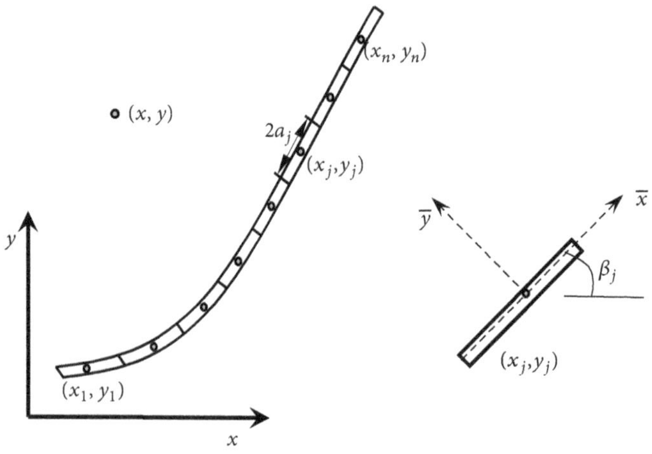

FIGURE 3: The local coordinate distribution after the discretion of natural fracture boundary.

isotropic elastic nonporous medium containing a finite small thin fracture with constant normal and shear displacement discontinuities. For an infinite elastic nonporous medium, the fractures are divided into N elemental segments with the displacement in each segment assumed to have a constant discontinuity. At any point (x, y), the influence of displacement discontinuities from all fractures in the system can be obtained by summing the effects of all N elements using the fundamental analytical solutions. The fracture length is $2a$ (a is the half length of fracture segment) and its center is located at (0, 0) as shown in Figure 3.

The deformation-induced stress in the field by the shear and dilation displacement of jth fracture unit is

$$\begin{aligned}
\sigma^{j}_{\overline{xx}} &= 2GD^{j}_{s}\left(2\overline{f}_{,\overline{xy}} + \overline{y}\overline{f}_{,\overline{xyy}}\right) + 2GD^{j}_{n}\left(\overline{f}_{,\overline{yy}} + \overline{y}\overline{f}_{,\overline{yyy}}\right) \\
\sigma^{j}_{\overline{yy}} &= 2GD^{j}_{s}\left(-\overline{y}\overline{f}_{,\overline{xyy}}\right) + 2GD^{j}_{n}\left(\overline{f}_{,\overline{yy}} - \overline{y}\overline{f}_{,\overline{yyy}}\right) \\
\sigma^{j}_{\overline{xy}} &= 2GD^{j}_{s}\left(\overline{f}_{,\overline{yy}} + \overline{y}\overline{f}_{,\overline{xyy}}\right) + 2GD^{j}_{n}\left(-\overline{y}\overline{f}_{,\overline{xyy}}\right)
\end{aligned} \tag{9}$$

where

$$\begin{aligned}
f\left(\overline{x}, \overline{y}\right) &= -\frac{1}{4\pi\left(1-\nu\right)}\left[\overline{y}\left(\arctan\frac{\overline{y}}{\overline{x}-a} - \arctan\frac{\overline{y}}{\overline{x}+a}\right)\right. \\
&\quad - \left(\overline{x}-a\right)\ln\sqrt{\left(\overline{x}-a\right)^{2}+\overline{y}^{2}} \\
&\quad \left. + \left(\overline{x}+a\right)\ln\sqrt{\left(\overline{x}+a\right)^{2}+\overline{y}^{2}}\right]
\end{aligned} \tag{10}$$

where the field point (x, y) must transform to the local jth fracture unit coordinate $(\overline{x}, \overline{y})$. The (x_j, y_j) is the midpoint of the jth fracture unit.

$$\begin{aligned}
\overline{x} &= \left(x - x_j\right)\cos\beta_j + \left(y - y_j\right)\sin\beta_j \\
\overline{y} &= -\left(x - x_j\right)\sin\beta_j + \left(y - y_j\right)\cos\beta_j
\end{aligned} \tag{11}$$

The deformation-induced stress is approximated as the sum of the all the displacement fracture unit, as shown in the following:

$$\begin{aligned}
\sigma_{xx} &= \sum_{j=1}^{n}\overset{j}{\sigma}_{xx} \\
\sigma_{yy} &= \sum_{j=1}^{n}\overset{j}{\sigma}_{yy} \\
\sigma_{xy} &= \sum_{j=1}^{n}\overset{j}{\sigma}_{xy}
\end{aligned} \tag{12}$$

2.3. Thermal-Induced Aperture and Stress Model. During the hydroshearing stimulation process, a large heat transfers from the reservoir matrix into the injected cold fluid. The temperature change causes thermal-induced stress and displacement within the rock and fracture. The thermal-induced stress model under the condition of the continuous heat source is deduced by using Green's function [25].

$$\begin{aligned}
&\widetilde{\sigma_{ij}}\left(x, y, s\right) \\
&\quad = -Q\rho_f c_f \int_0^{L_f} s\widetilde{\sigma_{ij}}^{cs}\left(R, s\right)\frac{\partial\widetilde{T}\left(x', 0, s\right)}{\partial x'}dx'
\end{aligned} \tag{13}$$

where

$$\begin{aligned}
\widetilde{\sigma}^{cs}_{rr} &= \frac{-E\alpha_T}{4\pi\left(1-\nu\right)K_{m,eff}}\frac{1}{s}\left[\frac{2}{\xi^2} - K_2\left(\xi\right) + K_0\left(\xi\right)\right] \\
\widetilde{\sigma}^{cs}_{\theta\theta} &= \frac{-E\alpha_T}{4\pi\left(1-\nu\right)K_{m,eff}}\frac{1}{s}\left[\frac{2}{\xi^2} - K_2\left(\xi\right) - K_0\left(\xi\right)\right]
\end{aligned} \tag{14}$$

$$\xi = R\sqrt{\frac{\rho_m c_m s}{K_{m,eff}}}$$

The displacement of the granite matrix can be given by the following Navier equation [20]:

$$G\nabla^2 u_y(x,y,t) + \frac{G}{1-2v}\nabla\left[\nabla \cdot u_y(x,y,t)\right] = \frac{2G(1+v)}{1-2v}\alpha_T \nabla T(x,y,t) \tag{15}$$

where G is the shear modulus, v is the Poisson's ratio, and u_y is the displacement of the fracture to the normal direction. Integrating (15) from the fracture surface to infinity, the aperture change of the fracture surface is shown as

$$-u_y(x,0,t) = \frac{(1+v)\alpha_T}{1-v}\int_0^{\infty} \Delta T(x,y,t)\,dy$$
$$\Delta T(x,y,t) = T(x,y,t) - T_{\infty} \tag{16}$$

where ΔT is the temperature difference. Differentiating (16) with respect to time and using the heat expression for heat conduction in the rock matrix,

$$\frac{\partial u_y}{\partial t} = \frac{1}{2}\frac{\partial w(x,t)}{\partial t} = \frac{(1+v)\alpha_T K_{m,eff}}{(1-v)\rho_m c_m}\left.\frac{\partial \Delta T(x,y,t)}{\partial y}\right|_{y=0} \tag{17}$$

The Bodvarsson analytical solution [26] was used to avoid the numerical oscillations and uncertainty in numerically calculated heat flux at the interface.

$$\frac{T_{m0} - T(x,y,t)}{T_{m0} - T_{in}} = efc\left[\frac{\sqrt{K_{m,eff}\rho_m c_m}}{Qc_f\rho_f\sqrt{t}}x + \sqrt{\frac{\rho_m c_m}{K_{m,eff}t}}\frac{y}{2}\right] \tag{18}$$

Assuming that the initial temperature of the granite rock is T_{m0} and T_{in} is the injection fluid temperature and C_w is the water specific heat. Differentiating (18) with respect to y,

$$\left.\frac{\partial T(x,y,t)}{\partial y}\right|_{y=0} = \frac{2\Delta T A_2}{\sqrt{\pi}}\exp\left[-(A_1)2\right] \tag{19}$$

Substitution (19) into (17) yields

$$\frac{\partial w(x,t)}{\partial t} = \frac{2(1+v)\alpha_T K_{m,eff}}{(1-v)\rho_m c_m}\left(\frac{2\Delta T A_2}{\sqrt{\pi}}\exp\left[-(A_1)^2\right]\right) \tag{20}$$

where A_1 and A_2 are

$$A_1 = \frac{x\sqrt{K_{m,eff}\rho_m c_m}}{Q\rho_f c_f\sqrt{t}},\quad A_2 = \frac{\sqrt{\rho_m c_m}}{2\sqrt{K_{m,eff}t}} \tag{21}$$

Table 2: Parameters used in numerical calculations.

Parameters	Value
Injected water density (kg/m^3)	1000
Injected water heat capacity (J/Kg/K)	4180
Hot dry rock fracture roughness	15
Hot dry rock Young's modulus (Pa)	4E10
Hot dry rock Poisson's ratio	0.28
Hot dry rock density (kg/m^3)	2650
Hot dry rock initial temperature (°C)	200
Hot dry rock specific heat capacity (J/Kg/K)	1070
Hot dry rock thermal conductivity(J/s/m/K)	2.60
Fracture begin shear expansion coefficient	0.3

2.4. Rock Matrix Stability Model. The induced stress will be generated through the deformation of single fracture, which will lead to redistribution of the stress state around the rock matrix. As a result, the rock matrix nodes and unconnected natural fractures may fail under the new stress environment, and the underground fracture network of hot rock is formed by the communication between the natural fractures.

Criterion of rock matrix failure is

$$\sigma_1 \geq \sigma_3 + \frac{2(c + \sigma_3\tan\varphi_b)}{(1-\tan\varphi_b\cot\alpha)\sin 2\alpha} \tag{22}$$

Criterion of unconnected fracture failure is

$$\sigma_1 \geq \sigma_3 + \frac{2\sigma_3\tan\varphi_b}{(1-\tan\varphi_b\cot\alpha)\sin 2\alpha} \tag{23}$$

where σ_1, σ_3 are maximum and minimum horizontal stress; c is rock cohesion, MPa; φ_b is basic friction angle; α is the angle between maximum horizontal stress and effective normal stress.

3. Numerical Solution and Verification

3.1. Model Parameters and Boundary Condition. The high temperature and fractured granite reservoir system is considered two sets of coupled partial differential equations for the natural fracture and the matrix. All the various input parameters used in the simulations are reported in Table 2.

The problem boundary condition of the fracture is prescribed along with the fracture direction and the rock matrix boundary is specified at the interface of the fracture-matrix. The initial and boundary conditions are (24)~(26) for the rock matrix and the within natural fracture, respectively, are

$$T_f(x,0,0) = T_m(x,y,0) = T_0$$
$$w(x,0,0) = w_0 \tag{24}$$
$$P_f(x,0,0) = P_{out}$$

$$T_f(0,0,t) = T_{in}$$
$$q(0,0,t) = q_{in} \tag{25}$$

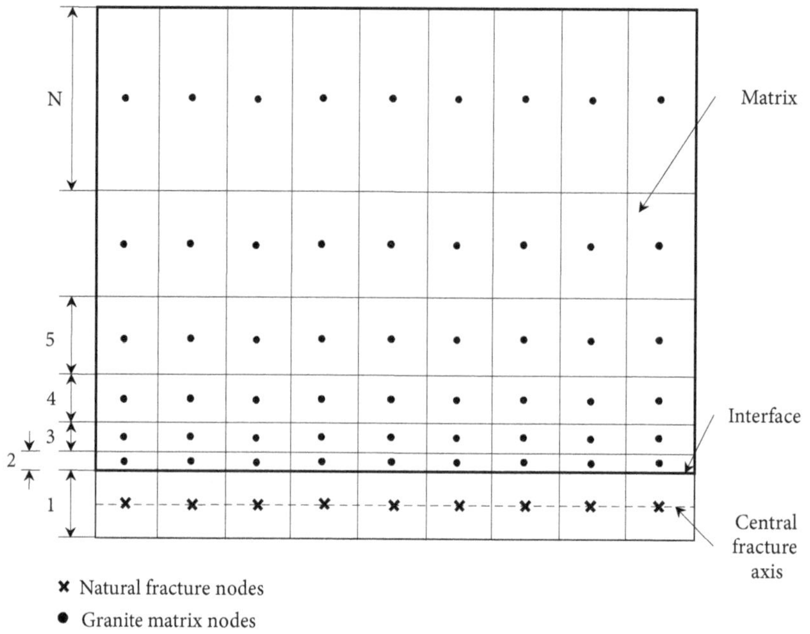

FIGURE 4: Discretization of the solution domain satisfying both the cell-based finite volume and the nodal-based finite-difference schemes [15].

$$\left.\frac{\partial T_m}{\partial y}\right|_{(x,L_m,t)} = 0$$
$$\left.\frac{\partial T_m}{\partial x}\right|_{(0,t)} = \left.\frac{\partial T_m}{\partial x}\right|_{(L_f,t)} = 0 \tag{26}$$

3.2. Numerical Solution. In this study, the system is solved numerically using a second-order central-difference finite-difference scheme. The solution is iterated in each time step to satisfy the continuity at the matrix-fracture interface. The grid size in the fracture is maintained uniform whereas a nonuniform size is adopted in the rock matrix; it is to be noted that all the fracture and matrix grid size does not change with time, shown as in Figure 4.

The heat transfer model is used to calculate the temperature profile through the coupling of the fracture and the matrix. The pressure distribution within the fracture is updated from the thermal-induced aperture change. The fracture deformation due to the effective pressure change will be analyzed through the fracture constitutive model. The coupling between thermal, deformation, and hydroflow is iterated at each time step. The convergence criterion for heat transfer is that the temperature difference is less than 0.01K. The convergence criterion for THM coupling is that the aperture difference is less than 0.00001mm. After the THM coupling, the new stress environment of each matrix nodes is updated from the calculation of induced stress. The stability analysis of matrix nodes is also completed. The coupling between THM, induced stress, and matrix stability is illustrated in Figure 5.

3.3. Verification. As previously described in this paper, the shear displacement and dilation behavior of a single fracture are compared with a numerical solution, as shown in Figure 6. It is to be noted that the prepeak behavior is not elastoplastic and it is simply a nonlinear (prepeak) model. The bottom of the rock is fixed in all three directions and the top displacement will be applied in one of the shear directions. In this comparative analysis, the confining normal stress is 2MPa, the joint's roughness is 15, and the joint's compressive strength is 150MPa, and other date has been shown in Table 2. The results indicate that the lab test and the numerical model agree very well.

During the verification of the heat transfer model, the temperature in both the rock matrix and natural fracture is a relative temperature which is defined as the ratio of current temperature to the initial reservoir temperature. The initial temperature of the matrix is 200°C and the injection water is 20°C. The water velocity between the inlet and outlet is maintained at a constant value of 50 m per day, and the other thermal parameters of the water/matrix are given in Table 2. The simulation results of the heat transfer and thermal-induced aperture for verification are shown in Figures 7 and 8. All comparative studies have shown that the numerical model and other researcher's results are in good agreement.

4. Case Study and Matrix Stability Analysis

4.1. Hydroshearing Stimulation and Pressure Fitting. RRG-9 ST1 well is located in Cassia County, southwestern Idaho. The open hole section of the well, from 5551 to 5900 ft MD, consists of granite and minor diabase. The tested initial minimum horizontal stress is 22MPa and the maximum is 32MPa [27, 28]. The hydroshearing stimulation target showed evidence of more than eighty natural fractures and the maximum temperature range from 130°C to 150°C. After cementing, a hydroshearing stimulation program was carried out in the open hole section in 2013. The objective of the

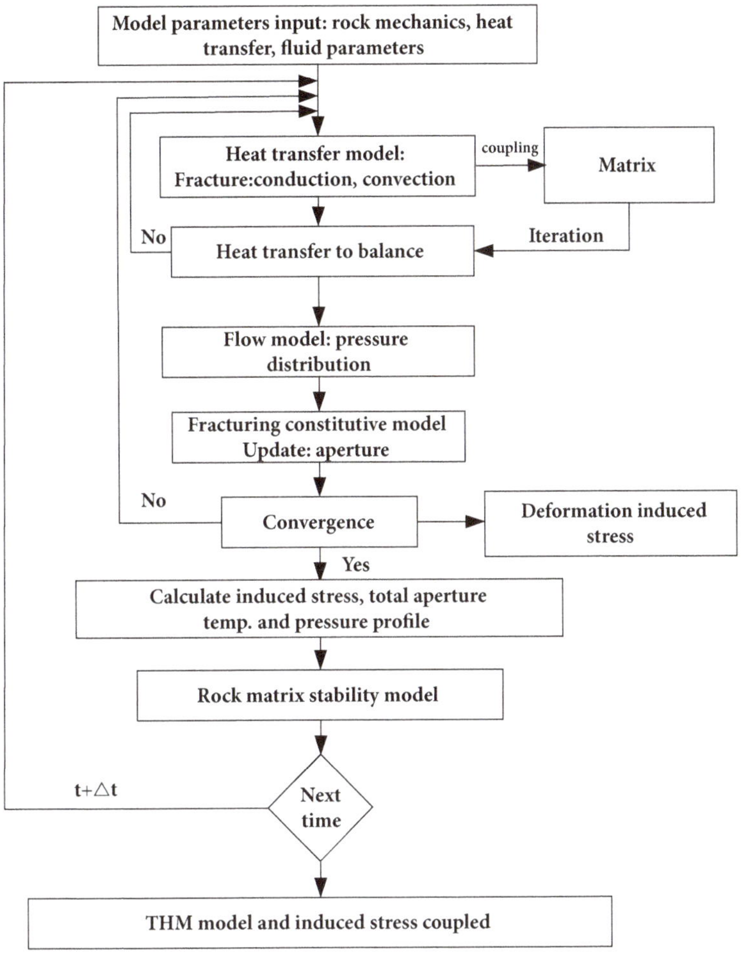

FIGURE 5: Coupled thermal-hydraulic-mechanical model and induced stress.

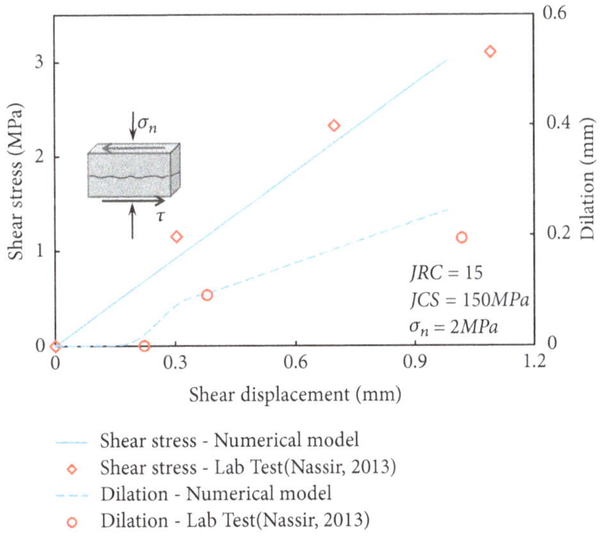

FIGURE 6: Comparison between the lab test and the numerical model for shear stress and displacement.

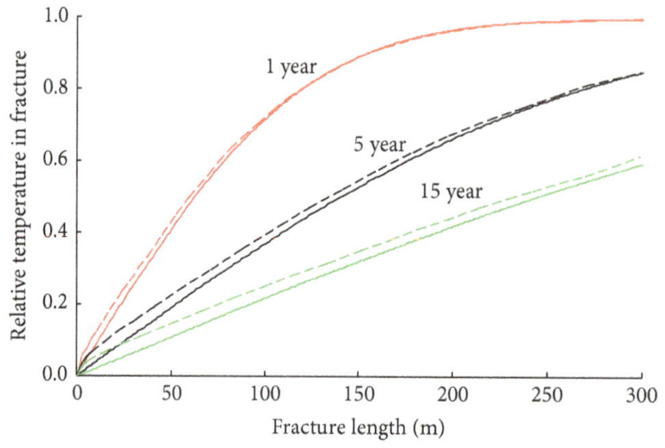

FIGURE 7: Comparison the normalized temperature distribution between numerical model and Cheng's solution in the fracture.

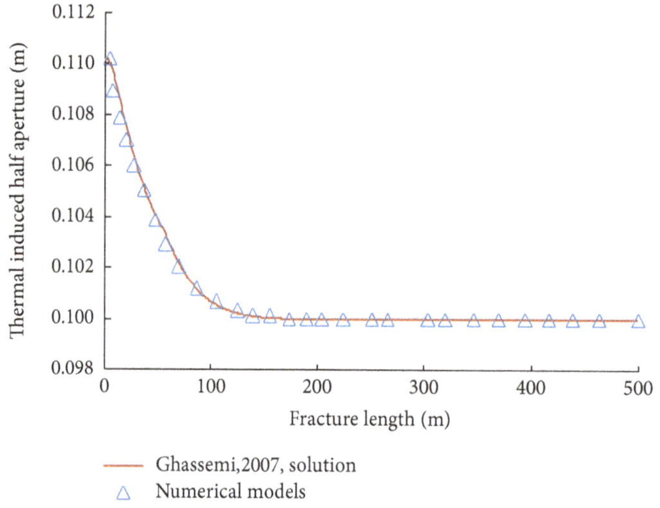

FIGURE 8: Comparison the thermal-induced aperture along the fracture between numerical model and Ghassemi's solution.

fracturing program is to improve the injectivity of the well. The entire stimulation can be divided into three stages: unstable pumping stage, stable pumping stage, and efficient pumping stage. In the unstable pumping stage, the injected water temperature from high (40°C) decreased to low (12°C), as shown in Table 3.

At the beginning of the fracturing, high temperature (40°C) water was injected into the fracture. The permeability of the fracture was very low and nearly equal to the initial conductivity while the surface pumping pressure increases with the injection rate. After unstable pumping stage, the permeability of the natural fracture is greatly improved; the ground injection pressure is maintained at a constant value. The pumping pressure fitting shows that the simulation results are in good agreement with the well site fracturing date, as shown in Figure 9. The coupled THM model to represent fracture aperture change and heat transfer worked well.

4.2. Fracture Aperture and Temperature Profile. Having verified the basic coupled fracture-matrix, heat transfer model, and fracture deformation equations, the total aperture change along the fracture is computed. Both models that affect the aperture change have been used in this work and the results are shown in Figure 10. It can be observed that both expressions yield similar trends at different pumping stages. Under the combined influence of thermal stress and decreasing net pressure, the fracture aperture increases with the shearing fracturing time. After 241 days of injection simulation, the change in fracture aperture in RRG-9 ST1 well occurs over a short zone near the injection well (the aperture change zone is about 10m). The fracture aperture is at a maximum (0.03m) near the injection well, and this maximum aperture decreases along the fracture axis as the fluid circulation continues between the inlet and outlet wells.

The combined influence of thermal stress and decrease in net pressure induced deformation in a fracture-matrix coupled system on fracture aperture is illustrated in Figure 11.

TABLE 3: Hydroshearing stimulation schedule of the RRG-9 ST1 well.

Hydro-shearing stimulation	Injected water temperature, °C	Injection rate, m^3/min	Pumping pressure, Psi	Time, day
Unstable pumping stage	40	0.16-0.53-0.99	270-540-810	35
	12	0.96	740-520	6
Stable pumping stage	29	0.5	270	90
Efficient pumping stage	45	0.8-1.9	260	110

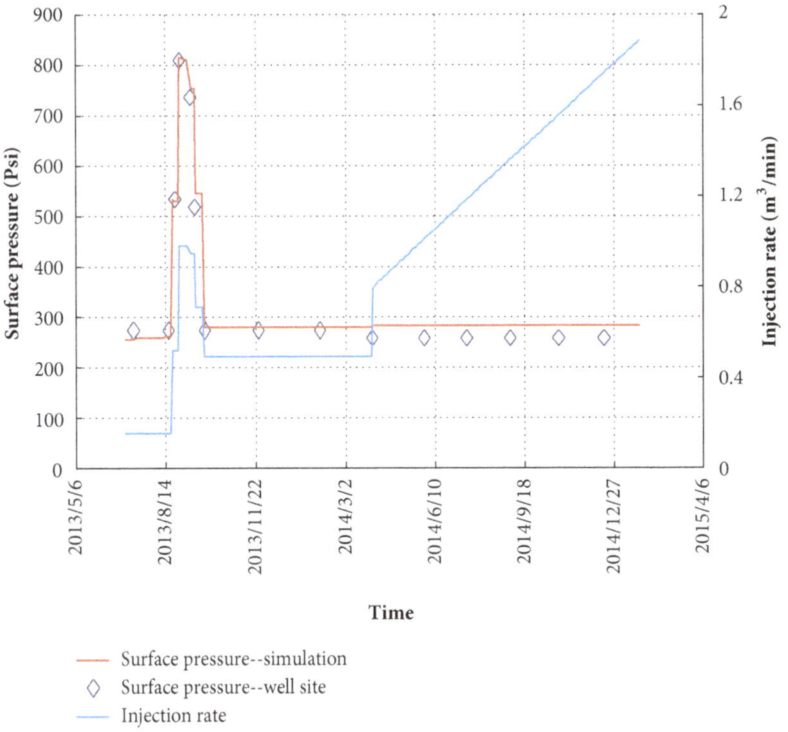

FIGURE 9: Pressure fitting on RRG-9 ST1 well hydroshearing simulation.

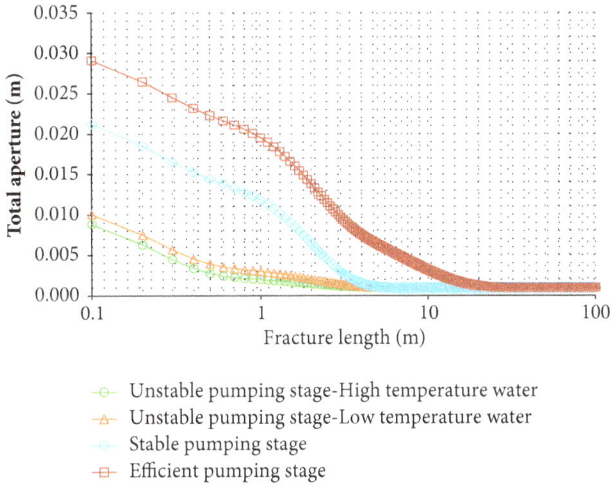

FIGURE 10: Change in fracture total aperture along the fracture due to coupling of thermal and mechanical.

Two distinct regimes in fracture aperture variation can be observed along the fracture between the inlet and outlet wells. The first regime occurs near the injection well, about 10 meters; the contribution of fracture conductivity is mainly from the thermal-induced aperture. The second regime, away from the injection well, the conductivity of the fracture is dominated by shear dilation. Therefore, the best way to enhance the conductivity of the underground

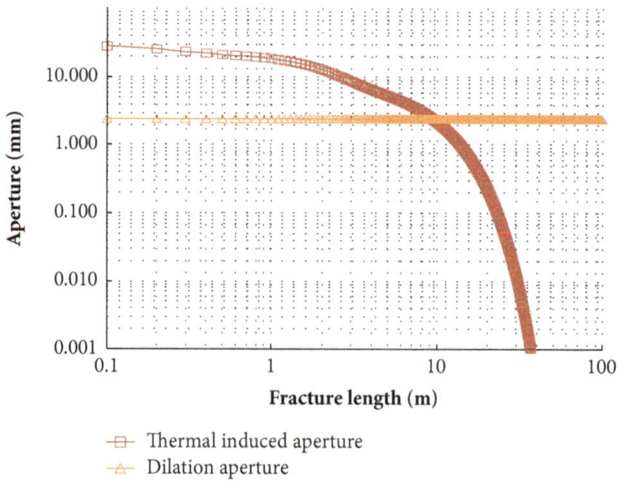

FIGURE 11: Change in thermal-induced aperture and dilation aperture along the fracture.

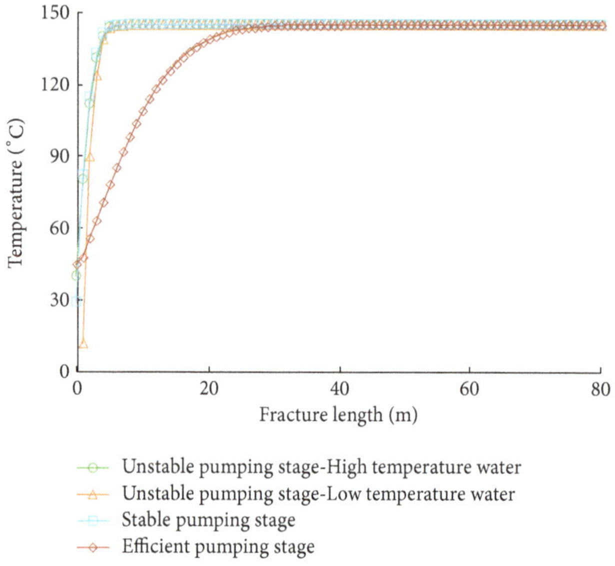

FIGURE 12: Change in temperature profile along the fracture in different pumping stage.

hundreds of meters' fracture area is the using of shear dilation.

The temperature profiles along the fracture axis after three different pumping stages are illustrated in Figure 12. The plot of the temperature profile is very sharp near the injection area, which indicates a large heat transfer from the reservoir into the fracture fluid. During long-term fracturing, it requires a greater distance from the injection well for the fluid to reach the matrix initial temperature. In this case, after 241 days' injection, the temperature of the circulating fluid nearly reaches the matrix initial temperature at 20 m from the injection well. That is why the thermal-induced aperture change regime is very small. It is indirectly proved that the hot reservoir has a strong thermal energy mining capacity.

4.3. Matrix Stability and Fracture Network Expansion. After the calculation of fracture total aperture and heat transfer in each time step, the deformation-induced stress and thermal-induced stress are calculated. Figure 13 shows the deformation-induced stress in minimum and maximum horizontal stress direction. The deformation-induced minimum horizontal stress and maximum horizontal stress have a similar change trend and have the maximum positive value at the injection point. In other words, the stress state of the matrix nodes is increased. In this case, after 241 days' injection, the maximum value of the deformation-induced stress in minimum and maximum horizontal stress direction are 35 MPa and 101 MPa, separately. The maximum value quickly reduced to a stable value in a very short range near the injection well (in this case, the affected range is about 1 meter).

The thermal-induced stress in both minimum and maximum horizontal stress directions are negative, which means that tensile stress is formed near the injection area, as shown in Figure 14. In this RRG-9 ST1 well hydroshearing simulation case, after 241 days' injection, the maximum value of the thermal-induced stress in minimum and maximum horizontal stress direction are -3 MPa and -4.5 MPa, separately. The thermal-induced stress affected range is about 2 meters. Therefore, the deformation-induced stress has a large value with a smaller affected range; on the contrary, thermal-induced stress has a small value with a larger affected range.

After the given of induced stress, the new stress state of the matrix's nodes can be updated. The new stress state consists of three parts: initial stress, deformation-induced stress, and thermal-induced stress. The new stress state will affect the stability of the rock matrix. In other words, when the new stress state of the matrix nodes is located above the strength envelope, the stability of the matrix node will be destroyed. Figure 15 shows the relationship between the new stress state and the strength envelope of the matrix (left), the failure modes, and their position in different pumping stages (right). In this simulation case, after three-stage fracturing, seven matrix nodes are located above the strength envelope. With the increase in fracturing pump time, the number of the failure nodes and the affected range also

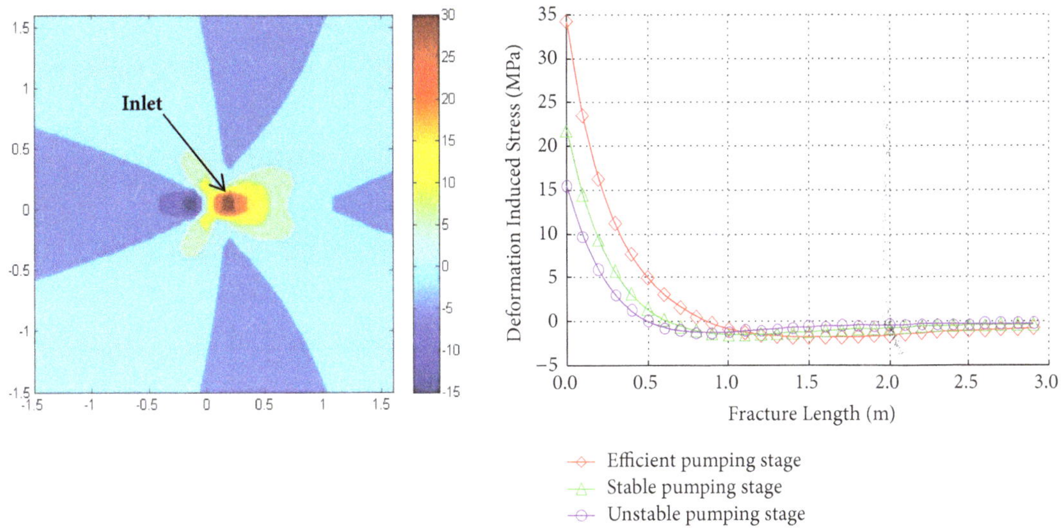

(a) Minimum horizontal stress direction

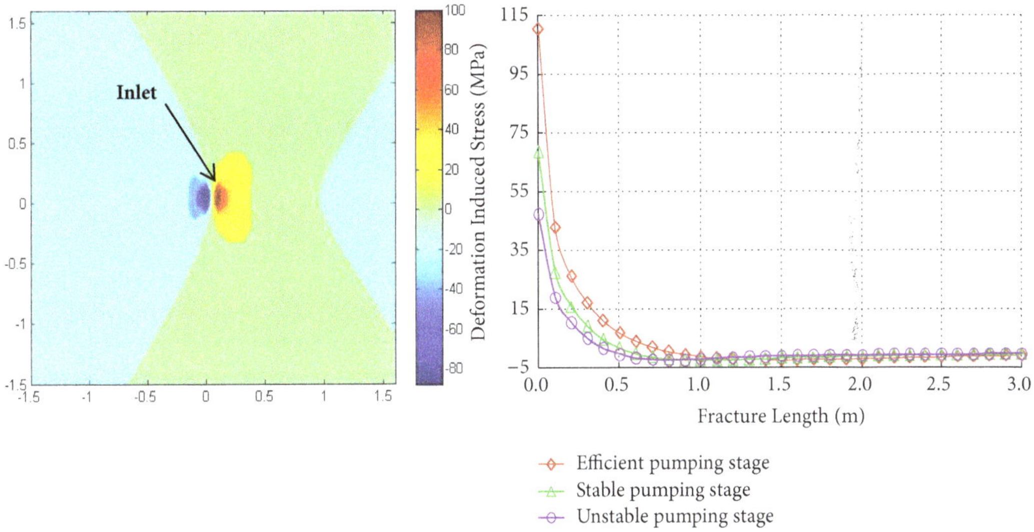

(b) Maximum horizontal stress direction

FIGURE 13: The variation of deformation-induced stress in minimum and maximum horizontal stress direction in different pumping stages.

increase. This mechanism is helpful to create complex fracture networks.

5. Conclusions

A coupled THM model is developed to simulate the combined effect of fracture fluid flow, heat transfer from the matrix to injected fluid, and shearing dilation behaviors in a coupled fracture-matrix hot volcanic reservoir system. In view of the expansion of the fracture network, deformation and thermal-induced stress models are added to the matrix node's in situ stress environment in each time step to analyze the stability of the matrix.

The coupled THM modeling results indicate that, in these case conditions, the change in fracture aperture due to thermoelastic stress occurs near the injection well, thus increasing the fracture opening and playing a dominant role. Away from the injection well, the conductivity of the fracture is contributed by shear dilation. Therefore, the best way to enhance the conductivity of the underground hundreds of meters' fracture area is the using of shear dilation.

The induced stress simulation results indicate that deformation-induced stress is a compressive stress with a positive and thermal-induced stress is tensile stress with a negative value. The induced stress has the maximum value at the injection point; the deformation-induced stress has a large value with smaller affected range; on the contrary, thermal-induced stress has a small value with larger affected range.

The new stress state of the matrix nodes is updated by initial stress, deformation-induced stress, and thermal-induced stress. The matrix stability simulation results indicate, in this case, some matrix nodes are located above the strength envelope after hydroshearing stimulation, which means the

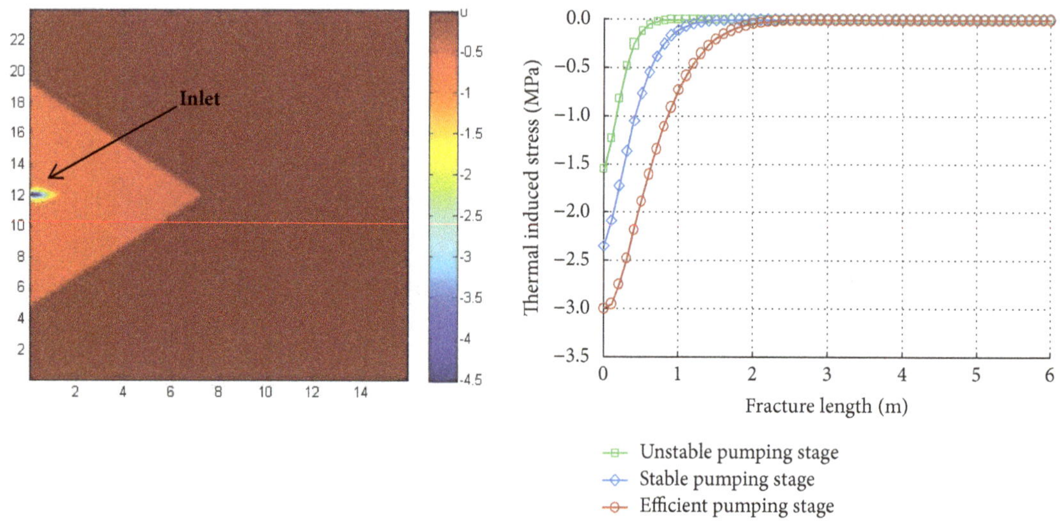

(a) Minimum horizontal stress direction

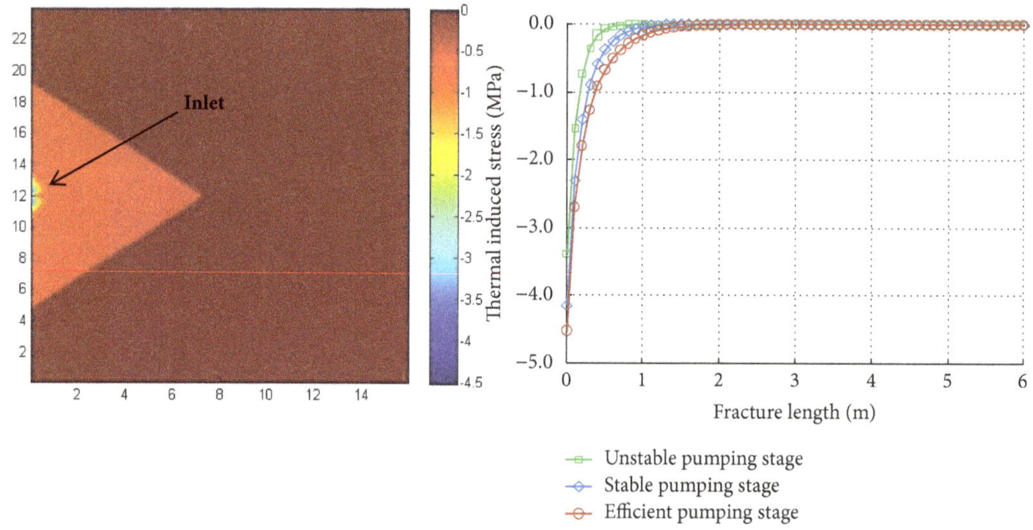

(b) Maximum horizontal stress direction

FIGURE 14: The variation of thermal-induced stress in minimum and maximum horizontal stress direction in different pumping stage.

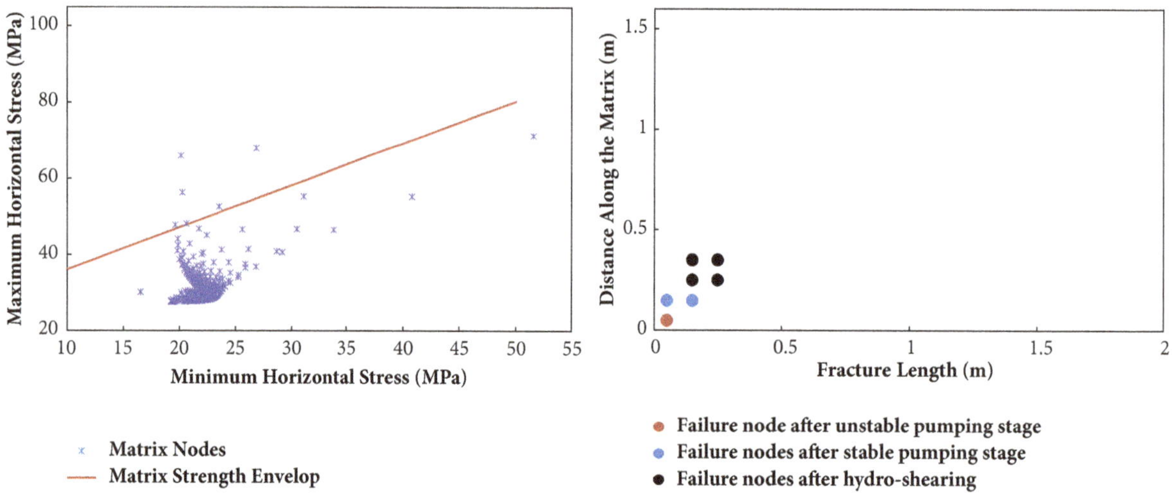

FIGURE 15: The relationship between the new stress state and the strength envelope of the matrix (left), the failure modes, and their position in different pumping stages (right).

stability of the matrix node will be destroyed. This mechanism is helpful to create complex fracture networks.

Conflicts of Interest

The authors declare that they have no conflicts of interest.

Acknowledgments

The authors would like to acknowledge the support of the National Science Fund for Distinguished Young Scholars (ID: 51525404) and Foundation of China Scholarship Council (File no. 201608510077).

References

[1] Z. Feng, J. Liu, P. Wang, S. Chen, and Y. Tong, "oil and gas exploration field: volcanic hydrocarbon reservoirs-Enlightenment from the discovery of the large gas field in Songliao basin," *Diqiu Wuli Xuebao*, vol. 54, no. 2, pp. 269–279, 2011.

[2] S. A. Holditch, "The Increasing Role of Unconventional Reservoirs in the Future of the Oil and Gas Business," *Journal of Petroleum Technology*, vol. 55, no. 11, pp. 34–79, 2003.

[3] Z.-Q. Feng, "Volcanic rocks as prolific gas reservoir: A case study from the Qingshen gas field in the Songliao Basin, NE China," *Marine and Petroleum Geology*, vol. 25, no. 4-5, pp. 416–432, 2008.

[4] J. Guo, B. Luo, C. Lu, J. Lai, and J. Ren, "Numerical investigation of hydraulic fracture propagation in a layered reservoir using the cohesive zone method," *Engineering Fracture Mechanics*, vol. 186, pp. 195–207, 2017.

[5] J. Guo, J. Wang, Y. Liu, Z. Chen, and H. Zhu, "Analytical analysis of fracture conductivity for sparse distribution of proppant packs," *Journal of Geophysics and Engineering*, vol. 14, no. 3, pp. 599–610, 2017.

[6] Y. Xiao, G. Jianchuan, and J. Mclennan, "Deformation and initiation in Fracture Due To Hydro-Shearing and Induced Stress in Enhanced Geothermal System," in *Paper presented at the 4th ISRM Young Scholars Symposium on Rock Mechanics*, 2017.

[7] Y. Xiao, G. Jianchuan, and B. Luo, "Characterization of Fracture Conductivity of Hydraulic Fracturing in Hot Dry Rock Exploitation," in *Paper presented at the 51st U.S. Rock Mechanics/ Geomechanics Symposium*, 2017.

[8] D. Vogler, F. Amann, P. Bayer, and D. Elsworth, "Permeability Evolution in Natural Fractures Subject to Cyclic Loading and Gouge Formation," *Rock Mechanics and Rock Engineering*, vol. 49, no. 9, pp. 3463–3479, 2016.

[9] A. P. Rinaldi, J. Rutqvist, E. L. Sonnenthal, and T. T. Cladouhos, "Coupled THM Modeling of Hydroshearing Stimulation in Tight Fractured Volcanic Rock," *Transport in Porous Media*, vol. 108, no. 1, pp. 131–150, 2015.

[10] J. Taron and D. Elsworth, "Thermal-hydrologic-mechanical-chemical processes in the evolution of engineered geothermal reservoirs," *International Journal of Rock Mechanics and Mining Sciences*, vol. 46, no. 5, pp. 855–864, 2009.

[11] J. Rutqvist, Y.-S. Wu, C.-F. Tsang, and G. Bodvarsson, "A modeling approach for analysis of coupled multiphase fluid flow, heat transfer, and deformation in fractured porous rock," *International Journal of Rock Mechanics and Mining Sciences*, vol. 39, no. 4, pp. 429–442, 2002.

[12] J. Taron, D. Elsworth, and K.-B. Min, "Numerical simulation of thermal-hydrologic-mechanical-chemical processes in deformable, fractured porous media," *International Journal of Rock Mechanics and Mining Sciences*, vol. 46, no. 5, pp. 842–854, 2009.

[13] Q. Gan and D. Elsworth, "Production optimization in fractured geothermal reservoirs by coupled discrete fracture network modeling," *Geothermics*, vol. 62, pp. 131–142, 2016.

[14] G. Izadi and D. Elsworth, "The influence of thermal-hydraulic-mechanical- and chemical effects on the evolution of permeability, seismicity and heat production in geothermal reservoirs," *Geothermics*, vol. 53, pp. 385–395, 2015.

[15] G. S. Kumar and A. Ghassemi, "Numerical modeling of non-isothermal quartz dissolution/precipitation in a coupled fracture-matrix system," *Geothermics*, vol. 34, no. 4, pp. 411–439, 2005.

[16] P. Olasolo, M. C. Juárez, M. P. Morales, S. Damico, and I. A. Liarte, "Enhanced geothermal systems (EGS): A review," *Renewable & Sustainable Energy Reviews*, vol. 56, pp. 133–144, 2015.

[17] S. L. Crouch, A. M. Starfield, and F. J. Rizzo, "Boundary element methods in solid mechanics," *Journal of Applied Mechanics*, vol. 50, p. 704, 1983.

[18] Y. Cheng, "Boundary element analysis of the stress distribution around multiple fractures: Implications for the spacing of perforation clusters of hydraulically fractured horizontal wells," in *Proceedings of the SPE Eastern Regional Meeting 2009: Limitless Potential / Formidable Challenges*, pp. 267–281, usa, September 2009.

[19] Q. Tao, *Numerical modeling of fracture permeability change in naturally fractured reservoirs using a fully coupled displacement discontinuity method*, Texas AM University, 2010.

[20] A. Ghassemi and G. Suresh Kumar, "Changes in fracture aperture and fluid pressure due to thermal stress and silica dissolution/precipitation induced by heat extraction from subsurface rocks," *Geothermics*, vol. 360, no. 2, pp. 115–140, 2007.

[21] E. Barbier, "Geothermal energy technology and current status: An overview," *Renewable & Sustainable Energy Reviews*, vol. 6, no. 1-2, pp. 3–65, 2002.

[22] J. Bradford, J. Moore, M. Ohren, J. McLennan, W. L. Osborn, and E. Majer, "Recent Thermal and Hydraulic Stimulation Results at Raft River," in *ID EGS Site*, Paper presented at the Fourtieth Workshop on Geothermal Reservoir Engineering, Stanford University, Stanford, California, 2015.

[23] N. Barton, S. Bandis, and K. Bakhtar, "Strength, deformation and conductivity coupling of rock joints," in *Proceedings of the Paper presented at the International Journal of Rock Mechanics and Mining Sciences & Geomechanics Abstracts*, 1985.

[24] N. Barton and V. Choubey, *The shear strength of rock joints in theory and practice*, Rock Mechanics amp; Rock Engineering, 1977.

[25] S. Tarasovs and A. Ghassemi, "Propagation of a system of cracks under thermal stress," in *Proceedings of the 45th US Rock Mechanics / Geomechanics Symposium*, usa, June 2011.

[26] G. Bodvarsson, "On the temperature of water flowing through fractures," *Journal of Geophysical Research*, vol. 74, no. 8, pp. 1987–1992, 1969.

[27] J. Bradford, J. McLennan, J. Moore, D. Glasby, D. Waters, and R. Kruwell, "Recent developments at the Raft River geothermal field," in *Proceedings of the Paper presented at the Proceedings*, Stanford University, Stanford, California, 2013.

[28] C. Jones, J. Moore, W. Teplow, and S. Craig, "Geology and hydrothermal alteration of the Raft River geothermal system," in *Proceedings of the Idaho. Paper presented at the Proceedings, Thirty-Sixth Workshop on Geothermal Reservoir Engineering*, Stanford University, Stanford, California, 2011.

13

Improved Model-Free Adaptive Sliding-Mode-Constrained Control for Linear Induction Motor considering End Effects

Xiaoqi Song,[1] Dezhi Xu,[1] Weilin Yang,[1] Yan Xia,[1] and Bin Jiang[2]

[1]*School of Internet of Things Engineering, Jiangnan University, Wuxi 214122, China*
[2]*College of Automation Engineering, Nanjing University of Aeronautics and Astronautics, Nanjing 211106, China*

Correspondence should be addressed to Dezhi Xu; xudezhi@jiangnan.edu.cn

Academic Editor: Tarek Ahmed-Ali

As a kind of special motors, linear induction motors (LIM) have been an important research field for researchers. However, it gives a great velocity control challenge due to the complex nonlinearity, high coupling, and unique end effects. In this article, an improved model-free adaptive sliding-mode-constrained control method is proposed to deal with this problem dispensing with internal parameters of the LIM. Firstly, an improved compact form dynamic linearization (CFDL) technique is used to simplify the LIM plant. Besides, an antiwindup compensator is applied to handle the problem of the actuator under saturations in case during the controller design. Furthermore, the stability of the closed system is proved by Lyapunov stability method theoretically. Finally, simulation results are given to demonstrate that the proposed controller has excellent dynamic performance and stronger robustness compared with traditional PID controller.

1. Introduction

In the past few decades, the LIM has been widely used in many fields, such as military, household appliances, industrial automation, and transportation [1–4]. Compared with the conventional rotary induction motors (RIM), the main advantages of LIM are as follows: (1) it does not have any converter, gear, or other intermediate conversion mechanism which can reduce mechanical loss; (2) it is only driven by magnetic force which makes the LIM have the features of high speed and low noise [5, 6]. Even though the driving principle of a LIM is similar to that of a RIM, the parameters of LIM are time-varying, such as end effects, slip frequency, dynamic air gap, three-phase imbalance, and track structure [7–9]. Among them, the end effects greatly affect the LIM control performance, and the faster the speed, the more significant the impact. Therefore, during the modeling of the LIM, the end effects must be considered.

With the quick development of science and technology, many model-based control methods are proposed to handle LIM control problems. In [10], an adaptive backstepping method is proposed to deal with the position tracking problem of the LIM. In [11], an optimized adaptive tracking control is applied for a LIM considering the uncertainties. In [12], the authors use input-output feedback linearization control technique with online model reference adaptive system (MRAS) method suiting the induction resistance to realize the velocity following goal, whereas the three mentioned methods are highly dependent on the accuracy of the model. Once the model is improperly defined or the system parameters cannot be accurately obtained, the dynamic response of the system will hardly be satisfied. Besides, some non-model-based control methods are also proposed for LIM control problems. In [13], the researchers present a real-time discrete neural control scheme based on a recurrent high order neural network trained online to a LIM. In [14, 15], some methods based on fuzzy control are also used to have the problem solved. However, even if we neglect the complexity of the selection of fuzzy rules and the uncertainty of the neural network nodes, these methods have not considered the input saturation problems which may result in system instability.

Model-free adaptive control (MFAC) was first proposed in 1994 and is a hot topic in the field of data-driven modeling [16–19]. It is a method that only relies on input/output (I/O) data and does not need any internal information of the plant. The main design steps of the MFAC are divided into three categories: (1) using CFDL technique to transfer the nonlinear system into self-designed linear model based on a parameter called pseudo-partial-derivative (PPD), (2) estimating the value of the PPD through a variety of methods, and (3) devising the controller based on self-designed linear model. For now, MFAC has been widely applied in all kinds of fields, such as multiagent systems [20], chemical process [19, 21], and intelligent transportation [22]. Moreover, due to the fact that sliding-mode control (SMC) is designed without object parameters and disturbance, it gets the merits of quick response and high fitness. SMC is also a hot topic and is applied in a variety of fields [23, 24] and has been used in combination with MFAC firstly in [25].

In this paper, an improved CFDL technique is used to linearize the LIM model considering end effects based on PPD estimation algorithm. And we design a model-free adaptive constrained sliding-mode control for the system considering input saturations. So as to avoid the instability caused by saturations, we design an antiwindup compensator to make the output continue to follow the given reference.

The rest of this paper is organized as follows. Section 2 briefly introduces the model of the LIM considering end effects. In Section 3, the main results of the proposed control strategy are given. The simulation results are shown in Section 4 to verify the effectiveness and robustness of the method. Finally, some conclusions are drawn in Section 5.

2. Problem Formulation for LIM

Similar to a RIM, a LIM is made up of primary and secondary components as shown in Figure 1. Besides, a LIM is obtained by a RIM that is opened longitudinally in a transverse direction. However, the biggest difference between a LIM and a RIM is that the LIM contains end effects which are caused by its structure. The end effects can be explained as follows: when the primary moves, eddy current occurs in the corresponding secondary conductor plate at the outlet and inlet terminals, and the direction of flow is opposite to the primary current, so that the air gap magnetic field will be distorted [9, 11]. Researchers generally use a parameter Q to express this phenomenon as

$$Q = \frac{l \cdot R_r}{v \cdot L_r}, \tag{1}$$

where l denotes the primary length, v denotes the speed of a LIM, and L_r and R_r denote the secondary inductance and resistance, respectively.

When the LIM is in a stationary state, we can consider its equivalent circuit as a RIM. Nevertheless, when the LIM is in a motion state, the model of a LIM in synchronously rotating reference frame should be improved as follows [8, 11]:

$$\begin{aligned} V_{sd} &= R_s i_{sd} + p\phi_{sd} - \omega_e \phi_{sq} \\ V_{sq} &= R_s i_{sq} + p\phi_{sq} - \omega_e \phi_{sd} \\ V_{rd} &= R_r i_{rd} + p\phi_{rd} - (\omega_e - \omega_r)\phi_{rq} \\ V_{rq} &= R_r i_{rq} + p\phi_{rq} + (\omega_e - \omega_r)\phi_{rd}, \end{aligned} \tag{2}$$

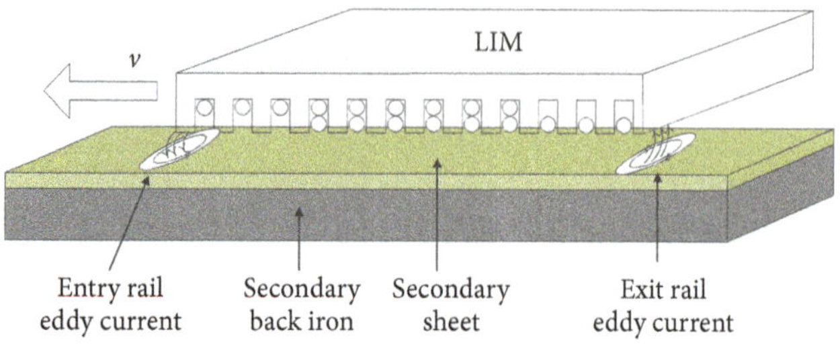

FIGURE 1: Structure of a LIM.

where (V_{sd}, V_{rd}), (i_{sd}, i_{rd}), and (ϕ_{sd}, ϕ_{rd}) denote the primary and secondary voltage, current, and flux linkage in d-axis; (V_{sq}, V_{rq}), (i_{sq}, i_{rq}), and (ϕ_{sq}, ϕ_{rq}) denote the corresponding parameters in q-axis; R_s denotes the primary resistance; ω_e and ω_r denote the angular frequency of stator and rotor; and p denotes the differential operator.

According to [8, 11], the flux linkage in dq-axis can be expressed as follows:

$$\begin{aligned} \phi_{sd} &= L_{sl} i_{sd} + L_m (1 - f(Q))(i_{sd} + i_{rd}) \\ \phi_{rq} &= L_{sl} i_{sq} + L_m (1 - f(Q))(i_{sq} + i_{rq}) \\ \phi_{rd} &= L_{rl} i_{rd} + L_m (1 - f(Q))(i_{sd} + i_{rd}) \\ \phi_{rq} &= L_{rl} i_{rq} + L_m (1 - f(Q))(i_{sq} + i_{rq}), \end{aligned} \tag{3}$$

where $f(Q) = (1 - e^{-Q})/Q$ is an important parameter during the process of modeling for a LIM, L_m is the magnetic inductance, and L_{sl} and L_{rl} are the primary and secondary leakage inductance. Meanwhile, the electromagnetic thrust force can be expressed as

$$F_{et} = K_f (\phi_{rd} \cdot i_{sq} - \phi_{rq} \cdot i_{sd}), \tag{4}$$

where $K_f = 3\pi P L_m/(2hL_r)$, P means the pole numbers, and h is the pole pitch.

By using the indirect vector control (IVC) technology, we can convert the linear induction motor model into a DC motor model which brings about great convenience to the control of the LIM. Thus, with IVC technology, orientate the rotor flux to the d-axis, and we get

$$\begin{aligned} \phi_{rq} &= \dot{\phi}_{rq} = 0 \\ V_{rd} &= V_{rq} = 0, \end{aligned} \tag{5}$$

where $\dot{\phi}_{rq}$ denotes the differential of ϕ_{rq}.

According to (2)–(5), the dynamic model of LIM considering end effects under IVC can be described as

$$\begin{aligned}
\frac{di_{sd}}{dt} &= -\frac{R_s}{L(Q)}i_{sd} + \frac{V_{sd}}{L(Q)} + \omega_e i_{sq} \\
\frac{di_{sq}}{dt} &= -\omega_e\left[i_{sd} + \frac{L_m(1-f(Q))}{L(Q)(L_r - L_m f(Q))}\phi_{dr}\right] \\
&\quad - \frac{R_s}{L(Q)}i_{sq} + \frac{V_{sq}}{L(Q)} \\
\frac{d\phi_{rd}}{dt} &= \frac{L_m[1-f(Q)]i_{sd} - \phi_{rd}}{T_r - L_m f(Q)/R_r} \\
\omega_{sl} &= \omega_e - \omega_r = \frac{L_m(1-f(Q))}{T_r - L_m f(Q)/R_r}\frac{i_{sq}}{\phi_{rd}} \\
F_{et} &= K_T i_{sq} = M\cdot\frac{dv}{dt} + D\cdot v + F_{\text{Load}},
\end{aligned} \tag{6}$$

where M denotes the total mass of the moving object, D denotes the viscosity coefficient, F_{Load} denotes the external force disturbance, ω_{sl} denotes the slip frequency, and

$$\begin{aligned}
L(Q) &= L_s - L_m f(Q) - \frac{[L_m(1-f(Q))]^2}{L_r - L_m f(Q)} \\
K_T &= \frac{3}{2}P\frac{\pi}{h}\frac{L_m(1-f(Q))}{L_r - L_m f(Q)}\phi_{dr} \\
T_r &= \frac{L_r}{R_r}.
\end{aligned} \tag{7}$$

Besides, according to (6), the acceleration of LIM can be redescribed as

$$\frac{dv}{dt} = \frac{K_T}{M}i_{sq} + A\cdot v + B, \tag{8}$$

where $A = -D/M$; $B = -F_{\text{Load}}/M$.

Remark 1. Taking into account the physical characteristics of the inverter structure and the safety of the system, the input saturation conditions must be considered. The control inputs are limited to

$$\begin{aligned}
u_{qs\min} &\leqslant u_{qs} \leqslant u_{qs\max}; \\
\dot{u}_{qs\min} &\leqslant \dot{u}_{qs} \leqslant \dot{u}_{qs\max},
\end{aligned} \tag{9}$$

where $\dot{u}_{qs}$ denotes the differential of u_{qs} and $(u_{qs\min}, u_{qs\max})$ and $(\dot{u}_{qs\min}, \dot{u}_{qs\max})$ denote the lower and upper bound of u_{qs} and $\dot{u}_{qs}$.

As speed is the most important performance parameter of motor control, we choose the velocity as our main control objective. Then, the model of a LIM considering end effects can be described in the following discrete-time unknown Nonlinear AutoRegressive with eXogenous input (NARX) model

$$\begin{aligned}
x(t+1) = g(&x(t),\ldots,x(t-t_x),u(t),\ldots,u(t-t_u), \\
&d(t),\ldots,d(t-t_d)),
\end{aligned} \tag{10}$$

where system output x denotes the speed of the LIM v, input u denotes the primary voltage in q-axis u_{qs}, and disturbance d denotes the external force disturbance F_{Load}. And t_x, t_u, and t_d mean the unknown orders, and $g(\cdot)$ is the unknown function. Apparently, the LIM satisfies the following two basic assumptions.

Assumption 2. The partial derivatives of $g(\cdot)$ for $u(t)$ and $d(t)$ are continuous.

Assumption 3. The plant (10) satisfies the condition of generalised Lipschitz, that is to say, $\forall t$, $|\Delta u(t-1)| \neq 0$ and $|\Delta d(t-1)| \neq 0$, satisfying $\Delta x(t) \leqslant \Lambda_1|\Delta u(t-1)|$ and $\Delta x(t) \leqslant \Lambda_2|\Delta u(t-1)|$, where $\Delta x(t) = x(t) - x(t-1)$, $\Delta u(t) = u(t) - u(t-1)$, and $\Delta d(t) = d(t) - d(t-1)$, and Λ_1, Λ_2 are unknown constants.

Remark 4. For general nonlinear systems, Assumption 2 is a common condition in the process of controller design. And Assumption 3 is a constrained condition that limits the changes of the outputs of the plant caused by system inputs and disturbance.

3. Main Results

In this section, an improved model-free adaptive SMC scheme is proposed for the LIM through the CFDL technology. The main contributions of this section are as follows:

(1) Transferring the LIM system into a data-based CFDL model considering the disturbance.

(2) Proposing the PPD estimation algorithm based on observers.

(3) Designing the model-free adaptive integral sliding-model controller via an antiwindup compensator.

(4) Proving the stability of the closed-loop system by Lyapunov stability theory.

3.1. Data-Driven Modeling for LIM and PPD Estimation Algorithm. Data-driven modeling method was originally proposed by HOU [17, 18, 26], and it is totally divided into three forms: CFDL, partial form dynamic linearization (PFDL), and full-form dynamic linearization (FFDL). In this paper, the CFDL technique is used to linearize the LIM system. When $|\Delta u(t)| \neq 0$, we can obtain the data-driven model as

$$\Delta x(t+1) = \theta_1\Delta u(t) + \theta_2\Delta d(t), \tag{11}$$

where $\theta_1 \leqslant \Lambda_1$ and $\theta_2 \leqslant \Lambda_2$ are the PPDs of the system.

The process of the proof is the same as that of [27].

To describe the system more conveniently, model (11) can be rewritten as follows:

$$x(t+1) = x(t) + Z^T(t)\Phi(t), \tag{12}$$

where $Z(t) = [\Delta u(t), \Delta d(t)]^T$ and $\Phi(t) = [\theta_1(t), \theta_2(t)]^T$.

The system output identification observer can be designed as

$$\hat{x}(t+1)=\hat{x}(t)+Z^T(t)\widehat{\Phi}(t)+Me_e(t), \tag{13}$$

where $\hat{x}(t)$ and $\widehat{\Phi}(t)$ mean the estimated value of output and PPDs of the system at time t, $e_e(t) = x(t) - \hat{x}(t)$ denotes the estimation error of the system output, and the gain K is chosen in the unit cycle. According to (12) and (13), the dynamic of the estimation error $e_e(t)$ can be described as

$$e_e(t+1)=Z^T(t)\widetilde{\Phi}(t)+Ne_e(t), \tag{14}$$

where $N = 1-M$ and $\widetilde{\Phi}(t) = \Phi(t)-\widehat{\Phi}(t)$ means the estimation error of the PPDs. The adaptive update PPD algorithm is given by

$$\widehat{\Phi}(t+1)=\widehat{\Phi}(t)+Z(t)\Gamma(t)\left(e_e(t+1)-Ne_e(t)\right), \tag{15}$$

where the gain function is chosen as

$$\Gamma(t)=\frac{2}{\|Z(t)\|^2+\partial} \tag{16}$$

Due to the fact that $\partial > 0$ is a chosen positive constant, it is for sure that $\Gamma(t)$ is positive. Besides, according to the practical assumption $\|Z(t)\| \leqslant \Omega$, $\Gamma(t)$ can be limited as

$$\Gamma(t)\geqslant\frac{2}{\Omega^2+\partial}=\upsilon>0 \tag{17}$$

In view of (14) and (15), the error dynamics of the system can be obtained as

$$\begin{aligned}e_e(t+1)&=Z^T(t)\widetilde{\Phi}(t)+Ne_e(t)\\ \widetilde{\Phi}(t+1)&=H\widetilde{\Phi}(t),\end{aligned} \tag{18}$$

where $H = I_{2\times 2}-Z(t)\Gamma(t)Z^T(t)$ and $I_{2\times 2}$ means the two-order unit matrix.

Theorem 5. *The equivalent of* $[e_e, \widetilde{\Phi}]$ *is globally uniformly stable. Furthermore, the estimation error of output* $e_e(t)$ *converges to 0; that is to say,* $\lim_{t\to\infty}|e_e(t)| = 0$

Proof. Consider the Lyapunov function as

$$V_A(t)=P_A{e_e}^2(t)+\lambda_A\widetilde{\Phi}^T(t)\widetilde{\Phi}(t), \tag{19}$$

where λ_A is a positive constant and P_A is also a positive constant figured by $P_A - F_A^2P_A = Q_A$ with Q_A being a positive constant. Then, the difference of $V_A(t)$ can be written as

$$\begin{aligned}\Delta V_A(t)&=V_A(t+1)-V_A(t)\\ &=P_AZ^T(t)\widetilde{\Phi}(t)\widetilde{\Phi}^T(t)Z(t)\\ &\quad-2P_AF_AZ^T(t)\widetilde{\Phi}(t)e_e(t)+P_AF_A^2e_e^2(t)\\ &\quad+\widetilde{\Phi}^T(t)\left(\lambda_AH^TH-\lambda_A\right)\widetilde{\Phi}(t)+P_Ae_e^2(t)\\ &=-\Theta_A^T(t)\left[\lambda_A\mu_A\Gamma^T(t)\Gamma(t)-P_A\right]\Theta_A(t)\\ &\quad-Q_Ae_e^2(t)+2P_AF_Ae_e(t)\Theta_A(t)\\ &\le-\left[\lambda_A\mu_A\Gamma^T(t)\Gamma(t)-P_A\right]\|\Theta_A(t)\|^2\\ &\quad+2P_AF_A|e_e(t)|\,\|\Theta_A(t)\|-Q_A|e_e(t)|^2\\ &\le-a_1|e_e(t)|^2-a_2\|\Theta_A(t)\|^2,\end{aligned} \tag{20}$$

where $\Theta_A(t) = Z^T(t)\widetilde{\Phi}(t)$, $a_2 = \lambda_A\mu_A\upsilon^2 - P_A - \iota P_A^2F_A^2$, and $a_1 = Q_A - (1/\iota)$. Thus, $\Delta V_A(t) \le 0$ can confirm that ι, Q_A, and λ_A satisfy the following inequalities:

$$\begin{aligned}Q_A&>\frac{1}{\iota},\\ \lambda_A\mu_A\upsilon^2-P_A-\iota P_A^2F_A^2&>0\end{aligned} \tag{21}$$

Since $V_A(t)$ is a nonnegative function and $\Delta V_A(t)$ is negative for sure, we can get the conclusion that when $t \to \infty$, $V_A(t) \to 0$. It is a signal where, for all k, $e_e(t)$ and $\widetilde{\Phi}(t)$ are bounded, and $\lim_{t\to\infty}e_e(t) = 0$.

From (13), we get the true value of the system output as follows:

$$x(t+1)=\hat{x}(t)+Z^T(t)\widehat{\Phi}(t)+Me_e(t)+e_e(t+1) \tag{22}$$

It is worth noting that $e_e(t + 1)$ is unknown in time t. So, we transfer $e_e(t + 1)$ into the following expression by two-step estimation technique:

$$e_e(t+1)\approx 2e_e(t)-e_e(t-1) \tag{23}$$

Therefore, (22) can be rewritten as

$$\begin{aligned}x(t+1)&=\hat{x}(t)+Z^T(t)\widehat{\Phi}(t)+(2-M)e_e(t)\\ &\quad-e_e(t-1)\end{aligned} \tag{24}$$

□

Remark 6. In order to make the parameter estimation law (15) have a strong capability in tracing time-varying parameters, a reset scheme should be considered as follows [17]:

$$\widehat{\Phi}(t)=\widehat{\Phi}(1),\quad \text{if } \widehat{\Phi}^T(t)\widehat{\Phi}(t)\le\vartheta \text{ or } \hat{\theta}_1(t)\le\vartheta, \tag{25}$$

where ϑ is a tiny positive constant and $\widehat{\Phi}(1)$ is the original value of $\widehat{\Phi}(t)$.

3.2. Model-Free Adaptive SMC Design and Stability Analysis. In order to eliminate the output non-following problem produced by the actuator saturation, an integral SMC based on antiwindup compensator is proposed [28]. Define the velocity tracking error as

$$e(t)=v^*(t)-v(t)-\xi(t), \tag{26}$$

where $v^*(t)$ means the given velocity reference value and $\xi(t)$ is the compensator signal which will be given later. To design the SMC, we choose an integral sliding surface as

$$s(t) = e(t) + \psi \sum_{i=1}^{t} T_s e(i), \tag{27}$$

where $\psi > 0$ and T_s denotes the sampling time of the control system. The closed-loop system stability can be guaranteed according to the following theorem.

Theorem 7. *When the integral sliding-mode surface is bounded, the tracking error of the control system is bounded, too. More specifically, for $|s(t)| \le \Omega$, the tracking error is bounded to a region as $\lim_{t\to\infty}|e(t)| \le 2\Omega/\psi T_s$.*

Proof. According to (27), we get

$$\begin{aligned} |e(t+1)| &= \frac{|e(t)|}{1+\psi T_s} + \frac{|s(t+1)-s(t)|}{1+\psi T_s} \\ &\le \frac{1}{1+\psi T_s}|e(t)| \\ &\quad + \frac{1}{1+\psi T_s}(|s(t+1)| + |s(t)|) \\ &\le \frac{1}{1+\psi T_s}|e(t)| + \frac{2\Omega}{1+\psi T_s} \end{aligned} \tag{28}$$

Due to the fact that $0 < 1/(1+\psi T_s) < 1$ and $2\Omega/(1+\psi T_s)$ is bounded, according to the stability criteria in [29], the tracking error can be bounded as

$$\lim_{t\to\infty} |e(t)| \le \frac{2\Omega}{\psi T_s} \tag{29}$$

The SMC law of the LIM can be designed based on observer (24) as

$$\begin{aligned} u^s(t) &= u(t-1) + u_f(t) + u_e(t) \\ u(t) &= \mathrm{Sat}\{(u(t-1) \\ &\quad + \mathrm{Sat}\{(u^s(t) - u(t-1)), T_s\dot{u}_{sq\min}, T_s\dot{u}_{sq\max}\}), \\ &\quad T_s u_{sq\min}, T_s u_{sq\max}\}, \end{aligned} \tag{30}$$

where $u_f(t)$ and $u_e(t)$ denote the feedback and equivalent laws and $u^s(t)$ and $u(t)$ denote the primary and actual control input signals, respectively. And Sat$(\cdot)$ function is defined as

$$\mathrm{Sat}(h, h_{\min}, h_{\max}) = \begin{cases} h_{\max} & h \ge h_{\max} \\ h & h_{\max} > h > h_{\min} \\ h_{\min} & h_{\min} \ge h, \end{cases} \tag{31}$$

where $h_{\max}$ and $h_{\min}$ mean the upper and lower bound of Sat$(\cdot)$. One important thing is that when the input signal is within saturation, the tracking performance cannot be guaranteed. Thus, we design an antiwindup compensator signal as follows:

$$\xi(t) = \gamma\xi(t-1) + \hat{\theta}_1(t)(u^s(t) - u(t-1)), \tag{32}$$

where γ is chosen in the unit disk. □

Remark 8. Since γ lies in the unit disk and assuming $u^s(t) - u(t-1)$ is bounded, the signal $\xi(t)$ is uniformly ultimately bounded (UUB) for all t according to [28].

Moreover, we concretely give the expressions of $u_f(t)$ and $u_e(t)$ as

$$\begin{aligned} u_f(t) &= -\frac{\hat{\theta}_2(t) K_s s(t)}{(\hat{\theta}_2^2(t)+\kappa)(1+\psi T_s)} \\ u_e(t) &= \frac{\hat{\theta}_2(t)}{\hat{\theta}_2^2(t)+\kappa}\Big(x^*(t+1) - \hat{\theta}_1(t)\Delta d(t) - \hat{x}(t) \\ &\quad - (2-M)e_e(t) + e_e(t-1) - \frac{e(t)}{1+\psi T_s} - \gamma\xi(t)\Big), \end{aligned} \tag{33}$$

where κ is also chosen in the unit disk, K_s is a negative constant chosen by $K_s^2/2 + K_s < 0$, and $x^*(t+1)$ means the reference signal value in time $t+1$.

Theorem 9. *For given $|\Delta x^*(t) - \Delta x^*(t-1)| \le \Delta x^*$, using control laws (31)–(33), the velocity tracking error of the LIM is UUB for all t with ultimate bound as $\lim_{t\to\infty}|e(t)| \le (q_1(t) + \sqrt{q_1^2(t) + 4q_0(t)q_2(t)})/q_0(t)\psi T_s$.*

Here, q_0 is a constant given by $q_0(t) \le -(K_s^2/2 + K_s)$, and

$$\begin{aligned} q_1(t) &= \frac{\kappa(K_s+1)(1+\psi T_s)q(t)}{\hat{\theta}_1^2(t)+\kappa} \\ q_2(t) &= \frac{\kappa^2(1+\psi T_s)^2 q^2(t)}{2(\hat{\theta}_1^2(t)+\kappa)} \\ q(t) &= -\frac{K_s s(t)}{1+\psi T_s} + x^*(t+1) - \hat{\theta}_2(t)\Delta d(t) - \hat{x}(t) \\ &\quad - (2-M)e_e(t) + e_e(t-1) - \frac{e(t)}{1+\psi T_s} \\ &\quad - \gamma\xi(t) \end{aligned} \tag{34}$$

Proof. Define the Lyapunov function $V_B(t) = (1/2)s^2(t)$; then, the difference of $V_B(t)$ can be figured by

$$\begin{aligned} \Delta V_B(t+1) &= V_B(t+1) - V_B(t) \\ &= \Delta s(t+1)\left[\frac{1}{2}\Delta s(t+1) + s(t)\right], \end{aligned} \tag{35}$$

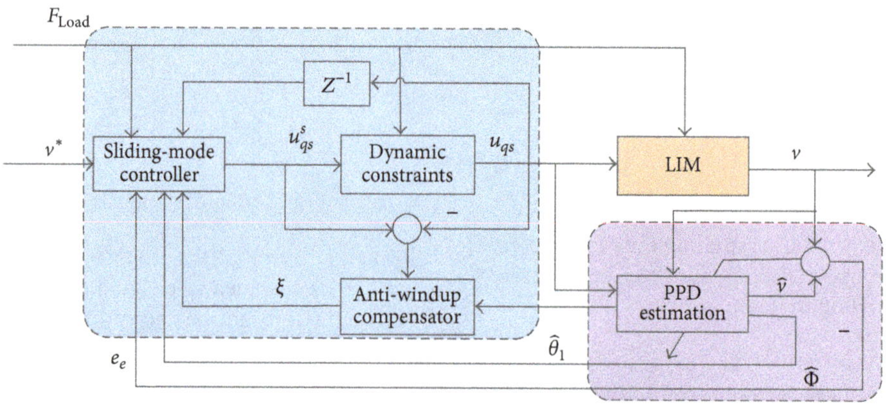

FIGURE 2: Diagram of LIM control systems.

where $\Delta s(t+1)$ is figured by

$$\begin{aligned}
\Delta s(t+1) &= s(t+1) - s(t) = (1+\psi T_s) e(t+1) \\
&- e(t) = (1+\psi T_s)\left(x^*(t+1) - \hat{x}(t)\right. \\
&- (2-M) e_e(t) - \hat{\Phi}(t) Z(t) + e_e(t-1) \\
&\left.- \xi(t+1)\right) - e(t) = (1+\psi T_s)\left(x^*(t+1) - \hat{x}(t)\right. \\
&- (2-M) e_e(t) + e_e(t-1) - \gamma\xi(t) - \hat{\theta}_2(t)\Delta d(t) \\
&\left.- \hat{\theta}_1(t)\left(u_f(t) + u_e(t)\right)\right) - e(t) = K_s s(t) \\
&+ \frac{\kappa(1+\psi T_s)}{\hat{\theta}_1^2(t)+\kappa}\left(-\frac{K_s s(t)}{1+\psi T_s} - \hat{x}(t) - \hat{\theta}_2(t)\Delta d(t)\right. \\
&+ x^*(t+1) - (2-M)e_e(t) + e_e(t-1) - \frac{e(t)}{1+\psi T_s} \\
&\left.- \gamma\xi(t)\right) = K_s s(t) + \frac{\kappa(1+\psi T_s)}{\hat{\theta}_1^2(t)+\kappa} q(t)
\end{aligned} \tag{36}$$

Besides, referring to (33), then we get

$$\begin{aligned}
\Delta V(t+1) &= \left(\frac{1}{2}K_s^2 + K_s\right) s^2(t) \\
&+ \frac{\kappa^2(1+\psi T_s)^2 q^2(t)}{2\left(\hat{\theta}_1^2(t)+\kappa\right)} \\
&+ \frac{\kappa(K_s+1)(1+\psi T_s) q(t)}{\hat{\theta}_1^2(t)+\kappa} s(t) \\
&\le -q_0(t) s^2(t) + q_1(t) s(t) + q_2(t),
\end{aligned} \tag{37}$$

where K_s is chosen to make $K_s^2/2 + K_s < 0$, and then $q_0(t) > 0$ is for sure. And when $s(t) > (q_1 + \sqrt{q_1^2 + 4q_0q_2})/2q_0$, $\Delta V(t+1) < 0$ can be guaranteed. Hence, the sliding surface is bounded as $\lim_{t\to\infty}|s(t)| \le (q_1 + \sqrt{q_1^2 + 4q_0q_2})/2q_0$. Finally, according to Theorem 9, we can get the conclusion that

$$\lim_{t\to\infty} |e(t)| \le \frac{\left(q_1 + \sqrt{q_1^2 + 4q_0q_2}\right)}{q_0\psi T_s} \tag{38}$$

□

TABLE 1: Parameters of LIM.

Parameters	Representation	Value
R_s (Ω)	Primary phase resistance	6.2689
R_r (Ω)	Secondary phase resistance	3.784
L_m (H)	Mutual inductance	0.0825
L_r (H)	Secondary phase inductance	0.1021
L_s (H)	Primary phase inductance	0.1021
M (kg)	Total mass of the object	3.25
D (kg/s)	Viscosity coefficient	40.95
h (m)	Polar distance	0.057

Remark 10. Because ψ and κ are tiny positive constants, respectively, the ultimate bound of tracking error is 0; i.e., $\lim_{t\to\infty}|e(t)| = 0$.

4. Simulation Results

In this section, a few simulation examples are given to testify the effectiveness of the designed controller compared to the classical PID controller. First of all, to clearly understand the control process of the LIM, a block diagram is given in Figure 2. Meanwhile, the parameters of the LIM are given in Table 1.

In order to obtain a satisfactory control effect, we choose the parameters of the controller as $N = 0.99$, $\vartheta = 300$, $\Phi(1) = [0.012, 0.005]^T$, $K_s = -1.5$, $\kappa = 0.0001$, $\psi = 10000$, $T_s = 0.001$, and $\gamma = 0.3$. Meanwhile, the parameters of the PID controller are $P = 300$, $I = 10000$, and $D = 0.1$. Two kinds of simulation experiments below are designed to prove the effectiveness of the proposed controller in this paper. By comparing the proposed controller with the PID

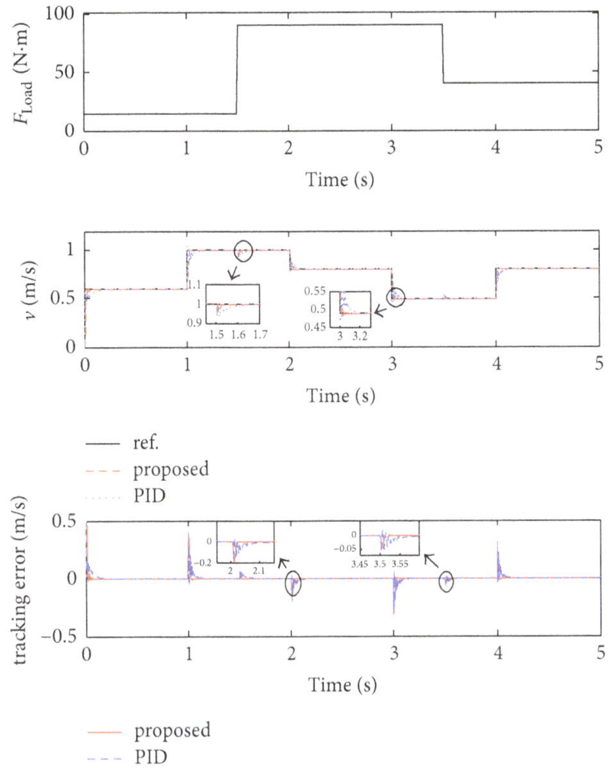

FIGURE 3: The reference tracking and tracking error curve of the proposed controller and PID controller (step signal).

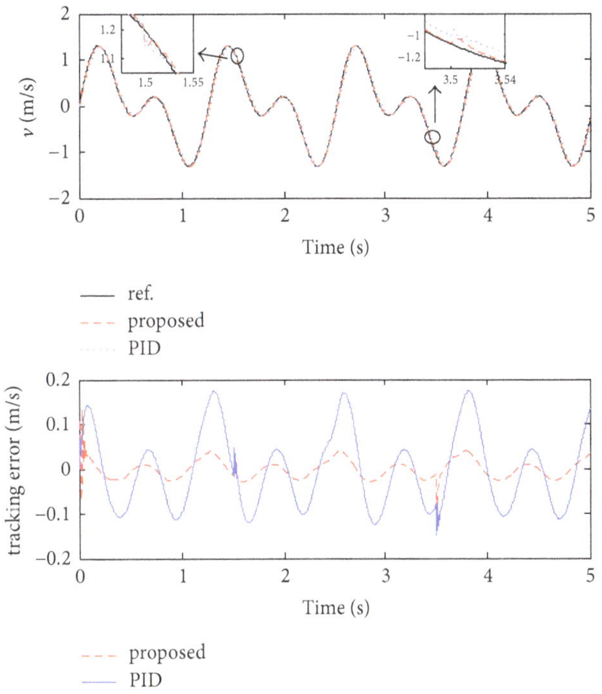

FIGURE 4: The reference tracking and tracking error curve of the proposed controller and PID controller (periodic signal).

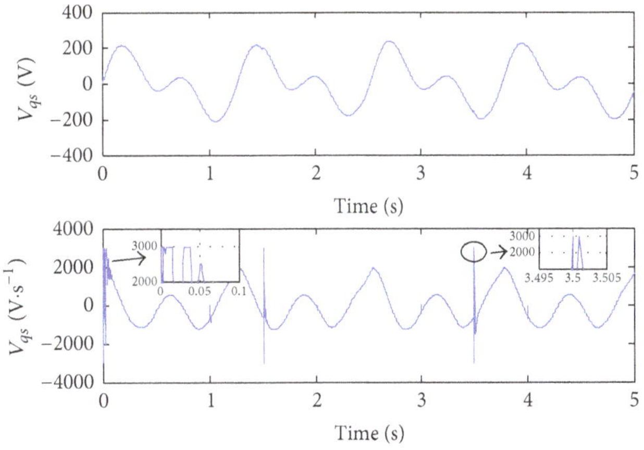

FIGURE 5: The value of input signals (periodic signal).

controller, we will analyze the control performance from the following aspects: dynamic performance, static performance, anti-interference, and robustness.

(1) To test the tracking performance and anti-interference, we select the step signal and time-varying periodic signal as our given velocity reference, respectively. Meanwhile, the load torque changes as shown in Figure 3. The velocity tracking performance and tracking error are also shown in Figures 3 and 4. As the figures show, it can be clearly known that both controllers can ensure that there is no steady-state error at steady state for step signal. However, the proposed control method enables the control system to enter steady state faster within 0.12 s (within is 0.3 s for the PID controller). Besides, when the load torque changes at 1.5 s and 3.5 s, the speed of the LIM under the proposed controller is still able to track the given signal quickly within 0.1 s after a small fluctuation (within 0.32 s for the PID controller). It can be seen more prominently in Figure 4 that the proposed controller can make the system output track the time-varying periodic signal perfectly with less than 0.05 m/s error. The input signal under time-varying periodic signal is shown in Figure 5. The compensator signal under time-varying periodic signal is shown in Figure 6. From Figures 5 and 6, we can get the information that, by adding the antiwindup compensator, the control system can quickly exit from saturation but still trace the reference quite well. The values of the PPDs are shown in Figure 7.

(2) To test the robustness of the proposed controller, we increase the mover mass to three times and five times the original, and this simulation is also under time-varying periodic signal. The tracking performance is shown in Figure 8. From Figure 8, we know that no matter how the mover mass changes, the speed of the LIM can always follow the given reference satisfactorily, and that is another merit of the model-free adaptive sliding-mode-constrained controller. Therefore, this simulation verifies the robustness of the proposed controller.

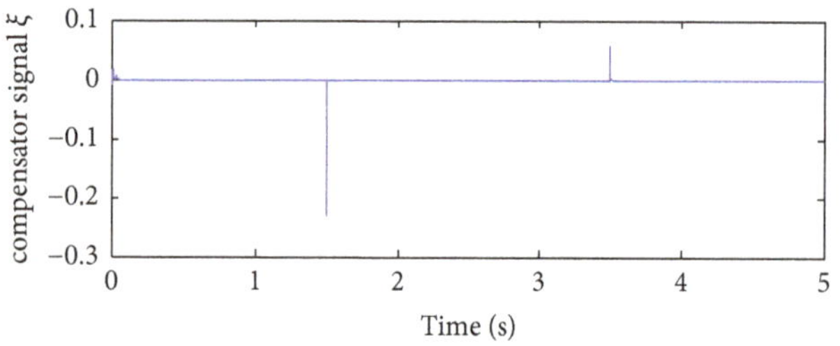

FIGURE 6: The value of compensator signals (periodic signal).

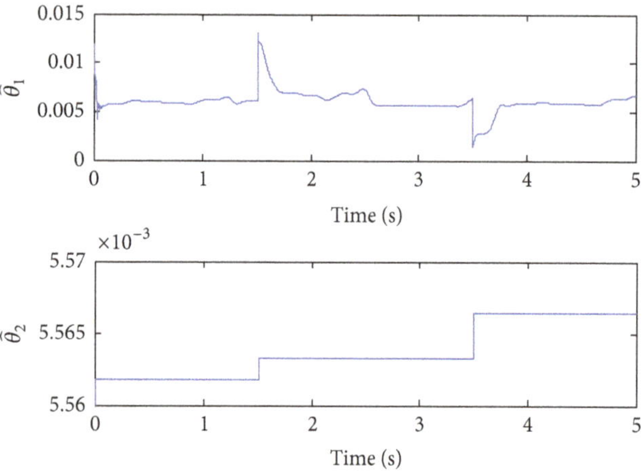

FIGURE 7: The value of PPDs (periodic signal).

5. Conclusion

In this paper, a model-free adaptive sliding-mode controller is proposed to deal with the problem of the speed tracking of the LIM considering end effects. First of all, the CFDL technique is applied to linearize the LIM model which has been transferred into a NARX form. Then, the controller is designed based on PPD estimation algorithm. Through the process of designing, an antiwindup compensator is designed to handle the problem of input saturation. Lyapunov stability theory proves the stability of the closed-loop system theoretically, and the simulation results verify the effectiveness of the proposed method to the LIM system.

Conflicts of Interest

The authors declare that there are no conflicts of interest regarding the publication of this paper.

Acknowledgments

This work was partially supported by the National Natural Science Foundation of China (61503156, 61403161, and 61473250) and the National Key Research and Development Program (2016YFD0400300).

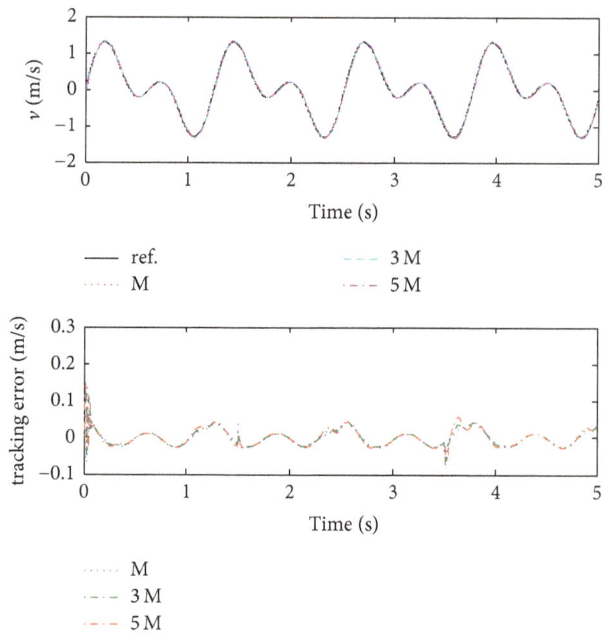

FIGURE 8: Velocity tracking curve under different mover mass (periodic signal).

References

[1] M. H. Ravanji and Z. Nasiri-Gheidari, "Design Optimization of a Ladder Secondary Single-Sided Linear Induction Motor for Improved Performance," *IEEE Transactions on Energy Conversion*, vol. 30, no. 4, pp. 1595–1603, 2015.

[2] X. Qiwei, S. Cui, Q. Zhang, L. Song, and X. Li, "Research on a new accurate thrust control strategy for linear induction motor," *IEEE Transactions on Plasma Sciences*, vol. 43, no. 5, pp. 1321–1325, 2015.

[3] H. Gurol, "General atomics linear motor applications: Moving towards deployment," *Proceedings of the IEEE*, vol. 97, no. 11, pp. 1864–1871, 2009.

[4] S. E. Abdollahi, M. Mirzayee, and M. Mirsalim, "Design and Analysis of a Double-Sided Linear Induction Motor for Transportation," *IEEE Transactions on Magnetics*, vol. 51, no. 7, 2015.

[5] A. Boucheta, I. K. Bousserhane, A. Hazzab, P. Sicard, and M. K. Fellah, "Speed control of linear induction motor using sliding mode controller considering the end effects," *Journal of Electrical Engineering & Technology*, vol. 7, no. 1, pp. 34–45, 2012.

[6] I. Boldea and S. A. Nasar, "Linear electric actuators and generators," *IEEE Transactions on Energy Conversion*, vol. 14, no. 3, pp. 712–717, 1999.

[7] A. Accetta, M. Cirrincione, M. Pucci, and G. Vitale, "Neural sensorless control of linear induction motors by a full-order luenberger observer considering the end effects," *IEEE Transactions on Industry Applications*, vol. 50, no. 3, pp. 1891–1904, 2014.

[8] G. Kang and K. Nam, "Field-oriented control scheme for linear induction motor with the end effect," *IEE Proceeding—Electrical Power Applications*, vol. 152, no. 6, pp. 1565–1572, 2005.

[9] F. E. Benmohamed, I. K. Bousserhane, A. Kechich, B. Bessaih, and A. Boucheta, "New MRAS secondary time constant tuning for vector control of linear induction motor considering the end-effects," *COMPEL - The International Journal for Computation and Mathematics in Electrical and Electronic Engineering*, vol. 35, no. 5, pp. 1685–1723, 2016.

[10] A. Boucheta, I. K. Bousserhane, A. Hazzab, B. Mazari, and M. K. Fellah, "Adaptive backstepping controller for linear induction motor position control," *COMPEL: The International Journal for Computation and Mathematics in Electrical and Electronic Engineering*, vol. 29, no. 3, pp. 789–810, 2010.

[11] H.-H. Chiang, K.-C. Hsu, and I.-H. Li, "Optimized adaptive motion control through an SoPC implementation for linear induction motor drives," *IEEE/ASME Transactions on Mechatronics*, vol. 20, no. 1, pp. 348–360, 2015.

[12] F. Alonge, M. Cirrincione, M. Pucci, and A. Sferlazza, "Input-output feedback linearization control with on-line MRAS-based inductor resistance estimation of linear induction motors including the dynamic end effects," *IEEE Transactions on Industry Applications*, vol. 52, no. 1, pp. 254–266, 2016.

[13] A. Y. Alanis, J. D. Rios, J. Rivera, N. Arana-Daniel, and C. Lopez-Franco, "Real-time discrete neural control applied to a Linear Induction Motor," *Neurocomputing*, vol. 164, pp. 240–251, 2015.

[14] A. Boucheta, I. K. Bousserhane, A. Hazzab, B. Mazari, and M. K. Fellah, "Fuzzy-sliding mode controller for linear induction motor control," *Revue Roumaine Des Sciences Techniques - Serie Electrotechnique Et Energetique*, vol. 54, no. 4, pp. 405–414, 2009.

[15] C.-Y. Hung, P. Liu, and K.-Y. Lian, "Fuzzy virtual reference model sensorless tracking control for linear induction motors," *IEEE Transactions on Cybernetics*, vol. 43, no. 3, pp. 970–981, 2013.

[16] Z. Hou, *The parameter identification, adaptive control and model free learning adaptive control for nonlinear systems [Ph.D. thesis]*, Northeastern Univerity, Shenyang, 1994.

[17] Z.-S. Hou and S.-T. Jin, "A novel data-driven control approach for a class of discrete-time nonlinear systems," *IEEE Transactions on Control Systems Technology*, vol. 19, no. 6, pp. 1549–1558, 2011.

[18] Z. Hou and S. Jin, "Data-driven model-free adaptive control for a class of MIMO nonlinear discrete-time systems," *IEEE Transactions on Neural Networks and Learning Systems*, vol. 22, no. 12, pp. 2173–2188, 2011.

[19] D. Xu, B. Jiang, and P. Shi, "A novel model-free adaptive control design for multivariable industrial processes," *IEEE Transactions on Industrial Electronics*, vol. 61, no. 11, pp. 6391–6398, 2014.

[20] X. Bu, Z. Hou, and H. Zhang, "Data-Driven Multiagent Systems Consensus Tracking Using Model Free Adaptive Control," *IEEE Transactions on Neural Networks Learning Systems*, vol. PP, no. 99, p. 11, 2017.

[21] D. Xu, B. Jiang, and P. Shi, "Adaptive observer based data-driven control for nonlinear discrete-time processes," *IEEE Transactions on Automation Science & Engineering*, vol. 11, no. 4, pp. 1549–1558, 2014.

[22] D. Xu, Y. Shi, and Z. Ji, "Model Free Adaptive Discrete-time Integral Sliding Mode Constrained Control for Autonomous 4WMV Parking Systems," *IEEE Transactions on Industrial Electronics*, 2017.

[23] Z. Wu, X. Wang, and X. Zhao, "Backstepping terminal sliding mode control of DFIG for maximal wind energy captured," *International Journal of Innovative Computing, Information and Control*, vol. 12, no. 5, pp. 1565–1579, 2016.

[24] X.-G. Yan and C. Edwards, "Adaptive sliding-mode-observer-based fault reconstruction for nonlinear systems with parametric uncertainties," *IEEE Transactions on Industrial Electronics*, vol. 55, no. 11, pp. 4029–4036, 2008.

[25] Z. Hou, W. Wang, and S. Jing, "Adaptive quasi-sliding-mode control for a class of nonlinear discretetime systems," *Control Theory Applications*, vol. 26, no. 5, pp. 505–509, 2009.

[26] Z.-S. Hou and Z. Wang, "From model-based control to data-driven control: survey, classification and perspective," *Information Sciences*, vol. 235, pp. 3–35, 2013.

[27] D. Xu, B. Jiang, and F. Liu, "Improved data driven model free adaptive constrained control for a solid oxide fuel cell," *IET Control Theory & Applications*, vol. 10, no. 12, pp. 1412–1419, 2016.

[28] N. Ji, D. Xu, and F. Liu, "A novel adaptive neural network constrained control for solid oxide fuel cells via dynamic anti-windup," *Neurocomputing*, vol. 214, pp. 134–142, 2016.

[29] J. Spooner, M. Maggiore, and K. Passino, *Stable adaptive control and estimation for nonlinear systems*, Wiley, New York, NY, USA.

14

Control System Based on Anode Offgas Recycle for Solid Oxide Fuel Cell System

Shuanghong Li, Chengjun Zhan, and Yupu Yang

Department of Automation, Key Laboratory of System Control and Information Processing, Ministry of Education, Shanghai Jiao Tong University, Shanghai, China

Correspondence should be addressed to Shuanghong Li; lishuanghong1989@163.com

Academic Editor: Alessandro Mauro

The conflicting operation objectives between rapid load following and the fuel depletion avoidance as well as the strong interactions between the thermal and electrical parameters make the SOFC system difficult to control. This study focuses on the design of the decoupling control for the thermal and electrical characteristics of the SOFC system through anode offgas recycling (AOR). The decoupling control system can independently manipulate the thermal and electrical parameters, which interact with one another in most cases, such as stack temperatures, burner temperature, system current, and system power. Under the decoupling control scheme, the AOR is taken as a manipulation variable. The burner controller maintains the burner temperature without being affected by abrupt power change. The stack temperature controller properly coordinates with the burner temperature controller to independently modulate the stack thermal parameters. For the electrical problems, the decoupling control scheme shows its superiority over the conventional controller in alleviating rapid load following and fuel depletion avoidance. System-level simulation under a power-changing case is performed to validate the control freedom between the thermal and electrical characteristics as well as the stability, efficiency, and robustness of the novel system control scheme.

1. Introduction

Hydrogen-fueled solid oxide fuel cells (SOFC) can directly generate electric power from hydrogen with numerous advantages, such as high electrical efficiency, reduced emissions, and quiet operation as compared with traditional power sources [1–4]. The SOFC system is especially suitable in the structuring of distributed alternative power stations, because it can efficiently produce electricity. Moreover, the quiet operation characteristic associated with the reduced emission feature makes the system appropriate in locating SOFC systems in downtown areas and densely populated regions for residual and commercial loads. Although the SOFC system possesses several advantages, the SOFC system has not reached the commercialization degree because of system durability and reliability problems. Therefore, a control method that not only ensures system safety but also efficiently operates the system must be developed [5–8].

Solid oxide fuel cell (SOFC) modeling is a low-cost method for studying and investigating fuel cells, optimizing and controlling their behavior, enhancing their efficiency and performance, and reducing high installation costs. Some review papers have examined the different aspects of SOFC and reviewed different mathematical modeling studies to aid future researchers in developing SOFC [9, 10]. In [11], a 0D mathematical model was introduced to analyze the performance of an SOFC-based microcogenerative power system that was fed by natural gas. The novelty of the proposed approach lies in its ability to accurately reproduce the logic of an on-board control system, which predicts different steady-state operating conditions by taking into account the actual operating ranges imposed by the manufacturer for the main parameters, such as stack temperature. Reference [12] investigated the 2D and 3D numerical modeling of SOFC by employing an accurate and stable fully matrix-inversion-free finite element algorithm. In [13], a new 3D finite element algorithm based on a detailed mathematical model for fuel cells and on the fully explicit artificial compressibility characteristic-based split scheme was employed to effectively and efficiently model the heat and

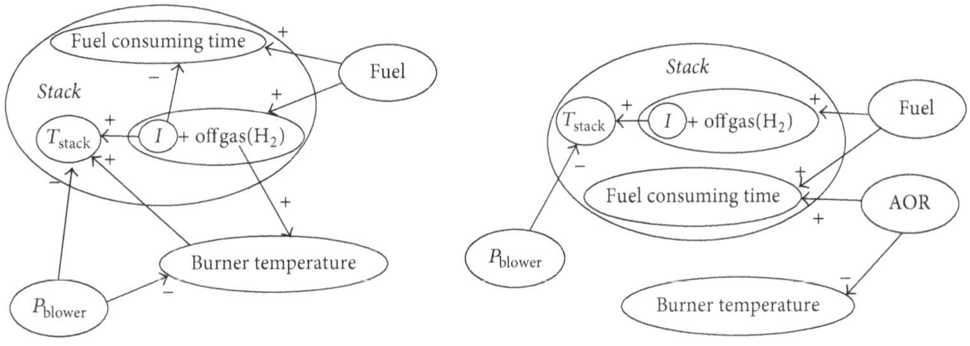

FIGURE 1: System coupling relationships before and after decoupling control.

mass transport phenomena coupled with electrochemical reactions in SOFC. The calculation results in [14] for the voltage and temperature distributions of a 3D computational fluid dynamics model were compared with the results of an experimental program with data for an 18-cell stack. This work is among the first to compare physical experiments with a comprehensive SOFC stack model. To perform SOFC measurements, [15] conducted an a priori uncertainty analysis of a cogenerative module based on HT-SOFC, on-board instruments, and meteorological characteristics declared by the manufacturer. Inspired by these studies, this paper examines the problem of an SOFC system based on a model.

In recent years, numerous studies on the control of SOFC systems have been published. Generally, these studies mainly attempted to address three control targets of the SOFC: (1) maintaining a rapid flow of power to address the power demand [1, 2, 6–8, 16–20]; (2) keeping the crucial components, such as the stack, burner, and exchanger temperature below the safety operation limits [5, 8, 16, 21–25]; and (3) ensuring high levels of fuel utilization and system efficiency [5, 8, 21, 23, 26, 27]. For the optimal operation of the SOFC system, an integral control scheme with rapidness, efficiency, easy application, and robustness must be designed to handle the control targets [28]. However, the strong couplings lead to the difficulty of the control design, which aims to optimally operate the SOFC system, as shown in Figure 1. As a multivariable system, the SOFC has several input and output variables that cross-couple or interact among one another. The characteristic, in which a change in one input affects several outputs, prevents control engineers from designing each input-output control loop independently, as adjusting one controller parameter affects the performance of another and may destabilize the entire system.

Therefore, by designing a decoupling control scheme, the multivariable SOFC system can be simplified to several single-variable systems that have no cross-coupling or interaction among variables, which is critical to SOFC control. Through the application of the decoupling control, each SOFC output variable is affected by only one reference input variable, and then each input-output pair can be controlled by a single-input single-output (SISO) controller, which is considerably easier for control engineering practice and demands less hardware cost as compared with the multivariable control. Wu et al. [29] designed a power decoupling controller for the SOFC-MGT hybrid system by self-tuning proportional-integral derivation (PID) to independently control the SOFC power and MGT power. Zhao et al. [30] developed a control solution based on dynamic disturbance decoupling control for a centrifugal compression system, which is used to supply compressed air to the fuel cell to manipulate the mass flow and pressure. However, to the best of the author's knowledge, no study has attempted to manage the decoupling control of the SOFC's thermal and electrical characteristics, thereby resulting in difficulty in achieving efficient system operation and guaranteeing safety.

The anode offgas recycle (AOR), through which the anode offgas is partially recycled to the anode inlet, has been used in recent research to increase SOFC system efficiency [31–34]. For the hydrogen-fueled SOFC system (Figure 2) decoupling controller designing, the plant output variables are difficult to manage independently because of the lack of control inputs needed to construct the input-output pairs. Therefore, to alleviate this problem in decoupling the interactions between the variables, this study adds the AOR as an independent manipulation variable to design the decoupling control architecture. The AOR is suitable in managing the system-level decoupling control of hydrogen-fueled SOFC because the hydrogen-fueled SOFC without reform reactions requires less heat generated by the burner for system heat self-sustenance, and the AOR can prevent the excessive hydrogen in the offgas from entering the burner to generate useless heat. Although many previous works have employed the AOR to improve system efficiency [31–34], its function in system-level control is first investigated in the current study.

This study aims to decouple the interactions among the thermal and electrical parameters in the hydrogen-fueled SOFC system based on the dynamic SOFC model to simplify the multiple-input multiple-output (MIMO) control system to several SISO control systems, which are suitable for controller development and implementation. By using the AOR-based method, we design the decoupling controller to solve the following problems.

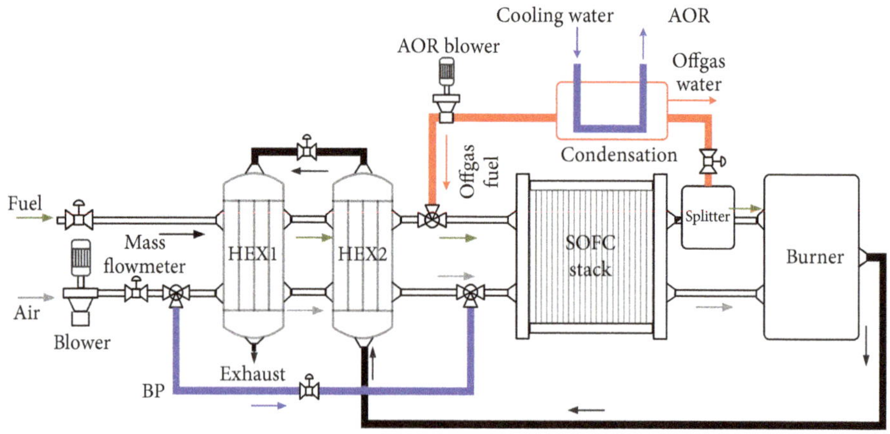

FIGURE 2: SOFC structure with anode offgas recycle.

(1) Decoupling Temperature from Power Changing. The burner temperature can be independently controlled by the AOR without being affected by the fuel flow increase caused by load power changing. The stack temperature can be independently managed without being significantly affected by power fluctuation.

(2) Rapid Load Following and Fuel Depletion Avoidance. The rapid loading capability of the system and the fuel depletion problem, which is limited and caused by the fuel supply time delay, are improved through proper handling of the AOR rate. Given that the AOR structure provides another part of fuel supply with high response speed to the stack, the fuel consumption time increases.

This study provides a helpful reference and example for the development of the control of the hydrogen-fueled SOFC system as well as SOFC with a MgH_2 tank, thereby allowing for both the robustness of controller and the feasibility of engineering implementation.

2. Description of the Studied AOR-SOFC

The kW-scale SOFC stand-alone system is developed as the platform for the decoupling control design. Special attention is drawn on the system-level thermal and electrical parameter simulation. The SOFC system is modeled based on transportation and conservation principles. The model has been built through many previous efforts [5, 27]. Given that the focus of this study is system controller design, the dynamic modeling method is briefly introduced. The model runs in a MATLAB/Simulink platform on a computer with 3 GHz and 12 G memory. The kW-scale SOFC stand-alone system model comprises a planar SOFC stack, a burner, and two heat exchangers (Figure 2), in which a special consideration for stack spatial temperature management is conducted by an air bypass manifold around heat exchangers. Particularly, an AOR structure is designed in this system. The SOFC system in this study has the following two characteristics:

(1) For stack temperature gradient along the gas flow direction to be minimized, the gas entering the stack should be preheated although the exchangers using the burner exhaust gas. In this study, another manifold bypassing the exchangers is added to the system whose cold air mixes with the preheated air passing through the main air manifold. Manipulating the flow rate of cold air is an effective way to manage the stack inlet temperature. Therefore, only by controlling both the bypass (BP) ratio and the air flow rate can the stack inlet and outlet temperature be efficiently managed.

(2) The AOR structure is used in the SOFC system: the AOR structure consists of a splitter, condensation, and an AOR blower and mixer. The condensation is the key element in the AOR because it is where the anode offgas is collected, condensed, stored, and transported to the stack by the AOR blower.

Some simplifying assumptions are made to obtain a computationally dynamic model.

(1) All of the gases in the system are ideal gases, and the pressure drop along the channels is neglected.

(2) The gas in the system is assumed to be incompressible.

(3) The system is assumed to be insulated from the system in which no heat is transferred to the environment.

In this section, the stack and thermal dynamic models are introduced first. The configuration and feasibility of the AOR structure is then discussed. Finally, some system-level operation parameters are defined for this special SOFC system.

Since the system model is used for control oriented analysis and design, some simplifications are considered for a balance between model details and computational burden. The simplifications consist of a quasi-two-dimensional approach for resolving geometrical features of the system components. This approach discretizes each component in the flow direction and resolves chemical and physical processes, such as electrochemical reaction, heat conduction, and convection. The discretized elements are called nodes. Each node includes two types of control volumes, gas phase and solid phase control volumes, representing the primary elements in the cross-wise direction [5, 16]. In each control volume, only the physical and chemical processes that affect the time scale of interest in the dynamic simulation are taken into account. For instance, electrochemical reaction and the dynamics of burn are assumed to occur at a time scale that is faster than that of interest to the dynamic model; those processes are considered quasi-steadily in the system model.

2.1. Stack Model. In this section, the electrical characteristics which are the most important component in planar type stack are conducted. The single cell voltage is obtained by calculating three polarization voltage losses from the irreversible open circuit voltage:

$$U_{\text{cell}} = U_{\text{OCV}} - U_{\text{loss}} = U_{\text{OCV}} - U_{\text{ohm}} - U_{\text{act}} - U_{\text{con}}, \quad (1)$$

where U_{cell} is the single cell output voltage, U_{OCV} is the irreversible open circuit voltage (OCV) that is also called Nernst voltage, U_{ohm} is the ohmic polarization voltage loss, U_{act} is the activation polarization voltage loss, U_{con} is the concentration voltage loss.

The irreversible open circuit voltage (OCV) is expressed as follows:

$$U_{\text{OCV}} = E^0 + \frac{RT_{\text{PEN}}}{2F} \ln\left(\frac{P_{H_2} P_{O_2}^{0.5}}{P_{H_2O}}\right). \quad (2)$$

According to the Ohm's law the ohmic polarization voltage loss can be calculated by current density and ohmic resistance:

$$U_{\text{ohm}} = iR_{\text{ohm}} = iT_{\text{PEN}} e^{a_1/T_{\text{PEN}} + a_0}. \quad (3)$$

The activation and concentration voltage loss can be described as follows:

$$U_{\text{act},a} = \frac{2RT}{n_e F} \sinh^{-1}\left(\frac{i}{2i_{oa}}\right),$$
$$U_{\text{act},c} = \frac{2RT}{n_e F} \sinh^{-1}\left(\frac{i}{2i_{oc}}\right), \quad (4)$$
$$U_{\text{con}} = \frac{RT}{n_e F} \ln\left[1 - \frac{i}{i_L A}\right].$$

Those polarization voltage losses are complicated chemical and physical process concerning with temperature, pressure, and current; consequently, the expression would not be of the same form. According to the Butler-Volmer equation, the current density is

$$i = i_0 \left\{\exp\left(\alpha \frac{n_e F U_{\text{act}}}{RT}\right) - \exp\left(-(1-\alpha)\frac{n_e F U_{\text{act}}}{RT}\right)\right\}, \quad (5)$$

where the α is the transfer coefficient, i_0 is the exchange current density, n is the electrons transferred per reaction, and U_{act} is the activation polarization.

2.2. Thermal Dynamic Equations. The model emulates the chemical and physical reactions in the SOFC system, such as electrochemical reaction, heat conduction, and convection. In particular, the most important part is the stack which will be individually described in the next section. The system variables such as temperature, molar flow rates, and mole fractions are described by energy, mass, and species conservation laws.

2.2.1. Solid Phase Control Volume. The dynamic solid-state energy conservation equation is in the general form:

$$\rho_s V_s C_s \frac{dT_s}{dt} = \sum \frac{dQ_{\text{in},s}}{dt} + \frac{dQ_{\text{react}}}{dt} - \frac{dW_{\text{out}}}{dt}, \quad (6)$$

where ρ_s, V_s, and C_s denote the density, volume, and specific heat capacity of each solid control volume, respectively.

Conduction and convection heat transfer between the solid phase control volumes are determined by Fourier's law and Newton's law. Fourier's law is utilized to capture conduction heat transfer among solid phase control volumes using the temperature of each control volume:

$$\frac{dQ_{\text{cond}}}{dt} = \frac{S_{\text{area}} \cdot k_{ss} \cdot (T_2 - T_1)}{L}. \quad (7)$$

Newton's law is used to calculate convection heat transfer between solid and gas phase control volumes:

$$\frac{dQ_{\text{conv}}}{dt} = S_{\text{area}} \cdot h_{gs} \cdot (T_2 - T_1). \quad (8)$$

2.2.2. Gas Phase Control Volume. The conservation of energy equation of the gas is as follows:

$$NC_v \frac{dT}{dt} = \frac{d(N_{\text{in}})}{dt} h_{\text{in}} - \frac{d(N_{\text{out}})}{dt} h_{\text{out}} + \sum \frac{d(Q_{\text{in}})}{dt}, \quad (9)$$

where N is control volume mole number, C_v is constant volume specific heat capacity of the gas mixture, N_{in} and N_{out} represent molar flow rate entering or exiting the control volume, h_{in} and h_{out} are enthalpy of the gas mixture entering or exiting the control volume, Q_{in} is heat transfer entering control volume.

The constant volume specific heat capacity of gas mixture C_v is calculated as follows:

$$C_v = \sum X_i C_{p,i}(T) - R, \quad i \in \{H_2, O_2, H_2O, N_2\}, \quad (10)$$

where R is the universal gas constant, X_i is mole fraction of species i, $C_{p,i}(T)$ is the constant pressure specific heat capacity.

The corresponding exit species mole fractions are calculated by the species conservation equation:

$$N\frac{d(X_i)}{dt} = \frac{d(N_{\text{in}})}{dt} X_{i,\text{in}} - \frac{d(N_{\text{out}})}{dt} X_{i,\text{out}} + R_i,$$
$$R_{H_2} = -\frac{iS_{\text{node}}}{F}, \quad (11)$$

where R_i is the reaction rate of individual species i and the R_{H_2} is the reaction rate of H_2. The exit molar flow rate is determined from the mass conservation equation:

$$\frac{d(N_{\text{out}})}{dt} = \frac{d(N_{\text{in}})}{dt} + \sum R_i. \quad (12)$$

Table 1: Comparison between the model and experiments.

	5000 W			3500 W		
	Model	Experiments		Model	Experiments	
Cell	130	24	22	130	24	22
P_{out} [W]	5000	923	846	3500	646	592
V_{cell} [V]	0.650	0.681	0.639	0.800	0.801	0.768
I_{stack} [A]	59.17	56.54	60.02	33.65	33.62	34.99
T_{Stack} [K]	1021	1023	1023	1006	1023	1023
T_{Air} [K]	912	923	923	937	923	923

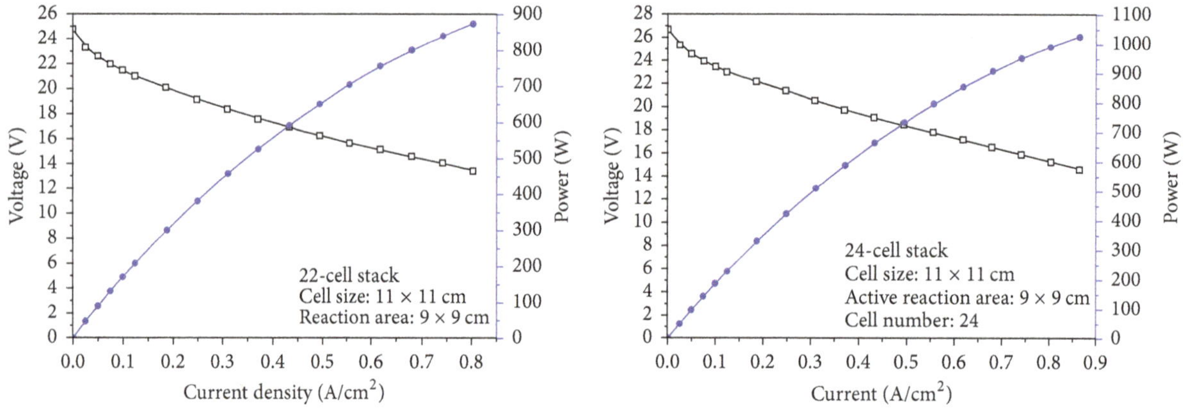

Figure 3: Stack testing data.

2.3. Model Validation. All the component submodels in the system model are built based on the mass and energy conservation law. All the model parameters are the same with the actual equipment to ensure the reliability of the SOFC system model. However, due to the influence of various factors and the existence of errors in the actual system, there will be a certain gap between the theoretical model and the actual system. In order to ensure the correctness of the system model building method, it is necessary to carry out the necessary experimental verification to fully demonstrate that the system model can describe the working characteristics of the real physical system. The accuracy of the model is of great significance to the development and design of practical system.

To collect experimental data on the stack, two units were assembled in our laboratory, including 22 cells and 24 cells of 11 $*$ 11 cm size battery with 9 $*$ 9 cm reaction area for 1 kW power supply. The stacks are tested with excess air and hydrogen in a constant temperature reheating furnace. The average operating temperature of the stack is 750°C and the gas temperature at the entrance of the stack is 650°C. The electrical properties of the two stacks are shown in Figure 3. Although the object of this study is the 5 kW SOFC system, the stack consists of 130 battery chips with the same performance, which can be regarded as five 1 kW stacks in series. The experimental data of 1 kW stack can be obtained and compared with the simulation data by scaling the operating points of 130 batteries to 22 and 24 batteries. The comparison between the model and the testing data are shown in Table 1.

2.4. Configuration and Feasibility of the AOR Structure. In this decoupling control-based study, three functions are demanded for the AOR structure to decouple the interactions and manage load-following problems:

(1) Limit the amount of fuel entering the burner.

(2) Safely and efficiently recycle excessive offgas fuel to the stack.

(3) Act as a buffer between the fuel consumption and fuel supply by storing some fuel in the AOR structure to supply the stack when needed.

2.4.1. AOR Configuration. Different recycle structures, such as gas recirculation ejector and blowers, have been used in recent studies to increase the power output and system efficiency. The reformer does not exist in the hydrogen-fueled SOFC system; therefore, a condensation is needed to remove the water gas from the offgas [31]. As shown in Figure 4, the AOR structure consists of a splitter, condensation, and an AOR blower and mixer. The condensation is the key element in the AOR where the anode offgas is collected, condensed, stored, and transported to the stack by the AOR blower. The splitter forces part of the offgas to enter the recirculation pipeline. During such time, the offgas contains H_2, which is needed to reenter the stack, and steam, which should be removed. Condensation is utilized to remove H_2O and store H_2 in the AOR structure. The function of the condensation is to remove the water gas down to 333 K (60°C) through the cooling water. The offgas only keeps the H_2, which is stored in the condensation. The mixer is utilized to mix the cold and hot fuel and maintain the anode pressure.

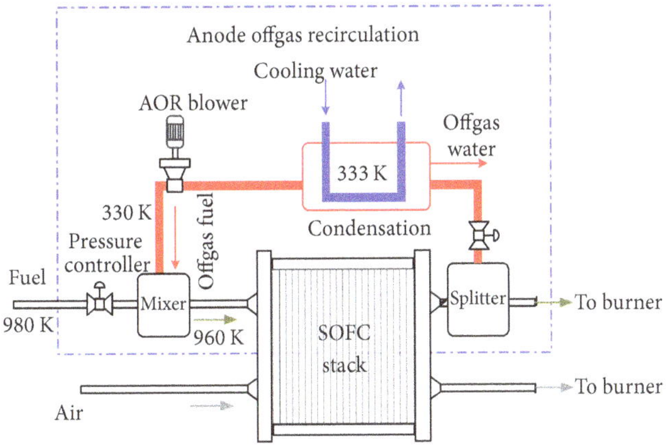

FIGURE 4: The anode offgas recycle structure.

2.4.2. AOR Feasibility. In this study, the AOR structure is designed to recycle the hydrogen to the stack inlet and limit the excessive fuel entering the burner, which generates useless high temperature. The feasibility of the proposed AOR structure aims to address four issues.

(1) Offgas Blower Power. The air supplied to the SOFC stack is significantly larger, usually more than five times that of the fuel entering the system [4, 27]. Thus, the air blower accounts for most of the parasitic loss, and the AOR blower power loss does not largely increase the power loss and decrease the system efficiency.

(2) Anode Pressure Controller. The offgas blower sends the offgas fuel to the mixer to mix with the primary fuel supply by a pressure controller. The pressure controller maintains the stack inlet pressure by manipulating the fuel flow valve.

(3) Condensation Volume. The condensation stores some hydrogen, thereby making the stack less sensitive to the abrupt load changes because the fuel in the condensation can be blown into the stack to supply the fuel supply delay period (in this study, the duration is 3 s). If we suppose that the largest load change is 5 kW and the fuel utilization is 80%, then the volume should be more than 6 L. At the system start-up period, the AOR should be open to fill the condensation with sufficient fuel used when the load abruptly increases, and the fuel can be replenished when the load decreases.

(4) Stack Temperature. The condensation cools the offgas down to 333 K to remove the H_2O from the gas. Before the offgas fuel enters the stack, the mixer is applied to prevent the cold fuel gas directly entering the stack from negatively affecting the stack performance. Otherwise, given that the air flow rate is considerably larger than the fuel flow rate, the stack temperature is mainly affected by the air flow rate, and the fuel temperature fluctuation caused by AOR can be regarded as external disturbance, which can be handled by the stack controller, which will be discussed in the next section.

2.5. Operation Parameters. To effectively manage the system-level parameters, some operation parameters should be defined in this section, including fuel utilization (FU), bypass ratio (BP), and system efficiency (SE) [4, 9, 20]. Particularly, the anode offgas recycle ratio (AORR) should be defined to effectively manage the SOFC thermal and electrical parameters, which will result in the SE decrease as compared with the SOFC system without AOR. The operating parameters are defined as follows:

$$\begin{aligned} \mathrm{FU} &= \frac{\text{fuel consumed in stack}}{\text{fuel entering the stack}} = \frac{nI}{2F\left(dN_{\mathrm{H}_2}/dt\right)}, \\ \mathrm{SE} &= \frac{\text{SOFC power}}{\text{stack power}} = \frac{P_{\text{stack}} - P_{\text{blower}} - P_{\text{offgas,blower}}}{P_{\text{stack}}}, \\ \mathrm{BP} &= \frac{\text{air through bypass manifold}}{\text{air through main manifold}} \\ &= \frac{d\left(N_{\text{air,bypass EX}}\right)/dt}{d\left(N_{\text{air}}\right)/dt}. \end{aligned} \tag{13}$$

The AORR is the most important variable in the AOR structure. The recycle rate is defined as the molar fraction of the anode offgas that is recycled.

$$\mathrm{AORR} = \frac{\text{fuel recycling rate}}{\text{anode offgas rate}} = \frac{d\left(N_{\text{recycle}}\right)/dt}{d\left(N_{\text{offgas}}\right)/dt}. \tag{14}$$

The AORR refers to the ratio of the recycled gas in the entire offgas.

3. Decoupling Control Scheme

In this section, the decoupling control scheme of the SOFC system is developed with the help of AOR to simplify the multivariable system to several single-variable systems. In each single-variable system, the output variables are controlled by one manipulation variable. Given that single-variable control is considerably easier than multivariable control, the decoupling control is especially appropriate for SOFC control engineering practice.

3.1. Control Objectives and Variable Parings. The proper control variable pairings for the MIMO system, SOFC system in this context, must be selected prior to the design and implementation of the decoupling controller. The selection of the proper control variable pairing will determine the input-output (IO) pair; then the output variable can be controlled by the input reference variable through the implementation of the SISO controller. All of the SISO controllers in the decoupling control scheme are PID-based controllers. Given its simple structure and good performance in engineering practice, the PID controller is suitable for handling complex SOFC systems with robustness and stability [9, 27].

In this study, the voltage is kept at a constant value, and the fuel flow and current are manipulated to control the power. For the electrical parameter, the power can be independently controlled by the current through the implementation of electronic devices, such as the DC/DC (DC/AC) converter:

TABLE 2: Control variable pairings.

Manipulate variables (MV)	Control variables (CV)
Current	Power
Air flow	Stack inlet T
Air bypass ratio (BP)	Stack outlet T
AORR	Burner T

(1) By manipulating the current, the system power can meet the demand [4–8, 16, 27].

(2) By modulating the fuel flow, the FU is maintained and the current is supplied [4, 27].

(3) Limiting the current ensures that the voltage remains at a safe value to prevent the fuel from depleting in the stack [9, 35].

For the electrical parameter, the stack inlet temperature, stack outlet temperature, and burner temperature can be independently controlled.

(4) The air flow rate and BP can be manipulated to maintain the stack operation temperature [4, 16, 27].

(5) The AOR acting as a manipulation variable is the key point to decoupling the interaction of the thermal and electrical parameters. The AORR is used to independently control the burner temperature without being affected by the power fluctuation. All the control variables can be found in Table 2.

The implementation of the selected control variable pairings and the decoupling design are addressed in the following sections.

3.2. Stack Temperature Control. The stack temperature should be kept as the highest priority because the purpose of the SOFC is to supply electricity through electrochemical reaction in the stack. Thus, the safety of the stack is crucial in fuel cell operation. For the SOFC to rapidly and efficiently supply current, its stack temperature must be maintained at a constant level, which will help in strictly controlling the stack thermal gradients and transients. However, several disturbances, including electrochemical reaction heat oscillation and exchanger preheat fluctuation, hamper the stack temperature stabilization.

In this section, the stack temperature controller is designed. The thermal response time has been discussed in [9], in which the thermal response, by manipulating the air flow rate, is more rapid than the inherent thermal response of the stack; thus, the controller can effectively and rapidly manage the stack inlet and outlet temperature. Given that the blower response is significantly faster than the stack thermal response, the air flow can effectively manage the stack temperature. According to the safety requirement, the stack inlet and outlet temperature should be closely maintained near the optimal value (960 and 1,050 K), respectively, during the transient operation period. The stack temperature fluctuation magnitude should be maintained at less than 10 K around the optimal value (OV).

A feedback control system is designed based on the selected control variable pairings, as shown in Figure 5. Two independent feedback loops are employed to achieve

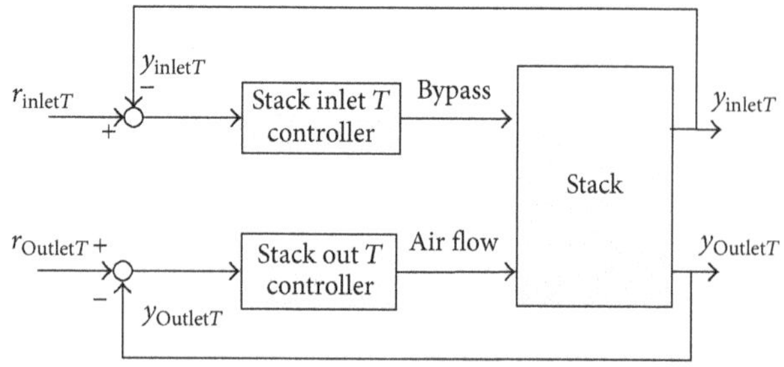

FIGURE 5: Stack temperature controller.

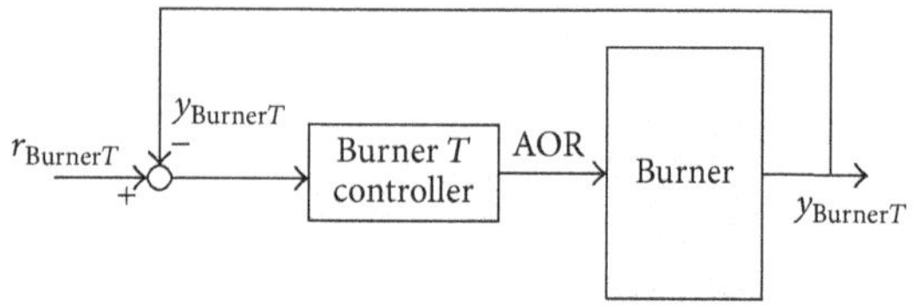

FIGURE 6: New control structure to decouple the burner temperature and the system power: AOR-based independent burner controller.

rapid temperature management. Only when the stack temperature is managed well, can the stack efficiently provide electricity. Given the coupling between the thermal and electrical parameters, the stack inlet temperature cannot be easily managed without affecting the system power. Thus, the stack temperature controller should cooperate with other controllers not only to maintain the system temperature but also to efficiently provide electricity.

3.3. Burner Temperature Control. This section discusses the implementation of the AOR to control the burner temperature without being influenced by electrical parameters. The burner temperature, especially the gas phase temperature, should be maintained because the exhausted gas entering the heat exchangers may negatively affect the stabilization of the stack temperature and the material durability of the burner.

The AOR-based controller contains two parts, namely, burner controller and fuel depletion controller, as shown in Figure 6. In this section, only the burner temperature control actuation is introduced. To maintain the burner temperature ($y_{\text{Burner}T}$), the AOR controls the fuel entering burner, and the excess fuel is recycled to the stack inlet. The fuel flow entering the stack must be decreased because a safe level of stack pressure needs to be maintained. The AOR control actuation in power-level control will be studied in the following section.

3.4. Power Control. For the power control, the voltage is maintained at a constant optimal value, and the power demand can be satisfied by manipulating the stack current using power electronic devices, such as a DC/DC or DC/AC converter. At the same time, as a manipulation variable, the current should be supplied by the fuel. Sufficient fuel must be fed to the stack to avoid fuel depletion and anode reoxidation. The fuel depletion problem should be avoided because it will oxidize the anode material and largely damage the ability to generate fuel cell electricity.

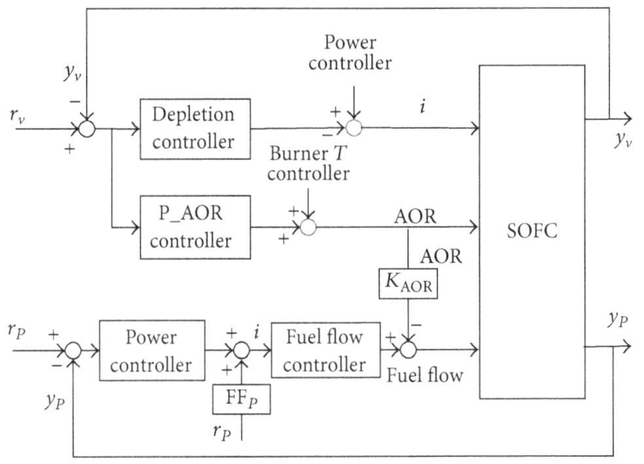

FIGURE 7: Electrical characteristics controller: depletion controller and power controller.

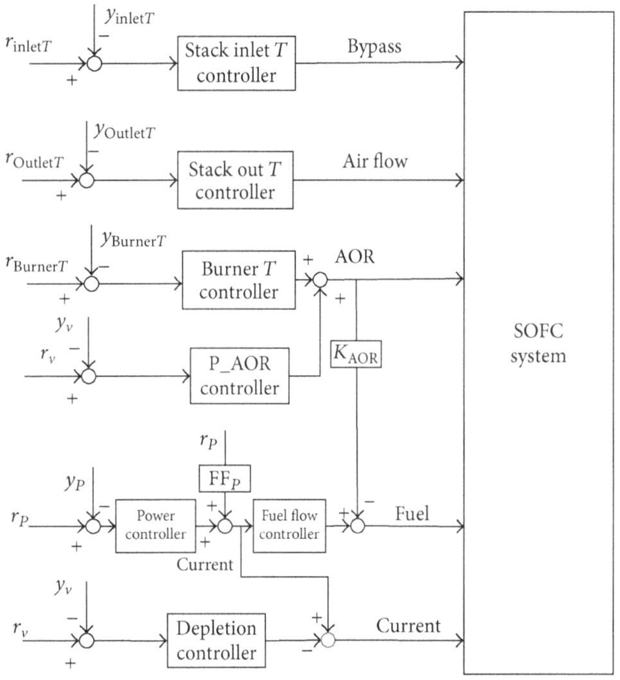

FIGURE 8: Integrate decoupling control.

The fuel supply delay results in the fuel depletion problem and leads to limitation of the load-following ability. The abrupt load increase will result in the stack fuel deficiency period because of the fuel supply delay. The fuel contained in the anode compartment will be consumed during that period. To avoid the fuel depletion, sufficient hydrogen must remain in the stack. As analyzed in Section 2, the fuel stored in the AOR condensation supplies the fuel deficiency period to the stack by the actuation of the AOR blower. Thus, the AOR structure can be regarded as an expansion of the stack compartment. Based on the analysis in [16], increasing the stack compartment size makes the stack less sensitive to the abrupt load change. The fuel contained in the condensation can be replenished when the load decreases. The excessive fuel, which previously enters the burner, is then recycled by the AOR to the condensation, which serves as a fuel supply buffer for the abrupt load changes.

The power control scheme consists of four controllers: depletion controller, AOR controller power actuator part (P_AOR controller), power controller, and fuel flow controller. For the depletion controller, to ensure that the hydrogen is not depleted in the stack, the current should be limited to keep the voltage from dropping. The fuel depletion controller is shown in Figure 7. The AOR controller power actuator part (P_AOR controller) is utilized to blow fuel contained in the AOR structure into the stack. The voltage decreases indicate that the fuel concentration in the stack decreases. For the power controller, the current is used to control the power demand, while the fuel flow should be sufficiently supplied by the fuel flow controller.

As shown in Figure 7, the fuel depletion control manages the voltage through the manipulation of the current by feedback (depletion controller). When the voltage (y_v) decreases and is smaller than the reference voltage (r_v), then the i should be lower to decrease the fuel consumption. At the same time, the AOR rate is increased (P_AOR controller) to blow the fuel contained in the AOR condensation tank to the stack satisfying the fuel demand as well as the load-following demand rapidly. The power is controlled by the power feedback (power controller) and the feedforward controller (FF_P) through the manipulation of the current (i). The AOR feedforward controller (K_{AOR}) helps in maintaining the stack fuel pressure by slightly decreasing the fuel flow. The AOR helps in preventing the fuel cell voltage from dropping to a low level, which ensures enough fuel in the anode compartment by enlarging the stack fuel consumption time. The fuel tank in the AOR structure can be regarded as one part of the stack aside from the second fuel supply element.

3.5. Decoupling Control Overview. To demonstrate the relationship of the control inputs and the outputs, the integrated decoupling control scheme is shown in Figure 8. The controller consists of six parts, namely, stack controller, burner temperature controller, P_AOR controller, power controller, depletion controller, and fuel flow controller. All the control inputs and outputs should be guaranteed within the operation range.

The stack controller manages the BP ratio and air flow to manipulate the stack inlet and outlet temperature, respectively. The burner temperature is managed by the AORR by controlling the fuel entering it. The P_AOR controller also manages the AORR to add the fuel to the stack when the voltage decreases. The fuel depletion controller decreases the current to prevent the voltage from dropping significantly. The power controller manages the system power following the demand rapidly by manipulating the fuel flow rate. The electrical characteristic controller simultaneously handles the rapid load following and the fuel depletion. Moreover, the system-level thermal parameters, such as the stack temperature and the burner temperature, are not influenced by the power changes. Therefore, under the control of the

TABLE 3: Control parameters.

(1) Stack inlet T control	
	K_p 2.1 K_i 0.61
(2) Stack outlet T control	
	K_p 0.3 K_i 0.001
(3) Burner T control	
	K_p 0.8 K_i 0 K_d 0.3
(4) Power control	
	K_p 0.000016 K_i 0.000002 K_d 0.000005 K_{AOR} 0.12 FF_p 0.00002
(5) Depletion control	
	K_P 1.2 K_i 0.01 FF_i 0.024
(6) Fuel flow control	
	K_P 0.21 K_i 0.012

decoupling controller, the interactions among the thermal and electrical characteristics can be decoupled.

4. Simulation Results

This section performs time-domain simulation of the studied AOR-SOFC and the performance of the decoupling control.

All of the SISO controllers in the decoupling control scheme are PID-based controllers due to their simple structure and good performance in engineering practice with robustness and stability. The control parameters of the decoupling control system, as shown in Table 3, are tuned by the Ziegler–Nichols tuning method, which is suitable for engineering practice. To evaluate the performance of the decoupling controller, especially the load-following ability and system temperature independent control, a power-changing case is set. This case assumes that the power demand rapidly changes from 3,500 W to 4,500 W and then decreases to 4,000 W. System power responses under the control of both the conventional controller and the decoupling controller are shown in the following sections. The decoupling performance is assessed from four aspects: power load-tracking characteristics, burner temperature, stack temperature, and system-level operation parameters, namely, fuel utilization (FU) and system efficiency (SE). In this study, the conventional control is the system-level control without the AOR control actuation, which consists of the stack controller, power controller, fuel flow controller, and depletion controller. The conventional control and the decoupling control parameters are the same, as shown in Table 2.

4.1. Power Load-Tracking Characteristics. The power load tracking characteristics are shown in Figure 8. When $0 < t < 100$ s, the power fluctuates because the SOFC needs time to start up from 0 to 3,500 W. When at $t = 1{,}000$ s, the power suddenly rises from 3,500 W to 4,500 W. Both the conventional and decoupling controller manipulate the system to follow the demand. Figure 9(b) shows the local information of the load-following increase. Figure 10 shows the control actuation of the power controller.

The AOR blows the fuel in the condensation into the stack to supply the abrupt load increase, as shown in Figure 10. However, when at $t = 1{,}020$ s, the system power stagnates because the depletion controller decreases the current to avoid stack fuel depletion. After several seconds, the fuel from the fuel source reaches the stack to supply sufficient fuel to the current, thereby causing the power to increase again. The conventional control makes the excessive fuel flow rate increase at the power increase, which may cause useless fuel to enter the burner. Figure 9(c) shows that when $t = 2{,}000$ s, the power demand decreases from 4,500 W to 4,000 W. The excessive fuel is blown by the AOR blower from the stack outlet into the AOR to fill the condensation. As a result, the power under the control of the decoupling controller decreases rapidly to the reference power.

Remark 1. The decoupling controller is better than the conventional controller in the power load following and fuel depletion avoidance because the AOR decreases the fuel deficiency period and makes the fuel supply less sensitive to the power variations. The convergence rate of the AOR controller is faster than the controller without AOR shown in Figures 9(b) and 9(c). As the power fluctuates, the AOR controller takes about 50 s to reach the demand value while the controller without AOR takes about 100 s. That is because when the power demand increases, the AOR recycles fuel in the offgas and transports it to stack again. The fuel intially contained in AOR will enter the stack quickly. Therefore, the power response speed of the AOR controller will be higher than the controller without AOR. For the power decrease condition, the AOR recycle blower helps to decrease the excessive fuel concentration in stack, so the power decrease speed is higher than the traditional controller.

4.2. Burner Temperature. The burner temperature dynamic response during system power changes is shown in Figure 11. The conventional controller causes a large temperature fluctuation, which may damage the burner and cause oscillation in the exchangers and stack temperatures. The AOR-based burner temperature controller maintains the burner temperature by limiting the fuel flow into the burner, as shown in Figure 12. Although the power and fuel flow rate change, the fuel to the burner is nearly maintained at a stable value by the decoupling controller, while the conventional control sends excessive fuel to the burner.

The temperature dynamic responses under the AOR controller are shown in Figure 13. When the power increases, more fuel is consumed in stack to generate current and the heat generated in stack will increase; then less heat will be generated in burner. Therefore, the heat transport to the heat exchanger will decrease. When the power increase, the excessive fuel entering the stack will be recycled, so the burner and the exchangers temperature will not largely change.

4.3. Stack Temperature. Figure 14 shows the stack inlet and outlet temperature under two controllers. The stack outlet temperatures are maintained within 10 K. For the stack inlet temperature, the decoupling controller causes a 10 K decrease because the AOR sends cold (333 K) fuel to the stack inlet.

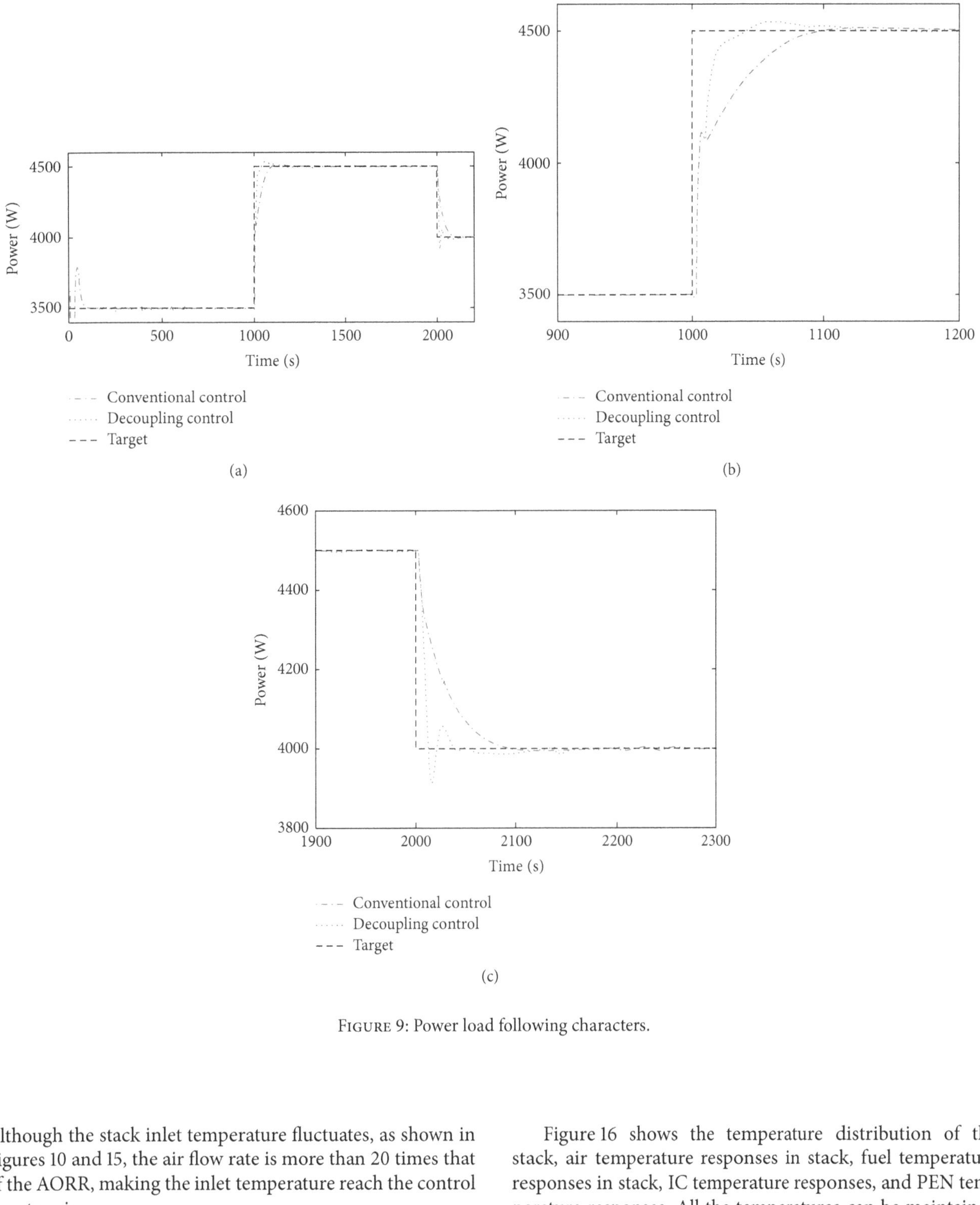

FIGURE 9: Power load following characters.

Although the stack inlet temperature fluctuates, as shown in Figures 10 and 15, the air flow rate is more than 20 times that of the AORR, making the inlet temperature reach the control target again.

Remark 2. Figure 15 shows that the air flow rate and BP rate of the decoupling control significantly decrease because the stabilization of the burner temperature makes it unnecessary for the stack temperature controller to utilize excessive control actuation. Decreasing the BP rate leads to a decrease in the cold air flow of the bypass heat exchangers, which positively influences the system efficiency.

Figure 16 shows the temperature distribution of the stack, air temperature responses in stack, fuel temperature responses in stack, IC temperature responses, and PEN temperature responses. All the temperatures can be maintained within 15 K level, despite the system power increase and decrease. According to Figure 16, the stack inlet and outlet temperatures can be kept at stable level.

Remark 3. The stack temperatures are mainly affected by the air temperatures because the air entering the stack is the main heat source for the stack operation temperature and the air flow rate is several times more than the fuel flow rate.

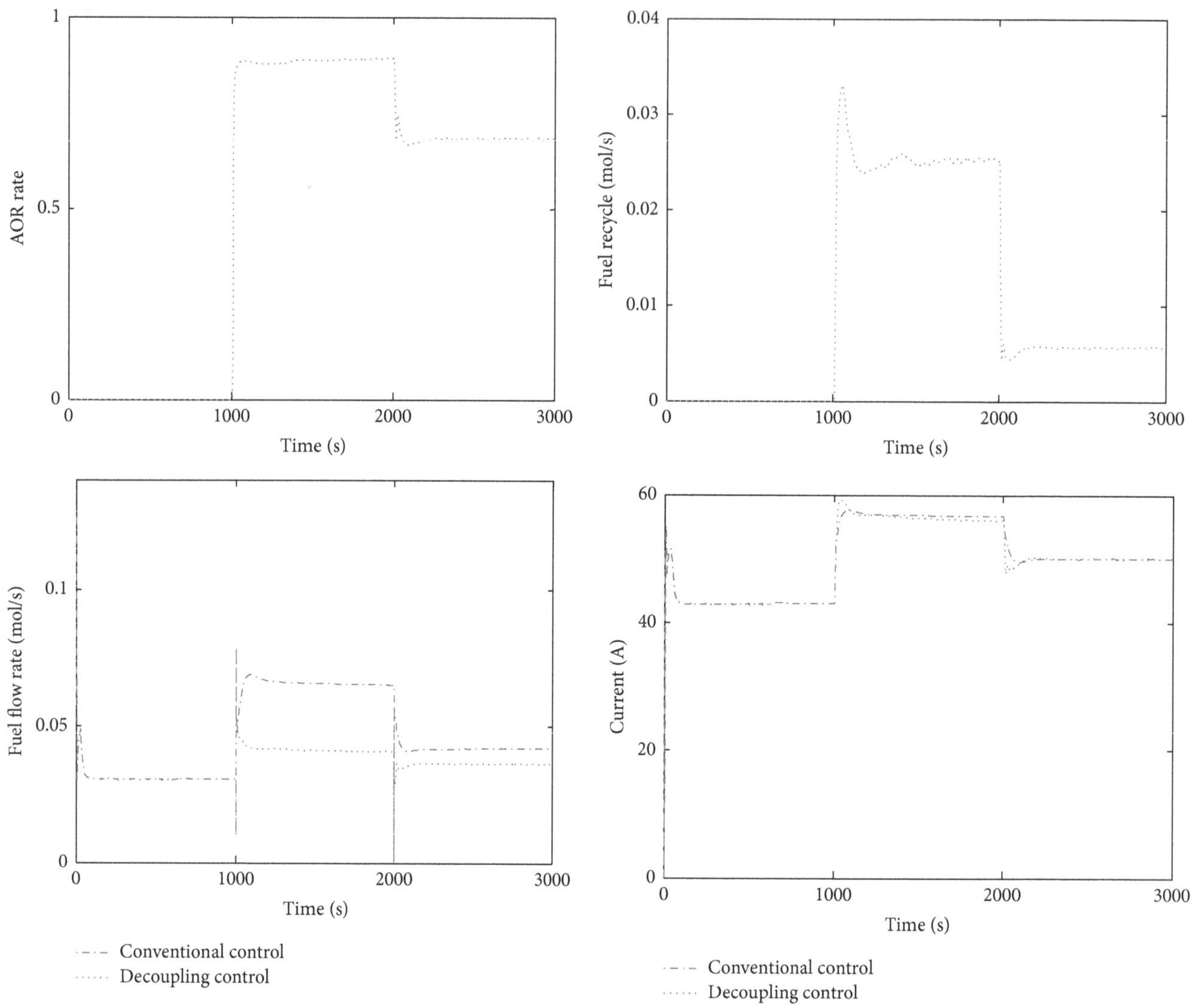

FIGURE 10: Control actuation of the decoupling power controller.

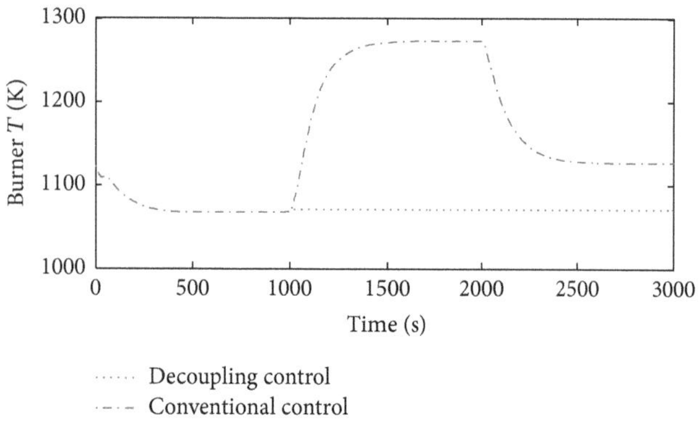

FIGURE 11: Burner temperature dynamic response.

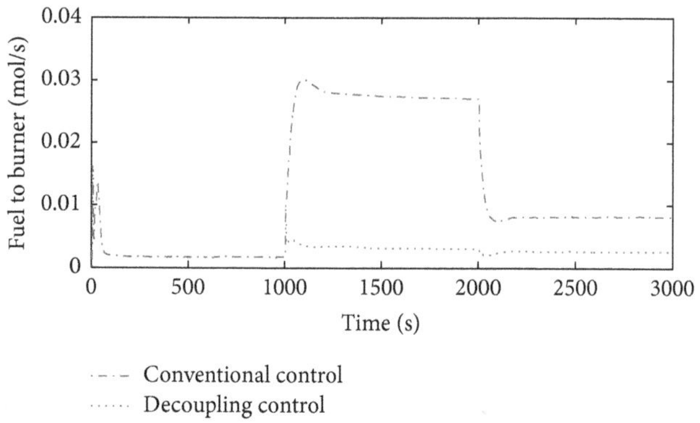

FIGURE 12: Fuel flow rate to burner.

According to Figure 16, the stack temperature distribution is nearly the linear growth along the gas flow direction. As a result, when the stack inlet temperature and the stack outlet temperature are controlled at the stable value, the whole stack temperature can be maintained at a constant value to some extent.

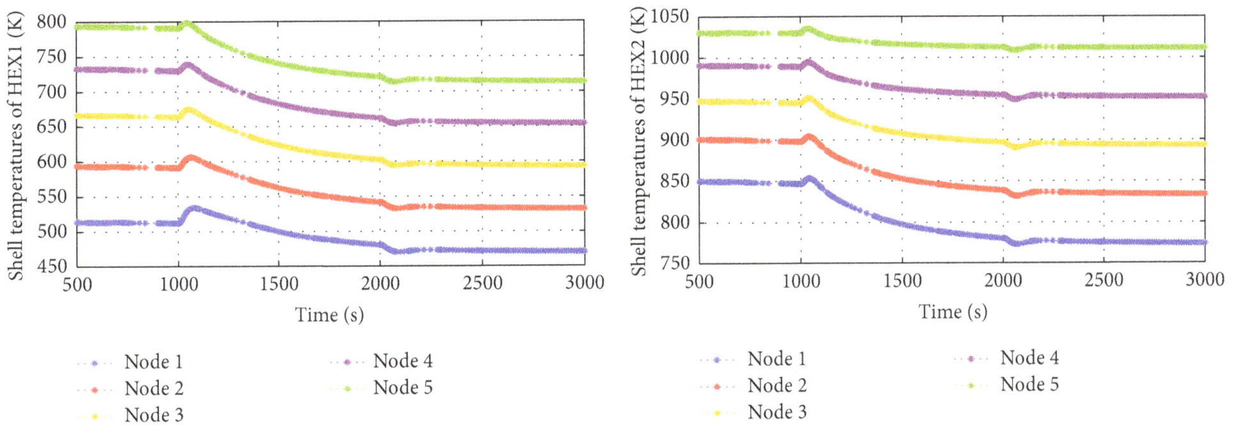

FIGURE 13: Heat exchangers temperature.

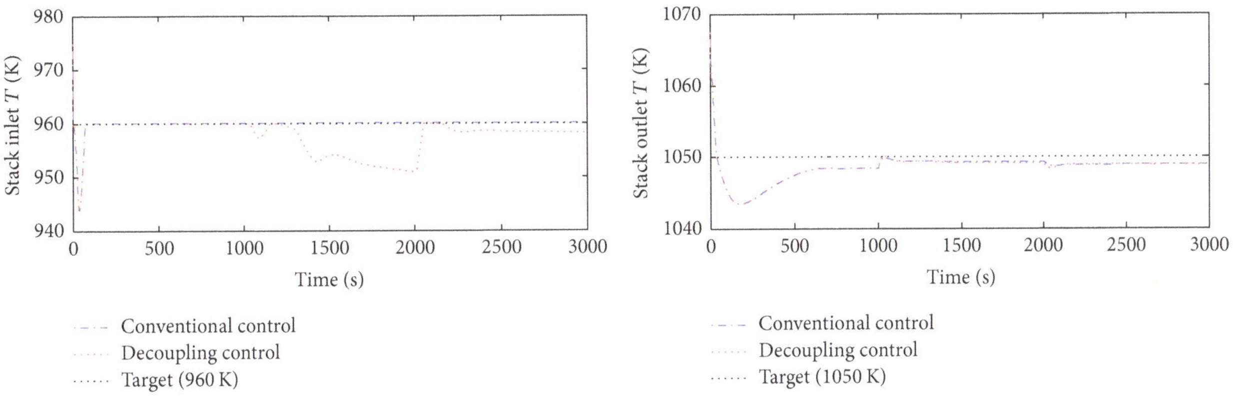

FIGURE 14: Stack temperature under conventional and decoupling controller.

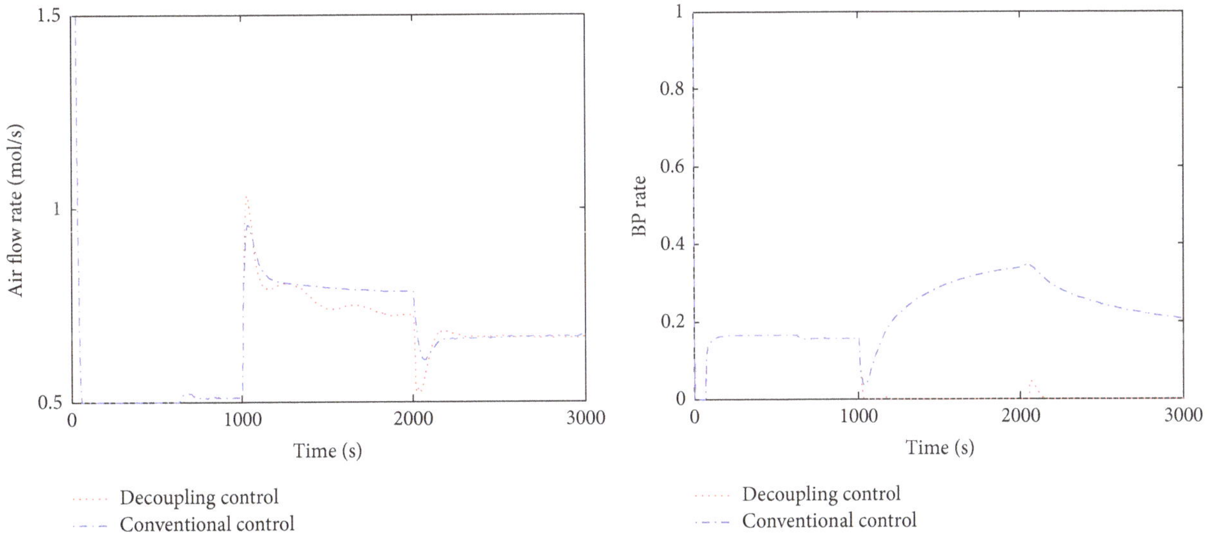

FIGURE 15: Stack temperature control actuation.

4.4. FU and SE. FU and SE are two important parameters for indicating the optimal working condition of the SOFC system. In Figure 17, the FU controlled by the conventional controller is influenced by the power changes. In other words, the interaction between the power and fuel flow rate prevents the SOFC from working at a high efficiency. Therefore, the FU and SE cannot remain steady under the control of the conventional scheme. However, the decoupling controller

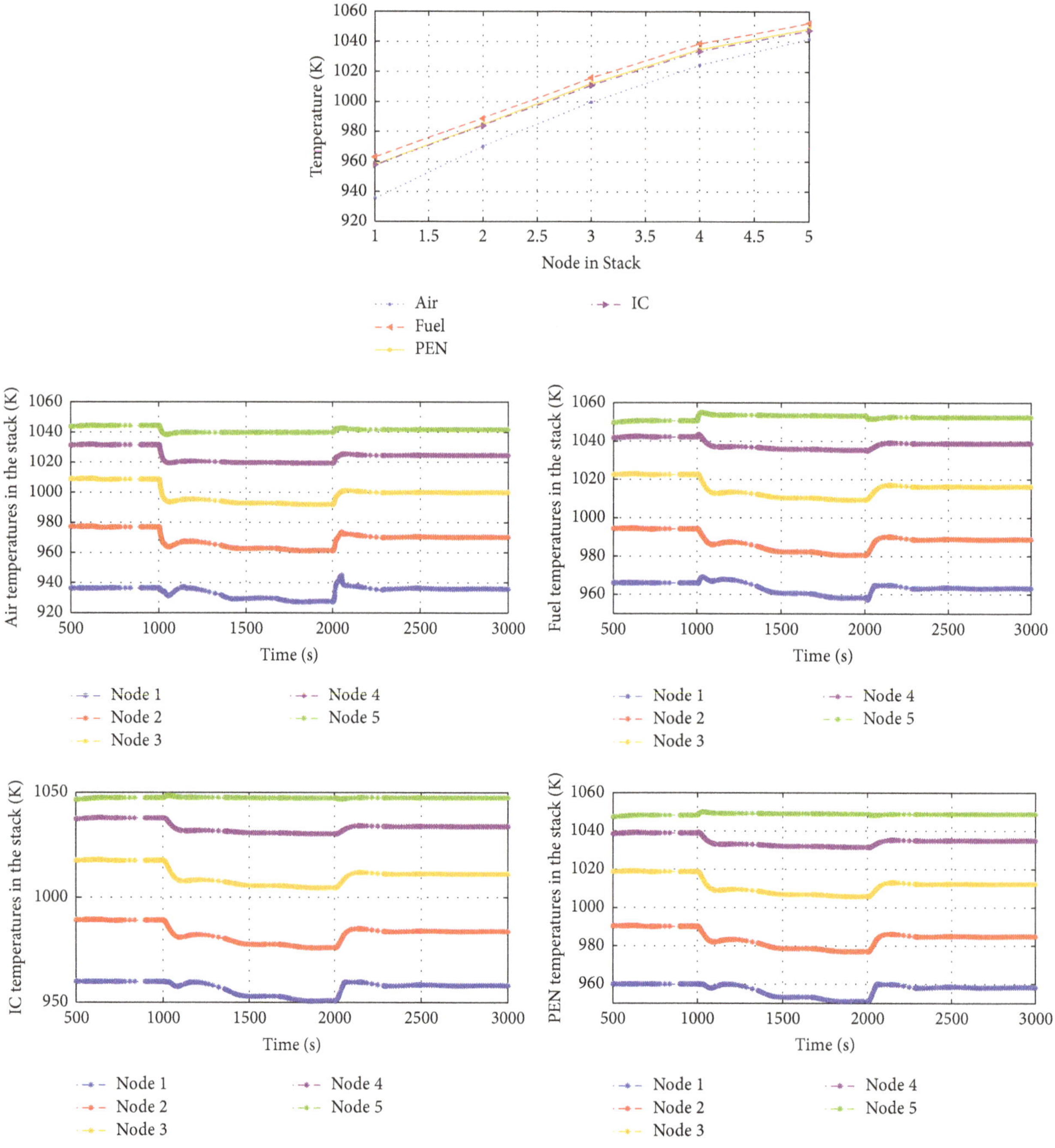

FIGURE 16: Stack temperatures.

independently manages the electrical characteristics without being affected by the fuel flow. As a result, as shown in Figure 17, the FU and SE are almost maintained at a high level even as the power changes.

Figure 18 shows that the power of the AOR blower does not play an important role in the SE compared with the air blower power.

Remark 4. The FU indicates the fuel consuming rate used in stack. If more fuel is consumed in stack, it means the electrochemical reaction ratio is at a high level. The SE indicates that the power ratio can be used by the load. Figure 17 shows that the AOR controller can maintain the FU and the SE at a high level even at the power-changing condition, while the controller without AOR will not perform well. The FU first decreases at the power increase, because when the power demand increases, more fuel will be added to stack that will cause the FU decrease. For the SE, when the system power demand increases, the fuel consumed in stack will increase as well as the heat generated in stack, so more air should be added to stack to cool the temperature down and the parasitic power increase; then the SE will decrease.

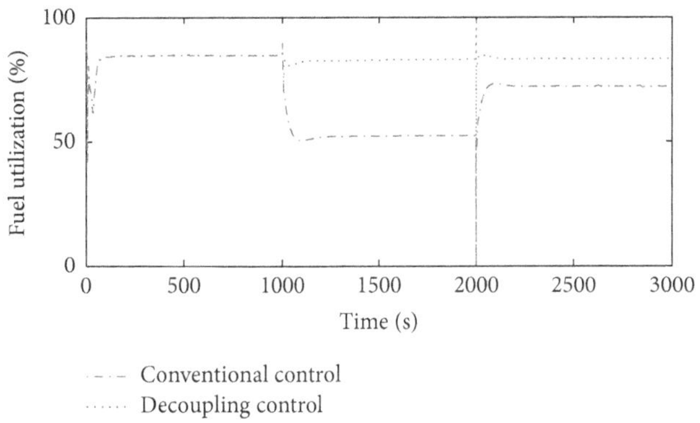

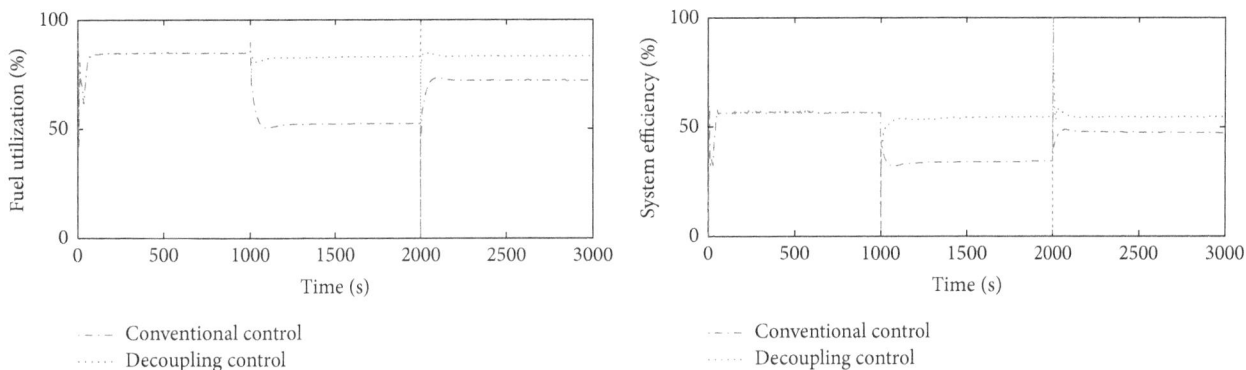

FIGURE 17: Fuel utilization and system efficiency under two controllers.

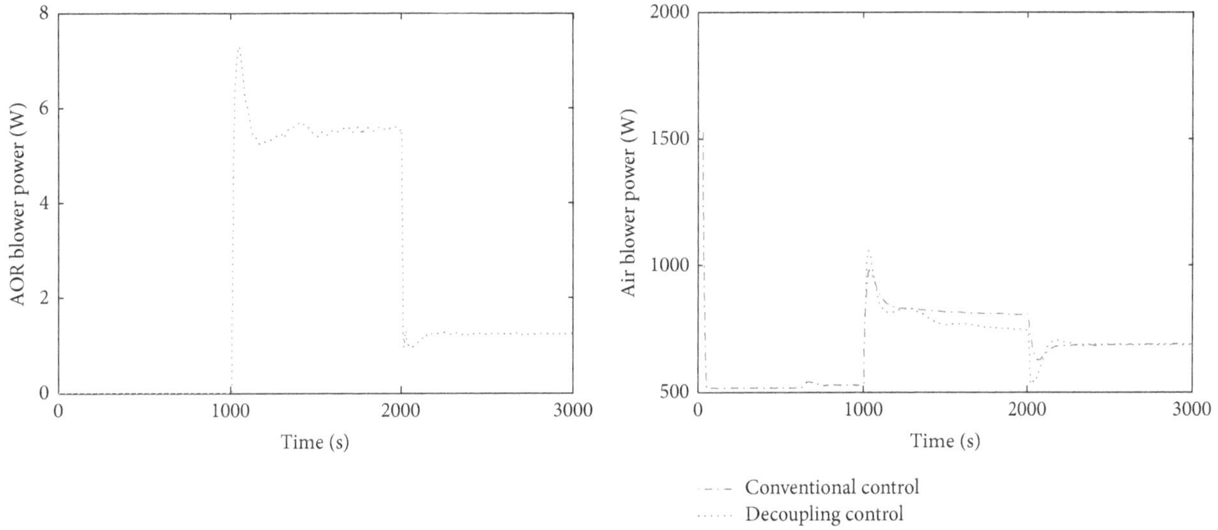

FIGURE 18: The power of AOR blower and the air blower.

Therefore, the AOR controller not only helps to control the system temperature but also controls the fuel utilization in stack and system efficiency.

5. Discussion and Conclusion

This study proposes an engineering-oriented decoupling controller to solve the following problems:

(1) The couplings between the electrical and thermal parameters are decoupled by the AOR-based decoupling control method. The system temperature can be independently controlled without being affected by the power changes.

(2) Fuel depletion avoidance and rapid load following. The fuel consumption time is significantly increased to avoid the fuel depletion problem. The stack fuel supply is less sensitive to the power variations.

(3) The decoupling control system can maintain the FU and SE at a high level even as the power changes.

Other problems, such as SOFC stack fault accommodation and fault reconfiguration control, can also be solved in the future through the decoupling-type modified control method based on the AOR. The control problems that SOFC combines with the MgH_2 tank can also be investigated.

Abbreviations

C: Specific heat capacity, kJ kg^{-1} K^{-1}
F: Faraday constant = 96485 C mol^{-1}
h_{gs}: Convection heat transfer coefficient, kW m^{-2} K^{-1}
I: Current, A
i: Current density, A cm^{-2}
N: Control volume mole number, mol
Q: Heat transfer, W
R_i: Reaction rate of species i, mol s^{-1}
S: Surface area of the heat transfer, m^2
T: Absolute temperature, K
W: Work, W
X_i: Species mole fractions i
α: Transfer coefficient
ρ: Density, kg m^{-3}.

Subscripts

act: Activation
con: Concentration
cond: Conduction
conv: Convection
H_2: Hydrogen
P: Pressure
S: Solid
OCV: Irreversible open circuit voltage
O_2: Oxygen
L: Limit.

Conflicts of Interest

The authors declare that they have no conflicts of interest.

Acknowledgments

This work was supported by the National Nature Science Foundation of China (nos. 61273161 and 51777122).

References

[1] J. Lee, J. Jo, S. Choi, and S.-B. Han, "A 10-kW SOFC low-voltage battery hybrid power conditioning system for residential use," *IEEE Transactions on Energy Conversion*, vol. 21, no. 2, pp. 575–585, 2006.

[2] C. M. Colson and M. H. Nehrir, "Evaluating the benefits of a hybrid solid oxide fuel cell combined heat and power plant for energy sustainability and emissions avoidance," *IEEE Transactions on Energy Conversion*, vol. 26, no. 1, pp. 140–148, 2011.

[3] L. Schlapbach and A. Züttel, "Hydrogen-storage materials for mobile applications," *Nature*, vol. 414, no. 6861, pp. 353–358, 2001.

[4] E. I. Gkanas and S. S. Makridis, "Effective thermal management of a cylindrical MgH2 tank including thermal coupling with an operating SOFC and the usage of extended surfaces during the dehydrogenation process," *International Journal of Hydrogen Energy*, vol. 41, no. 13, pp. 5693–5708, 2016.

[5] H. Cao, X. Li, Z. Deng, J. Li, and Y. Qin, "Thermal management oriented steady state analysis and optimization of a kW scale solid oxide fuel cell stand-alone system for maximum system efficiency," *International Journal of Hydrogen Energy*, vol. 38, no. 28, pp. 12404–12417, 2013.

[6] C. Wang and M. H. Nehrir, "Load transient mitigation for stand-alone fuel cell power generation systems," *IEEE Transactions on Energy Conversion*, vol. 22, no. 4, pp. 864–872, 2007.

[7] C. Wang and M. H. Nehrir, "Short-time overloading capability and distributed generation applications of solid oxide fuel cells," *IEEE Transactions on Energy Conversion*, vol. 22, no. 4, pp. 898–906, 2007.

[8] M. Fardadi, F. Mueller, and F. Jabbari, "Feedback control of solid oxide fuel cell spatial temperature variation," *Journal of Power Sources*, vol. 195, no. 13, pp. 4222–4233, 2010.

[9] S. A. Hajimolana, M. A. Hussain, W. M. A. W. Daud, M. Soroush, and A. Shamiri, "Mathematical modeling of solid oxide fuel cells: A review," *Renewable & Sustainable Energy Reviews*, vol. 15, no. 4, pp. 1893–1917, 2011.

[10] S. Kakaç, A. Pramuanjaroenkij, and X. Y. Zhou, "A review of numerical modeling of solid oxide fuel cells," *International Journal of Hydrogen Energy*, vol. 32, no. 7, pp. 761–786, 2007.

[11] F. Arpino, M. Dell'Isola, D. Maugeri, N. Massarotti, and A. Mauro, "A new model for the analysis of operating conditions of micro-cogenerative SOFC units," *International Journal of Hydrogen Energy*, vol. 38, no. 1, pp. 336–344, 2013.

[12] A. Mauro, F. Arpino, N. Massarotti, and P. Nithiarasu, "A novel single domain approach for numerical modelling solid oxide fuel cells," *International Journal of Numerical Methods for Heat & Fluid Flow*, vol. 20, no. 5, pp. 587–612, 2010.

[13] A. Mauro, F. Arpino, and N. Massarotti, "Three-dimensional simulation of heat and mass transport phenomena in planar SOFCs," *International Journal of Hydrogen Energy*, vol. 36, no. 16, pp. 10288–10301, 2011.

[14] R. Nishida, S. Beale, J. Pharoah, L. de Haart, and L. Blum, "Three-dimensional computational fluid dynamics modelling and experimental validation of the Jülich Mark-F solid oxide fuel cell stack," *Journal of Power Sources*, vol. 373, pp. 203–210, 2018.

[15] F. Arpino, N. Massarotti, A. Mauro, and L. Vanoli, "Metrological analysis of the measurement system for a micro-cogenerative SOFC module," *International Journal of Hydrogen Energy*, vol. 36, no. 16, pp. 10228–10234, 2011.

[16] F. Mueller, F. Jabbari, R. Gaynor, and J. Brouwer, "Novel solid oxide fuel cell system controller for rapid load following," *Journal of Power Sources*, vol. 172, no. 1, pp. 308–323, 2007.

[17] L. Wang and D.-J. Lee, "Load-tracking performance of an autonomous SOFC-based hybrid power generation/energy storage system," *IEEE Transactions on Energy Conversion*, vol. 25, no. 1, pp. 128–139, 2010.

[18] K. Sedghisigarchi and A. Feliachi, "Dynamic and transient analysis of power distribution systems with fuel cells - Part I: Fuel-cell dynamic model," *IEEE Transactions on Energy Conversion*, vol. 19, no. 2, pp. 423–428, 2004.

[19] K. Sedghisigarchi and A. Feliachi, "Dynamic and transient analysis of power distribution systems with fuel cells - Part II: Control and stability enhancement," *IEEE Transactions on Energy Conversion*, vol. 19, no. 2, pp. 429–434, 2004.

[20] T. Allag and T. Das, "Robust control of solid oxide fuel cell ultracapacitor hybrid system," *IEEE Transactions on Control Systems Technology*, vol. 20, no. 1, pp. 1–10, 2012.

[21] A. M. Murshed, B. Huang, and K. Nandakumar, "Estimation and control of solid oxide fuel cell system," *Computers & Chemical Engineering*, vol. 34, no. 1, pp. 96–111, 2010.

[22] C. Stiller, B. Thorud, O. Bolland, R. Kandepu, and L. Imsland, "Control strategy for a solid oxide fuel cell and gas turbine hybrid system," *Journal of Power Sources*, vol. 158, no. 1, pp. 303–315, 2006.

[23] Y. Inui, N. Ito, T. Nakajima, and A. Urata, "Analytical investigation on cell temperature control method of planar solid oxide fuel cell," *Energy Conversion and Management*, vol. 47, no. 15-16, pp. 2319–2328, 2006.

[24] M. Sorrentino and C. Pianese, "Model-based development of low-level control strategies for transient operation of solid oxide fuel cell systems," *Journal of Power Sources*, vol. 196, no. 21, pp. 9036–9045, 2011.

[25] A. Pohjoranta, M. Halinen, J. Pennanen, and J. Kiviaho, "Model predictive control of the solid oxide fuel cell stack temperature with models based on experimental data," *Journal of Power Sources*, vol. 277, pp. 239–250, 2015.

[26] X. W. Zhang, S. H. Chan, H. K. Ho, J. Li, G. Li, and Z. Feng, "Nonlinear model predictive control based on the moving horizon state estimation for the solid oxide fuel cell," *International Journal of Hydrogen Energy*, vol. 33, no. 9, pp. 2355–2366, 2008.

[27] H. Cao and X. Li, "Thermal Management-Oriented Multivariable Robust Control of a kW-Scale Solid Oxide Fuel Cell Stand-Alone System," *IEEE Transactions on Energy Conversion*, vol. 31, no. 2, pp. 596–605, 2016.

[28] B. Huang, Y. Qi, and M. Murshed, "Solid oxide fuel cell: Perspective of dynamic modeling and control," *Journal of Process Control*, vol. 21, no. 10, pp. 1426–1437, 2011.

[29] X.-J. Wu, Q. Huang, and X.-J. Zhu, "Power decoupling control of a solid oxide fuel cell and micro gas turbine hybrid power system," *Journal of Power Sources*, vol. 196, no. 3, pp. 1295–1302, 2011.

[30] D. Zhao, Q. Zheng, F. Gao, D. Bouquain, M. Dou, and A. Miraoui, "Disturbance decoupling control of an ultra-high speed centrifugal compressor for the air management of fuel cell systems," *International Journal of Hydrogen Energy*, vol. 39, no. 4, pp. 1788–1798, 2014.

[31] R. Peters, R. Deja, M. Engelbracht et al., "Efficiency analysis of a hydrogen-fueled solid oxide fuel cell system with anode off-gas recirculation," *Journal of Power Sources*, vol. 328, pp. 105–113, 2016.

[32] M. Carré, R. Brandenburger, W. Friede, F. Lapicque, U. Limbeck, and P. Da Silva, "Feed-forward control of a solid oxide fuel cell system with anode offgas recycle," *Journal of Power Sources*, vol. 282, pp. 498–510, 2015.

[33] R. Torii, Y. Tachikawa, K. Sasaki, and K. Ito, "Anode gas recirculation for improving the performance and cost of a 5-kW solid oxide fuel cell system," *Journal of Power Sources*, vol. 325, pp. 229–237, 2016.

[34] M. Engelbracht, R. Peters, L. Blum, and D. Stolten, "Comparison of a fuel-driven and steam-driven ejector in solid oxide fuel cell systems with anode off-gas recirculation: Part-load behavior," *Journal of Power Sources*, vol. 277, pp. 251–260, 2015.

[35] R. Gaynor, F. Mueller, F. Jabbari, and J. Brouwer, "On control concepts to prevent fuel starvation in solid oxide fuel cells," *Journal of Power Sources*, vol. 180, no. 1, pp. 330–342, 2008.

15

Fuzzy Fractional-Order PID Controller for Fractional Model of Pneumatic Pressure System

M. Al-Dhaifallah[1,2] **N. Kanagaraj,**[1] **and K. S. Nisar**[3]

[1]*Electrical Engineering Department, College of Engineering at Wadi Addawasir, Prince Sattam Bin Abdulaziz University, Wadi Addawasir, Saudi Arabia*
[2]*Systems Engineering Department, King Fahd University of Petroleum and Minerals, Dhahran, Saudi Arabia*
[3]*Department of Mathematics, College of Arts & Science at Wadi Addawasir, Prince Sattam Bin Abdulaziz University, AlKharj, Saudi Arabia*

Correspondence should be addressed to K. S. Nisar; ksnisar1@gmail.com

Academic Editor: Hung-Yuan Chung

This article presents a fuzzy fractional-order PID (FFOPID) controller scheme for a pneumatic pressure regulating system. The industrial pneumatic pressure systems are having strong dynamic and nonlinearity characteristics; further, these systems come across frequent load variations and external disturbances. Hence, for the smooth and trouble-free operation of the industrial pressure system, an effective control mechanism could be adopted. The objective of this work is to design an intelligent fuzzy-based fractional-order PID control scheme to ensure a robust performance with respect to load variation and external disturbances. A novel model of a pilot pressure regulating system is developed to validate the effectiveness of the proposed control scheme. Simulation studies are carried out in a delayed nonlinear pressure regulating system under different operating conditions using fractional-order PID (FOPID) controller with fuzzy online gain tuning mechanism. The results demonstrate the usefulness of the proposed strategy and confirm the performance improvement for the pneumatic pressure system. To highlight the advantages of the proposed scheme a comparative study with conventional PID and FOPID control schemes is made.

1. Introduction

Pneumatic pressure is one among the vital variables used in industries like power plants, chemical reaction control, pneumatic position servo systems, well drilling, heating, ventilating and air conditioning systems, automobile, and so on. The dynamic characteristics of pneumatic pressure plants are highly nonlinear due to the compressibility of air, load variations, and external disturbances. Further, the industrial pneumatic pressure plants are usually interconnected and operating at different pressure level. Therefore, the precise control of pressure plant is complex due to the presence of uncertainties and nonlinearity. Hence, an efficient control strategy is needed for trouble-free operation of the pneumatic system in industries. The classical PI and PID controllers are widely used in industrial applications in the past because of its advantages. PID control is a simple and effective control method and it can be easily implemented for industrial control applications. However, the PID control algorithm is not advisable for the complex and nonlinear system. On the other hand, the fractional calculus is getting much more attention in the field of control system engineering due to its potential and significant importance [1–3]. The controllers, making use of fractional-order derivatives and integrals, give improved results compared to the classical controllers in terms of robustness [4–6]. Fractional-order (FO) controllers are usually expressed by fractional-order differential equations. The FO controllers are derived from the integer order by adding the fractional powers in integral and derivative terms. For example, in addition to the proportional (K_P), integral (K_I), and derivative (K_D) parameters which comprise the integer-order PID, the FOPID controller has two more parameters an integrator order (λ) and a differentiator order (μ). Adding the integral and derivative terms of fractional order will improve system frequency response to be better and leads to design an improved control system [7–10]. The FOPID control scheme has certain merits whereby it offers five parameters to be tuned. However, this control scheme has its own demerits as it makes the system more complex than the classical one.

Recent research trends in FO control are looking towards using fuzzy with FO control scheme to improve the control performances. The rule base fuzzy set theory provides more flexibility in designing complex industrial control system. In fuzzy set theory, linguistic notations are used to express the observations easily to form a control structure. The fuzzy logic controller (FLC) design is becoming simple, even for more complex and nonlinear industrial process without knowledge of the exact mathematical description of the system [11]. Further, FLC is combined with the FO controller for fine-tuning parametric gains and guarantee optimal performance owing to nonlinearities, load disturbances, and plant parameters variations [12–14]. The adaptive method provided by the fuzzy system will improve the dynamic performances of the FO controller through which the controller may respond quickly to parameter variation.

By considering all these aspects, a fuzzy-based FOPID control scheme is proposed for faster response and better control performances. To demonstrate the performances of proposed control technique, a novel pneumatic pressure system model is developed and the system performances are studied under load disturbances and changes in set-point conditions. This paper is organized as follows. In Section 2, the mathematical background of fractional calculus is discussed. Section 3 gives an overview of the experimental setup for pneumatic pressure control system. The modeling of the proposed system is discussed in Section 4. Section 5 describes the controller design for the proposed system. Section 6 presents the simulation results. The findings are given as a conclusion in Section 7, followed by the references.

2. Mathematical Background of Fractional Calculus

Fractional-order calculus (FOC) is one of the popular and emerging mathematics branches that deals with differentiation and integration of real or complex order [1, 2, 15]. The fractional-order calculus provides efficient tools for many situations related to the fractal dimension, infinite memory, and chaotic behavior. Recently, FOC attracted various researchers due to its application in electronics, Bioengineering, control theory, and many more areas [7–10]. Fractional-order mathematical phenomena are very useful to describe and model real-time system more accurately than the conventional integer methods. The fractional-order differentiator can be denoted as a continuous differintegral operator [1, 2, 16–18], which is given by

$$aD_t^\gamma = \begin{cases} \dfrac{d^\gamma}{dt^\gamma}, & \Re(\gamma) > 0 \\ 1, & \Re(\gamma) = 0 \\ \displaystyle\int_a^t (d\tau)^{-\gamma}, & \Re(\gamma) < 0, \end{cases} \qquad (1)$$

where γ is the order of the differintegration and a is constant related to initial conditions.

The most commonly used definitions in FOC are Riemann-Liouville (e.g., in calculus), Caputo (e.g., in numerical integration and physics), and Grunwald-Letnikov (e.g., communications and control).

The definitions due to Riemann-Liouville and Caputo are, respectively, given by

$$aD_t^\gamma f(t) = \frac{1}{\Gamma(n-1)} \frac{d^n}{dt^n} \int_a^t \frac{f(\tau)}{f(t-\tau)^{\gamma-n+1}} d\tau,$$
$$aD_t^\gamma f(t) = \frac{1}{\Gamma(\gamma-n)} \int_a^t \frac{f^{(n)}(\tau)}{f(t-\tau)^{\gamma-n+1}} d\tau, \qquad (2)$$

where $(n-1 < \gamma < n)$ and $\Gamma(\gamma)$ is the familiar gamma function defined by

$$\Gamma(\gamma) = \int_0^\infty e^{-u} u^{\gamma-1} du, \quad \Re(\gamma) > 0. \qquad (3)$$

Here, in this paper the Grunwald-Letnikov definition is used and is expressed by

$${}_aD_\gamma^t f(t) = \lim_{h\to 0} h^{-\gamma} \sum_{j=0}^{[(t-a)/h]} (-1)^j \binom{\gamma}{j} f(t-jh), \qquad (4)$$

where $(n-1 < \gamma < n)$, $[(t-a)/h]$ is integer part, and a, t are the limits of operator. The binomial coefficient is evaluated by the gamma function that generalizes the factorial operator:

$$\binom{\gamma}{j} = \frac{\gamma!}{j!(\gamma-j)!} = \frac{\Gamma(\gamma+j)}{\Gamma(j+1)\Gamma(\gamma-j+1)}. \qquad (5)$$

The equation given in (4) is very useful to obtain a numerical solution of fractional differential equation [19].

FIGURE 1: Photograph of the pneumatic pressure system.

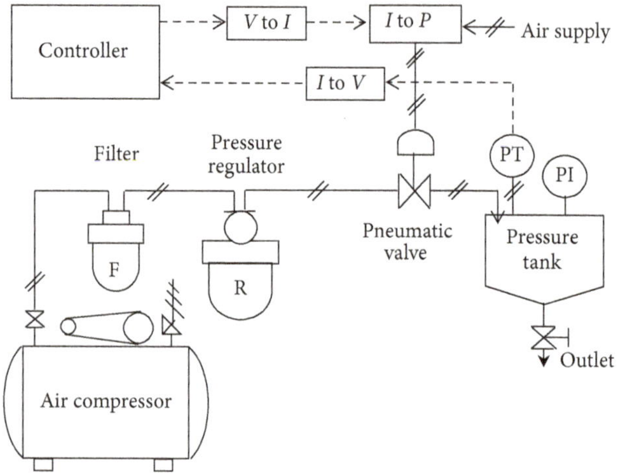

PT: pressure transmitter
PI: pressure indicator
V to I: voltage to current converter
I to V: current to voltage converter
I to P: current to pressure converter

FIGURE 2: Schematic diagram of pneumatic pressure control system.

3. Pneumatic Pressure Control Experimental Setup

The experimental setup of the pneumatic pressure control system is shown in Figure 1. It consists of an air compressor, pressure transmitter, pressure regulator, electropneumatic control valve, pressure indicator, and controller interfacing units. Figure 2 depicts the scheme of the pressure control. The air compressor is attached to the pressure regulator, which supplies the air at constant pressure. An equal percentage electropneumatic control valve of 50 mm size at inlet regulates air flow to the pressure tank. A precise pressure transmitter (PT) attached to the pressure tank measures pressure at each sampling period and gives output as a current signal of 4 to 20 mA. The current signal is converted into a voltage signal in the range of 0 to 5 volts by current to voltage (V to I) converter. The controller block of this setup computes tank pressure using input voltage signal and by applying control algorithm the position of the control valve will be manipulated to keep the tank pressure at the desired level. A 32-bit advanced RISC architecture ARM7 (AT91M55800A) microcontroller is used in the controller part. The microcontroller output voltage magnitude is based on the measured and reference values. The controller output voltage is first converted into a current signal of 4 to 20 mA and then to pressure signal suitably to manipulate the control valve. A pressure indicator (PI) fixed at the top of the pressure tank is used to read the tank pressure manually. Based on the valve characteristics, it is obvious that the pneumatic system possesses inherent nonlinearity and parameters uncertainty because of air compressibility.

4. Modeling of Pneumatic Pressure System

The transient method is commonly used in the industrial process to identify the dynamic model of the system. To obtain the system transient response, the process is initially allowed to run for sufficient time to reach the steady state at the normal operating condition. Then, the controller is disconnected from the loop and an open-loop transient is introduced by a step change input signal and the system response is plotted against time. In case of the proposed pneumatic system, an open-loop transient is introduced by a step change input signal to the control valve. The step change to the control valve is conveniently provided from the controller. The input-output data are recorded at a uniform time interval over a period of time and expressed as

$$\begin{aligned} u &= [u(t), u(2t), u(3t), \ldots, u(Nt)] \\ y &= [y(t), y(2t), y(3t), \ldots, y(Nt)], \end{aligned} \tag{6}$$

where u and y are the input and output measured values, t is sampling time, and Nt is the time of the final measurement. The time-domain data are measured with a sampling period of 0.1 sec. By using the measured time-domain data, the integer-order model of the proposed system is identified by linear ARX regressive system identification method [20]. The identified system model is given in

$$G(s) = \frac{2.87}{44.6s^2 + 105.6s + 1} e^{-1.4s}. \tag{7}$$

In most of the industrial processes, the standard input-output (integer-order) model will be available for experimental study. The system FO model could be obtained easily from the integer-order model. Several tools are available to identify the FO model from the integer-order model. Among them, the Fractional-Order Modeling and Control (FOMCON) toolbox of MATLAB [21] is useful for FO model identification and controller design and optimization. By using FOMCON toolbox, the fractional-order model of the proposed system is identified with Grunwald-Letnikov definition [22] and using the approach described in [23]. To identify the FO model, the fractional-order differential equation is transformed into a fractional-order integral equation. By expanding integral term a least-squares expression is created that allows the implicit time delay term to have an explicit appearance in the parameter vector. This supports for simple estimation of model parameters simultaneously with time delay. The resulting model is given in

$$G(s) = \frac{3.96s^{0.12}}{133s^{1.13} + 105.6s + 1} e^{-1.4s}. \quad (8)$$

5. Controller Design

5.1. Integer-Order PID Controller. The selection of the controller parameters and their optimal values is essential to obtain good control. A good controller is supposed to have minimum overshoot, settling time, and robustness to load disturbances. The Ziegler-Nichols (ZN) controller tuning [24] is applied in the proposed system to identify the controller parameters values. The ZN method is considered to be a better choice for the process of pneumatic pressure control having dead time. In the controller tuning process, three variables, namely, process gain (g_p), dead time (t_d), and time constant (τ), are calculated from the open-loop time-domain plot. From the calculated values, the controller gain, integral time, and derivative time could be obtained for PID controller using the following expressions:

$$K_c = \frac{1.2\tau}{g_p t_d}$$
$$T_i = 2t_d \quad (9)$$
$$T_d = \frac{T_i}{4}.$$

The calculated values from open-loop experimental data are time delay $t_d = 1.4$ s, process gain $g_p = 2.87$, and time constant $\tau = 52$ s. Using (9), the computed values of integer-order PID controller parameters, namely, gain (K_c), integral time (T_i), and derivate time (T_d), are 15.5, 2.8, and 0.7, respectively.

5.2. Fractional-Order PID Controller. Recent research studies have shown that FO controllers could perform better than conventional (integer-order) controllers in terms of system performance and robustness [25, 26]. The application of FO calculus for the dynamic system has been started in the year 1960 [27]. Since then the research on FO control was extended to various fields of engineering. The fractional-order PID controller is a sum of fractional operators along with controller gains. The FOPID controller transfer function representation is expressed as

$$G_c(s) = \frac{u(s)}{e(s)} = K_P + K_I s^{-\lambda} + K_D s^{\mu}, \quad (10)$$

where $G_c(s)$ is controller transfer function, $e(s)$ is error, and $u(s)$ is the output. K_P, K_I, and K_D are the gains for proportional, integral, and derivative terms. The term λ is the fractional component of integral parts and μ is the fractional component of derivative parts. The FOPID controller time-domain representation is given in

$$u(t) = K_P e(t) + K_I D^{-\lambda} e(t) + K_D D^{\mu} e(t). \quad (11)$$

Table 1: Controller parameters of fractional-order PID controller.

K_P	K_I	λ	K_D	μ
6.22	0.09	0.91	0.33	1.33

It is evident that, in FOPID controller, apart from the usual three parameters K_P, K_I, and K_D, the parameters of integral-order λ and derivative-order μ should be considered. Hence, the FOPID controller design procedure consists of solving five nonlinear equations with five unknowns K_P, K_I, K_D, λ, and μ related to the system. On the other hand, the complexity of the five nonlinear equations is very significant, mainly because of the fractional order. By considering the difficulties, the MATLAB with the suitable tool could be a better choice to design the controller. Further, the MATLAB optimization toolbox gives the best solutions with minimum error. The controller design for the proposed system is made using FOMCON toolbox with Oustaloup's rational approximation technique described in [28, 29]. Further, Nelder-Mead optimization technique [30] with integral of square error (ISE) performance metric is applied to optimize controller parameters. For optimization, the following design specifications are selected.

Gain margin = 10 dB.

Phase margin = 60 deg.

Sensitivity function: $|S(j\omega)|_{\text{dB}} \leq -20$ dB, for $\omega \leq \omega_s = 0.01$ rad/s.

Noise rejection: $|T(j\omega)|_{\text{dB}} \leq -20$ dB, for $\omega \leq \omega_t = 10$ rad/s.

With the help of the optimization indices, the calculated fractional-order controller parameters' values are given in Table 1.

5.3. Fuzzy Fractional-Order PID Controller. The rule base fuzzy set theory gives more flexibility in designing systems and expressing the observations in easy-to-follow linguistic notation. Further, the fuzzy logic system performs better in tuning of controller parameters in closed-loop control system, particularly system with nonlinearity between its inputs and outputs. The classical controllers, including fractional-order controller work on the basis of the inputs of errors with a fixed gain value for the proportional, integral, and derivative terms. Hence, the controller performance is not up to the expected level for nonlinear and complex system. An attempt can be made to incorporate dynamic gain value for the proportional, integral, and derivative terms instead of a fixed gain. Dynamically modifying the gain in a FOPID control structure will enhance controller performance and bring the system output quickly to stable condition during load variation and external disturbances. By considering these aspects, a fuzzy logic combined with fractional-order control scheme is proposed in this work. The FFOPID controller is a combination of rule base fuzzy control with FOPID controller. In this control strategy, the FLC is designed to use system error and derivative error inputs to obtain the scaling factor of the proportional, integral, and derivative terms.

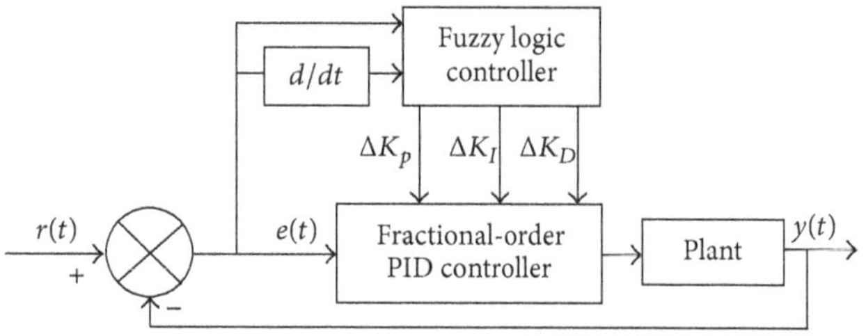

FIGURE 3: Block diagram of the fuzzy fractional-order PID controller.

Using these scaling factors, the controller gain magnitude will be updated at each sampling period. The frame of a typical FFOPID control structure is shown in Figure 3.

In the proposed control structure, the FLC uses the error and derivative error inputs and computes the scaling factor for proportional, integral, and derivative terms and then these values are used to update the gain parameters of FOPID controller. So, the final gain values of k_P, k_I, and k_D for FOPID controller are computed from the following expression:

$$k_P = K_P + \Delta K_P$$
$$k_I = K_I + \Delta K_I \quad (12)$$
$$k_D = K_D + \Delta K_D,$$

where K_P, K_I, and K_D are the initial gain value of FOPID controller and ΔK_P, ΔK_I, and ΔK_D are the scaling factors computed from FLC. The triangle membership functions are used for input and output fuzzy sets and Mamdani-type fuzzy inference system is applied. Three membership functions are selected for the inputs and to produce precise output; five membership functions are chosen for the outputs. The membership functions used in the FLC design are shown in Figure 4; the membership functions are designated with linguistic variables NB (negative big), NS (negative small), N (negative), Z (zero), P (positive), PS (positive small), and PB (positive big). The rule base is the vital part of the FLC design and it relates the input and output linguistic variables based on the current input. Mudi and Pal 1999 [30] recommended the methods for framing rule base using intuitive logic. The rule base used for each output is shown in Table 2. The relationship between inputs and output of each case is shown in surface view, in Figure 5. The surface view specifies how the scaling factor value could vary based on the input error and change-in-error magnitude. From the surface view, it is observed that the input and output have a nonlinear relationship, particularly for the proportional and derivative scaling factors. Further, the surface obviously shows that the values of the proportional, integral, and derivative scaling factors are more for larger amplitude of error and change-in-error. Also, these values are gradually reduced for the smaller error and change-in-error values. The center of gravity defuzzification method is selected to determine the crisp output.

TABLE 2: Fuzzy linguistic rule-base and surface view.

(a)

ΔK_P		Δe	
	NE	Z	P
e			
NE	NB	NS	Z
Z	NB	NS	NS
P	Z	PS	PS

(b)

ΔK_I		Δe	
	NE	Z	P
e			
NE	NB	NS	NS
Z	NS	Z	PS
P	Z	PS	PB

(c)

ΔK_D		Δe	
	NE	Z	P
e			
NE	NS	NS	Z
Z	Z	Z	PS
P	Z	PS	PS

6. Simulation Results

The closed-loop pneumatic pressure system performance under various control schemes is studied using step input, load disturbances, and set-point change. The unit step response of the pressure control system with conventional PID controller, FOPID controller, and FFOPID controller is shown in Figure 6. From the step response result, the system with conventional controller takes much time to reach the desired output. Also, the output has a reasonable overshoot before reaching the steady-state condition. In case of the system with FOPID controller, the system output reaches desired level faster than conventional control scheme but with a small overshoot. On the other hand, the system with fuzzy logic based fractional controller makes system output settle quicker than conventional PID and FOPID control scheme. Due to the online gain modification in FOPID control scheme using fuzzy logic, the gain factor of proportional, integral, and derivate terms is updated at each sampling time which makes the controller perform better. From the step input simulation results, one can easily say that the FFOPID controller is more suitable for pneumatic pressure regulating system with dead time.

To validate the robustness of controller, load disturbance is introduced at the steady-state condition. Figure 7 shows the system performance under three different control schemes with load disturbance at steady-state condition. It is noticed that the system using FFOPID controller reaches steady state faster than the other two methods because the fuzzy system tracks the load disturbances and the related parameter variation easily and updates the gain parameters in FOPID

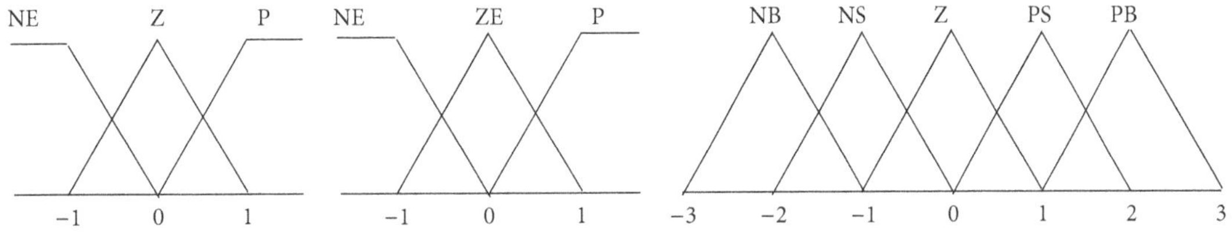

(a) Input (error) (b) Input (change-in-error) (c) Outputs

FIGURE 4: Fuzzy membership functions.

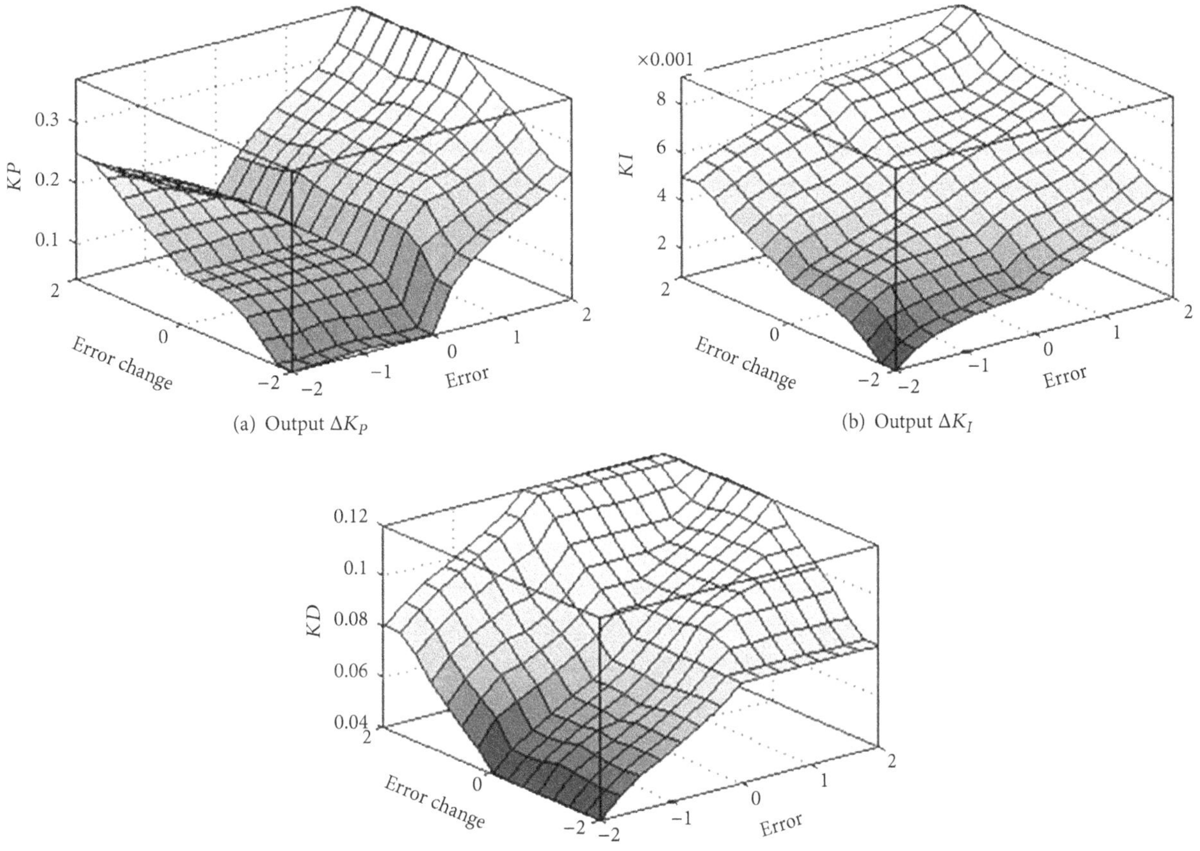

(a) Output ΔK_P

(b) Output ΔK_I

(c) Output ΔK_D

FIGURE 5: Surface views for input versus output.

controller block. Figure 8 depicts the relevant control signals corresponding to load disturbance shown in Figure 7.

The controllers' performance for set-point change is also studied for the proposed pneumatic system. The system response and related control signals for three different control structures are shown in Figures 9 and 10. From the results, it is observed that FFOPID control scheme effectively identify the changes in the set point and modify the controller parameters accordingly to make system output reach a new level with the short span of time as compared to other controllers. Numerical comparative analysis of three different control schemes using performance measures such as settling time, overshoot, integral square error (ISE), and integral absolute error (IAE) are given in Table 3. The performance measures in Table 3 evidently depict that the fuzzy logic combined fractional-order controller outperforms the FOPID and conventional PID controllers in all aspects.

7. Conclusion

In this paper, a rule base fuzzy fractional-order PID controller was designed with online gain changing strategy. To demonstrate the proposed control scheme, a pneumatic pressure tank experimental system model has been developed using open-loop experimental data. Using the system model, the controller performances have been demonstrated for step input, under load variation and set-point change conditions. The results evidently showed that the proposed control

Table 3: Numerical comparison of controllers' performance.

Controller	Settling Time	Over shoot	ISE	IAE
Fuzzy fractional order PID controller	7	0%	2.19	3.20
Fractional order PID controller	12	2%	2.62	3.47
Conventional PID controller	14	6.5%	3.12	4.07

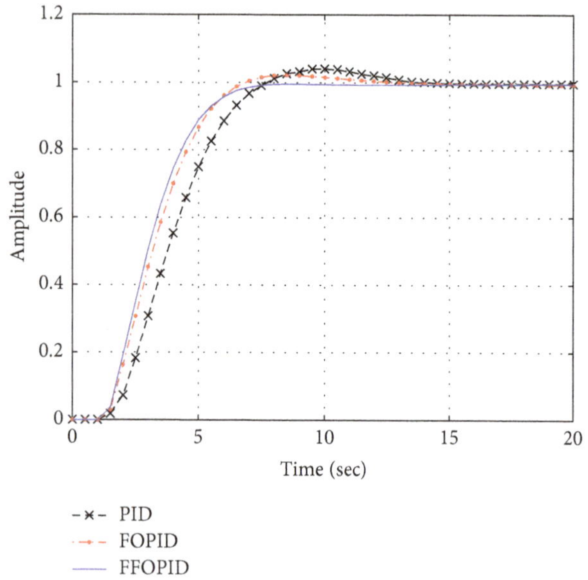

Figure 6: Step response of various controllers for pneumatic pressure control system.

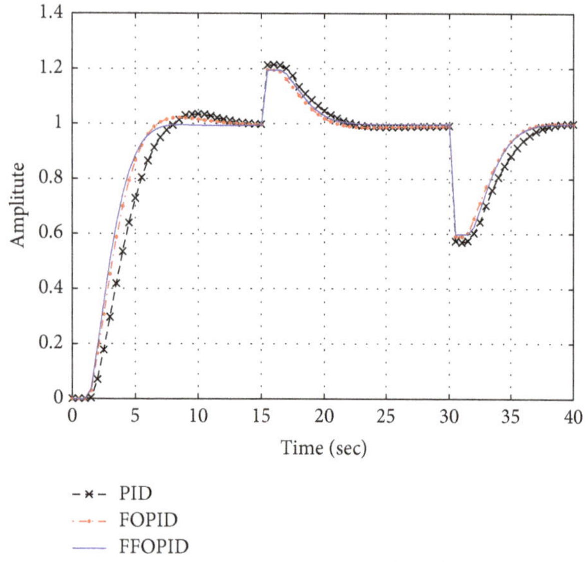

Figure 7: Load disturbances response of various controllers.

structure outperforms other two controllers at different test conditions for the pneumatic pressure system. The fuzzy logic system of FFOPID control structure enhances the controller

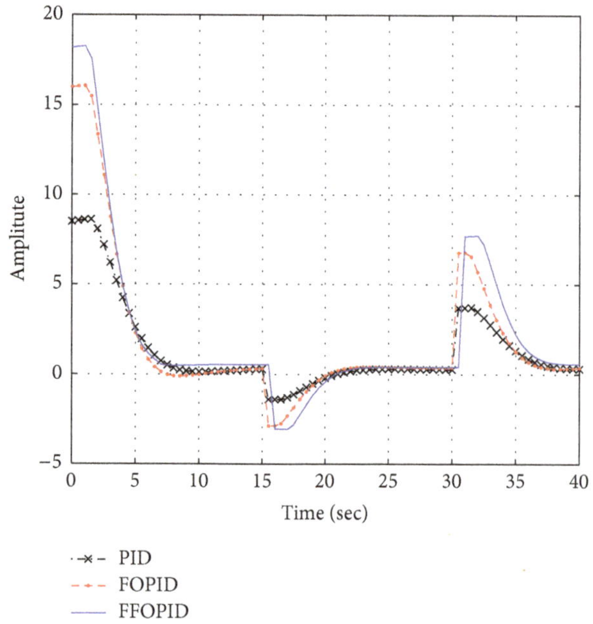

Figure 8: Control signal for load disturbances of various controllers.

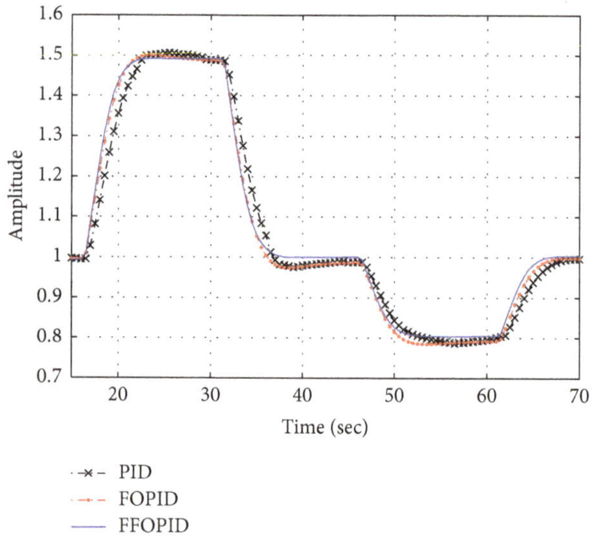

Figure 9: Change in set-point response of various controllers.

performance through online gain tuning mechanism. The robustness and adaptability of the proposed control scheme were also studied under load disturbance circumstances. From the results, it is observed that the FFOPID control scheme is better for pressure control application. Comparative studies are also made with conventional PID and FOPID control scheme to validate the proposed method. Moreover, this control scheme is simple, is effective, and could be a better alternative to the existing conventional control scheme. This study can be extended in future to digital algorithm and implemented using microprocessor or microcontroller.

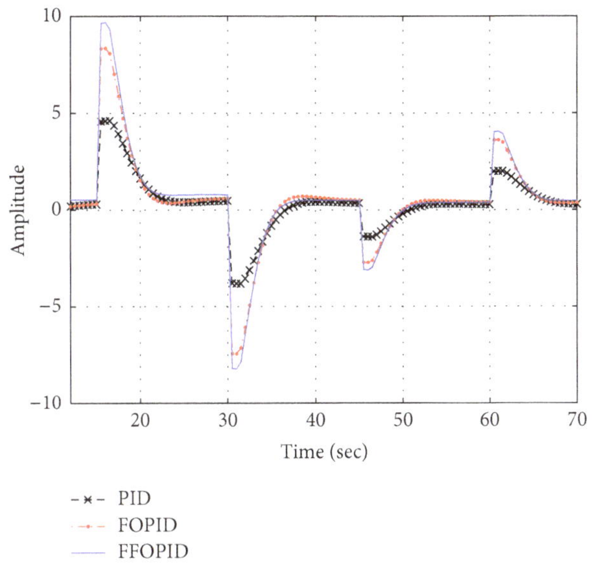

FIGURE 10: Control signal for change in set point of various controllers.

Conflicts of Interest

The authors declare that they have no conflicts of interest.

Acknowledgments

This project was supported by the Deanship of Scientific Research at Prince Sattam Bin Abdulaziz University under the Research Project no. 2016/01/6637.

References

[1] K. B. Oldham and J. Spanier, *Fractional Calculus: Theory and Applications of Differentiation and Integration to Arbitrary Order*, Academic Press, New York, NY, USA, 1974.

[2] K. S. Miller and B. Ross, *An Introduction to the Fractional Calculus and Fractional Differential Equations*, A Wiley-Interscience Publication, John Wiley & Sons, New York, NY, USA, 1993.

[3] A. Atangana and D. Baleanu, "New fractional derivatives with non-local and non-singular kernel: Theory and application to heat transfer model," *THERMAL SCIENCE*, vol. 20, no. 2, pp. 763–769, 2016.

[4] I. Petras, L. Dorcak, and I. Kostial, "Control quality enhancement by fractional order controllers," *Acta Montanistica Slovaca*, vol. 3, pp. 143–148, 1998.

[5] N. M. F. Ferreira and J. A. T. Machado, "Fractional-order hybrid control of robotic manipulators," in *Proceedings of the 11th International Conference on Advanced Robotics*, pp. 393–398, Coimbra, June 2003.

[6] A. Kailil, N. Mrani, M. M. Touati, S. Choukri, and N. Elalami, "Low Earth-orbit satellite attitude stabilization with fractional regulators," *International Journal of Systems Science*, vol. 35, no. 10, pp. 559–568, 2004.

[7] C. Ma and Y. Hori, "The application of fractional order control to backlash vibration suppression," in *Proceedings of the 2004 American Control Conference (AAC)*, pp. 2901–2906, usa, July 2004.

[8] D. Xue, C. Zhao, and Y. Chen, "Fractional order PID control of A DC-motor with elastic shaft: A Case Study," in *Proceedings of the American Control Conference*, pp. 3182–3187, Minneapolis, Minn, USA, June 2006.

[9] K. Erenturk, "Fractional-order $PI^{\lambda}D^{\mu}$ and active disturbance rejection control of nonlinear two-mass drive system," *IEEE Transactions on Industrial Electronics*, vol. 60, no. 9, pp. 3806–3813, 2013.

[10] P. Rastogi, S. Chatterji, and D. S. Karanjkar, "Performance analysis of fractional-order controller for pH neutralization process," in *Proceedings of the 2012 2nd International Conference on Power, Control and Embedded Systems, ICPCES 2012*, India, December 2012.

[11] K. M. Passino and S. Yurkovich, *Fuzzy-Control*, Wesley Longman, California, USA, 1998.

[12] M. Vahedpour, A. R. Noei, and H. A. Kholerdi, "Comparison between performance of conventional, fuzzy and fractional order PID controllers in practical speed control of induction motor," in *Proceedings of the 2nd International Conference on Knowledge-Based Engineering and Innovation, KBEI 2015*, pp. 912–916, Iran, November 2015.

[13] N. Bouarroudj, B. Djamel, and F. Boudjema, "Tuning fuzzy fractional order PID sliding-mode controller using PSO algorithm for nonlinear systems," in *Proceedings of the 2013 3rd International Conference on Systems and Control, ICSC 2013*, pp. 797–803, Algeria, October 2013.

[14] G. Mann, B. Hu, and R. Gosine, "Analysis and performance evaluation of linear-like fuzzy PI and PID controllers," in *Proceedings of the 6th International Fuzzy Systems Conference*, pp. 383–390, Barcelona, Spain, 1997.

[15] I. Podlubny, *Fractional Differential Equations*, vol. 198 of *Mathematics in Science and Engineering*, Academic Press, San Diego, Calif, USA, 1999.

[16] Y. Q. Chen, I. Petráš, and D. Y. Xue, "Fractional order control—a tutorial," in *Proceedings of the American Control Conference (ACC '09)*, pp. 1397–1411, June 2009.

[17] R. E. Gutiérrez, J. M. Rosário, and J. T. MacHado, "Fractional order calculus: basic concepts and engineering applications," *Mathematical Problems in Engineering*, vol. 2010, Article ID 375858, 19 pages, 2010.

[18] I. Petráš, *Stability of Fractional-Order Systems with Rational Orders: A Survey Fractional Calculus Applied Analysis12*, 2009, pp. 269-298.

[19] I. Petráš, "Tuning and implementation methods for fractional-order controllers," *Fractional Calculus and Applied Analysis*, vol. 15, no. 2, pp. 282–303, 2012.

[20] L. Ljung, *System Identification Toolbox TM Users Guide*, MathWorks Co. Ltd, 2015.

[21] A. Tepljakov, E. Petlenkov, and J. Belikov, "FOMCON: a MATLAB toolbox for fractional-order system identification and control," *International Journal of Microelectronics and Computer Science*, vol. 2, no. 2, pp. 51–62, 2011.

[22] A. T. Azar, S. Vaidyanathan, and A. Ouannas, *Fractional Order Control and Synchronization of Chaotic Systems*, Springer, 2017.

[23] S. Ahmed, "Parameter and delay estimation of fractional order models from step response," in *Proceedings of the IFAC 9th International Symposium on Advanced Control of Chemical Processes*, vol. 48, pp. 942–947, Whistler, British Columbia, Canada, June 7-10, 2015.

[24] J. G. Ziegler and N. B. Nichols, "Optimum settings for automatic controllers," *Transactions of the ASME*, vol. 64, pp. 759–768, 1942.

[25] C. Y. Quan, "Applied Fractional Calculus," in *Proceedings of the American Control Conference-ACC2009*, St. Louis, Missouri, USA, 2009.

[26] C. Junyi and C. Binggang, "Fractional-order control of pneumatic position servosystems," *Mathematical Problems in Engineering*, vol. 2011, Article ID 287565, 14 pages, 2011.

[27] S. Manabe, "The non-integer integral and its application to control systems," *JIEE (Japanese Institute of Electrical Engineers) Journal*, vol. 6, pp. 83–87, 1961.

[28] A. Oustaloup, F. Levron, B. Mathieu, and F. M. Nanot, "Frequency-band complex noninteger differentiator: characterization and synthesis," *IEEE Transactions on Circuits and Systems I: Fundamental Theory and Applications*, vol. 47, no. 1, pp. 25–39, 2000.

[29] M. A. Luersen and R. le Riche, "Globalized nelder-mead method for engineering optimization," *Computers & Structures*, vol. 82, no. 23–26, pp. 2251–2260, 2004.

[30] R. K. Mudi and N. R. Pal, "A robust self-tuning scheme for PI- and PD-type fuzzy controllers," *IEEE Transactions on Fuzzy Systems*, vol. 7, no. 1, pp. 2–16, 1999.

16

Tracking Control of Chaotic Systems via Optimized Active Disturbance Rejection Control

Fayiz Abu Khadra

Mechanical Engineering Department, King Abdulaziz University, Rabigh, Saudi Arabia

Correspondence should be addressed to Fayiz Abu Khadra; fabukhadra@kau.edu.sa

Academic Editor: Anna Vila

For tracking control of chaotic systems, we develop an active disturbance rejection (ADR) control method. Using the first state of the system as the only available state, a time-varying bandwidth extended state observer reconstructs the remaining states and the total disturbance. A time-varying bandwidth feedback controller forces all the states of the system to follow exactly the reference signal and its derivative. The parameters of the ADR controller are optimized using a genetic algorithm. As the objective function, we chose the weighted sum of the integral of the absolute error and the integral of the absolute control signal. Two chaotic systems—the Duffing system and the Genesio–Tesi system—are considered in computer simulation tests. Results of these simulations are presented to demonstrate the effectiveness of the ADRC method in controlling chaotic systems.

1. Introduction

Chaos is the complex, unpredictable, and irregular behavior of systems. The behavior is very sensitive to a slight change in the initial conditions. Chaotic systems are very interesting nonlinear systems and have been intensively studied in numerous fields. Indeed, chaos can be found in many engineering systems such as oscillators, chemical reactions, robotics, lasers, and secure communications. Controlling a chaotic system requires designing a controller that stabilizes the system towards equilibrium points or forces the system to follow a time-varying reference signal. Transforming the system's response into useful signals is important in practice. Towards this end, a variety of control methods have been proposed, ever since the pioneering work reported in [1]. Examples include sliding mode control [2–8], adaptive control [9–18], and backstepping control [19, 20].

The control methods cited above are examples from two categories. The first assumes that, for a system model described accurately by a mathematical model, all states are available, neglecting uncertainties and external disturbances. This leads to a simple control method. The second assumes that the mathematical description of the system is partially known. This leads to a complex control method. In most cases in practice, only one state is available; the parameters of the system are uncertain and the system is subject to external disturbances. These uncertainties and external disturbances affect the performance of the controller. Therefore, it is essential to develop techniques that provide reliable estimates of the other states and the external disturbances, from the available single output. This problem can be solved if an observer is used. The observer design is a branch of control theory, for which solutions exist for linear and nonlinear systems. An observer is designed to reconstruct the unmeasured states and the external disturbance.

The active disturbance rejection (ADR) control method [21–23] is a nonlinear control method that is efficient and easy to implement. In addition, it is an effective method in dealing with uncertain nonlinear systems subject to external disturbances. In the (ADR) control method the unknown dynamics of the system and the external disturbances are considered the "total disturbance". The extended state observer (ESO) reconstructs the missing states and the total disturbance. These outputs of the ESO are used by a feedback controller. The procedure keeps the controller relatively simple, as the plant is reduced to a chain of integrators.

The objective of this paper is as follows:

(i) To control chaotic systems with uncertain parameters, in the presence of external disturbances using ADRC method.
(ii) To optimize the ADRC controller parameters so that the system performs well by rejecting the total disturbances acting on the system and minimizing the tracking error to improve the overall system performance.
(iii) To suppress the peaking phenomena and improve the controller performance by using the time-varying observer and controller bandwidths.

As an objective function, a weighted sum of the integral of the absolute error and the integral of the absolute control signal is used, so that good tracking and concomitantly a smaller control signal can be produced. Two well-known examples are considered: the Duffing system (DS) as a typical second-order chaotic system and the Genesio–Tesi system (GTS) as a typical third-order chaotic system.

The organization of this article is as follows. Section 2 states the problem at hand in mathematical terms. Section 3 describes the ADRC method. In Section 4, simulation experiments are performed to show the effectiveness of the proposed scheme in controlling the Duffing and Genesio–Tesi chaotic systems. Section 5 concludes the paper.

2. Problem Formulation

In this study, the chaotic systems to be controlled are a class of n-th order single-input and single-output (SISO) continuous nonlinear systems, described by the dynamic equation:

$$\begin{aligned} \dot{x}_1 &= x_2 \\ \dot{x}_2 &= x_3 \\ &\vdots \\ \dot{x}_n &= f(X,t) + \Delta f(X,t) + d(t) + u(t) \\ y &= x_1(t) \end{aligned} \tag{1}$$

where $X = [x, \dot{x}, \ldots, x^{n-1}]^T = [x_1, x_2, \ldots, x_n]^T \in \mathbb{R}^n$ is the state vector, $f(X,t)$ is function of the states and time, not exactly known but assumed to be continuous and bounded, $y \in R$ is the output, and $u(t) \in R$ is the control signal. The functions $\Delta f(X,t), d(t)$ are the bounded uncertain term and the disturbance term, respectively. In mathematical form,

$$\begin{aligned} |\Delta f(X,t)| &\le \alpha, \\ |d(t)| &\le \beta \end{aligned} \tag{2}$$

where α *and* β are positive constants. The given reference vector R(t) is assumed to be bounded and has up to (n−1) bounded derivatives; that is,

$$R(t) = \left[r, \dot{r}, \ldots, r^{(n-1)}\right]^T = \left[r_1, r_2, \ldots, r_n\right]^T \tag{3}$$

The aim of a control problem is to design a controller $u(t)$ to force the system output $y(t)$ to follow the given bounded reference signal $R(t)$ and its derivatives, that is, to minimize the tracking error defined as $E(t) = X(t) - R(t)$,

$$\lim_{t \to +\infty} \|E(t)\| = \|X(t) - R(t)\| \longrightarrow 0 \tag{4}$$

where $\|\cdot\|$ is the Euclidian norm (or 2-norm) of a vector.

3. ADR Controller

In practice, only partial information of the mathematical model is known for most control systems; that is, the system has uncertainties and external disturbance. The main part of ADR controller is the ESO [18]. The main idea of the ESO is to use an additional state $x_{n+1} = f(X,t) + \Delta f(X,t) + d(t)$ so that system (1) can be written in state equation form as

$$\begin{aligned} \dot{x}_1 &= x_2 \\ \dot{x}_2 &= x_3 \\ &\vdots \\ \dot{x}_n &= x_{n+1} + u(t) \\ \dot{x}_{n+1} &= \dot{F}(X,t) \\ y &= x_1(t) \end{aligned} \tag{5}$$

Here $x_{n+1} = F(X,t)$ is the total disturbance or the extended state of the system. Note that the extended state can be considered as total unknown factor including linear/nonlinear function of state and the disturbance part.

The ESO in its nonlinear version takes the form:

$$\begin{aligned} \dot{\hat{x}}_1 &= \hat{x}_2 + \alpha_1 \omega_o(t)(y - \hat{x}_1) \\ \dot{\hat{x}}_2 &= \hat{x}_3 + \alpha_2 \omega_o^2(t)(y - \hat{x}_1) \\ &\vdots \\ \dot{\hat{x}}_n &= \hat{x}_{n+1} + \alpha_n \omega_o^n(t)(y - \hat{x}_1) + u(t) \\ \dot{\hat{x}}_{n+1} &= \alpha_{n+1} \omega_o^{n+1}(y - \hat{x}_1) \end{aligned} \tag{6}$$

With suitable values of $\beta_1, \beta_2, \ldots, \beta_n, \beta_{n+1}$the observer can track the states so that $\hat{x}_1 \approx x_1, \hat{x}_2 \approx x_2, \ldots, \hat{x}_n \approx x_n, \hat{x}_{n+1} \approx x_{n+1} = F(X,t)$.

Note that a third-order observer is required when implementing an ADR controller for a second-order system. With a good estimation of all states of the system and the additional state, the control signal in the ADR controller is defined as

$$u(t) = -\hat{x}_{n+1} + k_1(t)\hat{e}_1 + k_2(t)\hat{e}_2 + \cdots + k_n(t)\hat{e}_n \tag{7}$$

where $k_1(t), k_2(t), \ldots, k_n(t)$ are the time-varying gains of the controller. By substituting (7) into (5), the uncertain and disturbed system (1) can be reduced to a simple disturbance-free system represented as set of linear integrators:

$$\begin{aligned}\dot{x}_1 &= x_2 \\ \dot{x}_2 &= x_3 \\ &\vdots \\ \dot{x}_n &= k_1\hat{e}_1 + k_2\hat{e}_2 + \cdots + k_n\hat{e}_n\end{aligned} \tag{8}$$

The objective in any control is to obtain an accurate estimation and then a cancellation of the total disturbance along with an accurate tracking of the reference. The speed at which the observer estimates the total disturbance $F(X,t)$ to be cancelled by the controller in real time is crucial. The procedure to determine the parameters of the ADR controller for the tracking control problem of the chaotic system considered in this study is explained in the next section.

4. Parameter Optimization of ADR Controller

4.1. Selecting Parameters. Parameter tuning in the ADR controller is usually a manual process. It is based on experience in the control field and is relatively difficult. Moreover, the numerous parameters requiring adjustments by manual methods weaken the control performance. The parameters of the ADR controller are both observer-related and controller-related. For the observer, one set of parameters that are required are the gains. With the parameterization technique proposed in [21], the roots of the characteristic polynomial

$$\lambda(s) = s^{n+1} + \beta_1 s^n + \beta_2 s^{n-1} + \cdots + \beta_{n-1}s^2 + \beta_n s + \beta_{n+1} \tag{9}$$

are compared with polynomial

$$G_o(s) = (s + \omega_o(t))^{n+1} \tag{10}$$

thence placing all of the observers poles at $-\omega_0$ and reducing the characteristic polynomial to Hurwitz-type. The same can be done to determine the controller gains $k_1, k_2, \ldots k_n$. The roots of the characteristic polynomial

$$\lambda_c(s) = s^n + k_n s^{n-1} + \cdots + k_2 s + k_1 \tag{11}$$

are compared with the following polynomial:

$$G_c(s) = (s + \omega_c(t))^n \tag{12}$$

so that all the poles are placed at $-\omega_c$, where ω_c is the bandwidth of the feedback controller.

A higher bandwidth results in better reference signal tracking and disturbance rejection. Nevertheless, the sensor noise and dynamic uncertainties are two factors that limit the maximum closed-loop bandwidth [19].

If the initial value of the extended observer is different from the initial value of the plant, the phenomenon of peaking appears for very large ω_0. This can affect significantly the convergence of the extended observer. A possible way to mitigate peaking is to let the observer and the controller bandwidth vary with time as, for example, [24]

$$\boldsymbol{\omega_o}(\boldsymbol{t}) = A\frac{1 - e^{-Bt}}{1 + e^{-Bt}} \tag{13}$$

where A and B are two positive constants. If $\omega_c(t) = \omega_o(t)/C$, with constant C that can be chosen from the range [2–10], then, the parameters of the ADR controller, A, B, and C, can be optimized via an appropriate selection of an objective function.

4.2. Objective Function. The selection of the objective function is important for accuracy during parameter setting. To achieve good tracking performance, and at the same time an acceptable control signal, a fitness function J is introduced that is a combination of the integral of the absolute error (IAE) and the integral of the absolute control signal (IAU). The IAE performance index is adopted to obtain stability and accuracy of the tracking process; the IAU is included to minimize the control signal so that the need to limit the control signal using saturation function is eliminated. In practical applications, there are limiting conditions on the control energy that leads to a failure to achieve the desired performance. The minimization problem is formulated as

$$J(A,B,C) = \lambda\int_0^t |e_1(t)|\,dt + \int_0^t |u(t)|\,dt \tag{14}$$

where λ is a positive constant. Setting λ large leads to a more accurate tracking and a larger control signal.

4.3. Optimization Method. Genetic algorithms (GAs) [22, 23] are optimization methodologies based on Darwinian principles of evolutionary biology. They are often used to solve nonlinear or nondifferentiable optimization problems. As shown in Figure 1, a basic genetic algorithm consists of the following steps:

1. Initial population creation: randomly generate n chromosomes that are possible solutions of the problem.

(2) Calculating the fitness: by running the simulation obtain the fitness of the n chromosomes using the objective function.

(3) Generating the new population: by applying the GA operators until a new population is complete.

(4) Replacing: combine the old with newly generated solutions.

(5) Termination: if the termination criterion is satisfied, stop.

6. Loop: go to Step (2) for fitness evaluation.

The GA operators in step (3) are (a) Selection: based on their fitness, select two parent chromosomes, (b) Crossover: Cross over the parents to form new children, (c) Mutation: mutate new offspring at a randomly selected position in the chromosome, and (d) Acceptance: place new children in the

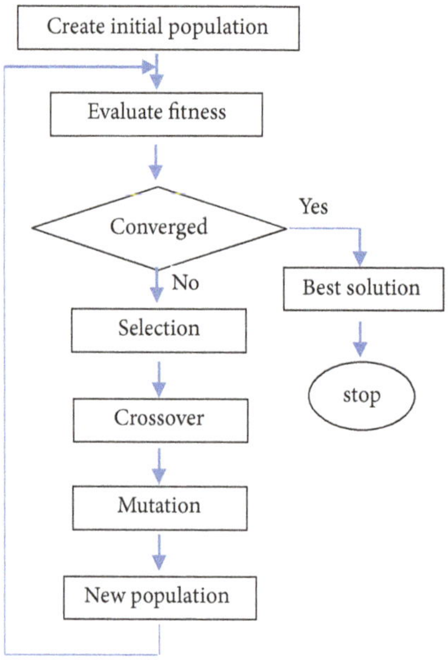

FIGURE 1: Basic genetic algorithm.

new population. Selection methods include, for example, the roulette-wheel selection, random selection, and tournament selection. The crossover operation can be single-point, two-point, or uniform. Mutation prevents the algorithm from being stuck in a local minimum. The mutation can be performed using procedures such as Flipping, Interchanging, and Reversing. The algorithm is terminated when the population converges through a variety of convergence criteria. A couple of popular ones are the fixed number of generations, and the best objective value is no longer changing.

5. Simulation Results

In this section, two well-known chaotic systems, the Duffing and Genesio–Tesi systems, are considered to demonstrate by numerical experiments the effectiveness of the proposed ADR control scheme.

5.1. Duffing System (DS). A nonlinear oscillator with a cubic stiffness term to describe the hardening spring effect observed in many mechanical problems was introduced by Duffing. Duffing's equation has the form

$$\ddot{x} + a_1\dot{x} + a_2x + a_3x^3 = u(t) + q\cos\omega t \tag{15}$$

where x is the oscillation displacement, a_1 the damping constant, a_2 the linear stiffness constant, a_3 the cubic stiffness constant, q the excitation amplitude, and ω the excitation frequency. To establish the dynamic behavior of the chaotic DS, we select parameter settings: $a_1 = -1.1, a_2 = 0.4, a_3 = -1, q = 1.3,$ and $\omega = 1.8$ rad/s. From the DS response for the initial conditions $x = 1, \dot{x} = 2$ (Figure 2), the system develops chaotic behavior when no control signal is applied.

To test the performance of the ADR controller in its tracking task, the DS is controlled to follow a reference signal and its derivative given by

$$\begin{aligned} r_1(t) &= 10\sin(t) \\ r_2(t) &= 10\cos(t) \end{aligned} \tag{16}$$

By defining the states of (15), $x_1 = x$ and $x_2 = \dot{x}$, it can be rewritten as two first-order ordinary differential equations,

$$\begin{aligned} \dot{x}_1(t) &= x_2(t) \\ \dot{x}_2(t) &= a_1x_1(t) + a_2x_2(t) + a_3x_1^3(t) + q\cos(\omega t) \\ &\quad + u(t) \\ y(t) &= x_1(t) \end{aligned} \tag{17}$$

Taking $F(x_1, x_2, \omega, t) = a_1x_1(t) + a_2x_2(t) + a_3x_1^3(t) + q\cos(\omega t)$ as the extended state x_3, (17) can be rewritten as

$$\begin{aligned} \dot{x}_1 &= x_2 \\ \dot{x}_2 &= x_3 + u(t) \\ \dot{x}_3 &= \dot{F}(x_1, x_2, \omega, t) \end{aligned} \tag{18}$$

The ESO for a second-order system is as follows:

$$\begin{aligned} \dot{\hat{x}}_1(t) &= \hat{x}_2(t) + 3\omega_o(t)(y(t) - \hat{x}_1(t)) \\ \dot{\hat{x}}_2(t) &= \hat{x}_3(t) + 3\omega_o^2(t)(y(t) - \hat{x}_1(t)) \\ \dot{\hat{x}}_3(t) &= \omega_o^3(t)(y(t) - \hat{x}_1(t)) \end{aligned} \tag{19}$$

The feedback controller is described by

$$u(t) = -\hat{x}_3 + \omega_c^2(t)\hat{e}_1 + 2\omega_c\hat{e}_2 \tag{20}$$

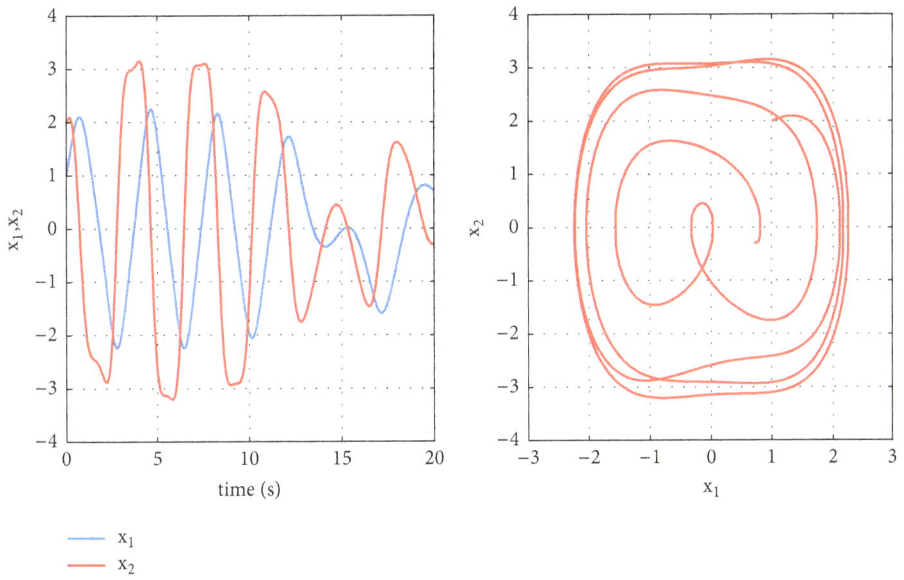

FIGURE 2: Chaotic behavior of DS without the control input in 20 seconds.

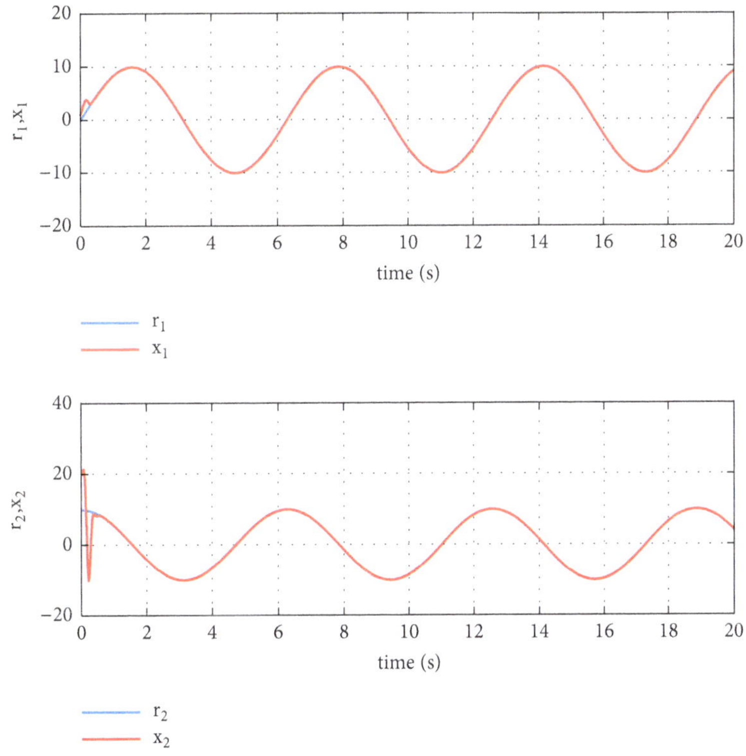

FIGURE 3: Controlled time response of the states x_1, x_2.

To optimize the three parameters A, B, and C, the lower and upper bounds of the parameters are selected as (after a number of experimental simulations) $A = [5, 200], B = [0.1, 1]$, and $C = [2, 10]$. The parameter settings of the GA are as follows: (i) a population size of 20—the tournament selection is applied to all chromosomes as the number of chromosomes in the population is small; (ii) a two-point crossover function and a uniform mutation function; (iii) during the optimization process, the stopping condition is invoked if the maximum number of generations reaches 100; and (iv) to increase the accuracy of tracking, a value of $\lambda = 100$ is set in objective function.

The optimized controller parameters are then $A = 170.0221, B = 0.9985$, and $C = 1.300$. The simulation results are presented in Figures 3–6. The time-dependent responses of the states of the DS (Figure 2) show the states exactly

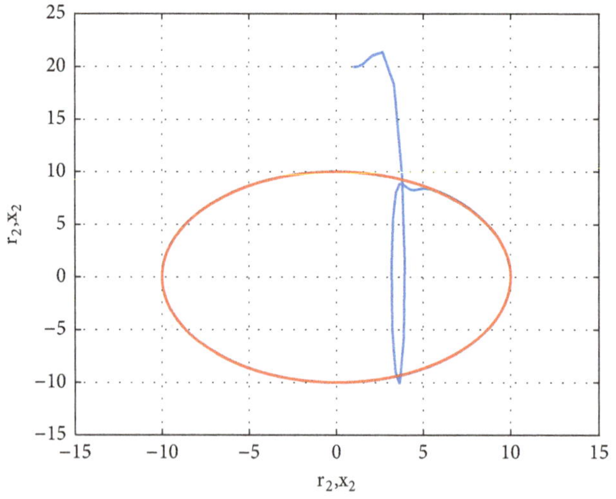

FIGURE 4: The controlled trajectory of the DS system.

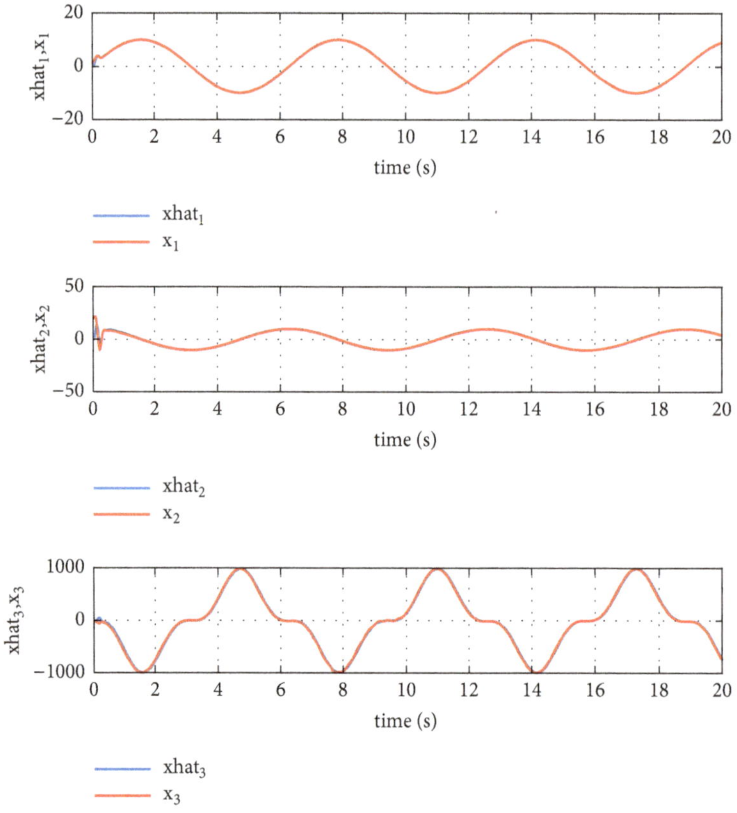

FIGURE 5: The performance of the ESO in estimating the states.

following the reference signal and its derivative after a short time. A good view of the tracking of the DS of the reference signal is seen in the phase-plane plot (Figure 4). The ESO performs well in estimating the two states of the system and the extended state in less than one second (Figure 5). The control effort required to force the DS follows the reference signal (Figure 6); note that the control signal required to control the DS is limited because the optimization of the ADR controller parameters based on the objective function (14) includes the minimization of the control signal as part of it. Therefore, the need to saturate the control signal is eliminated. The ADR controller parameter settings obtained from the minimization process leads to a very fast control of the system's chaotic behavior.

5.2. Genesio–Tesi System (GST). As our second example, the GTS is considered having the form,

$$\dot{x}_1 = x_2$$

$$\dot{x}_2 = x_3$$

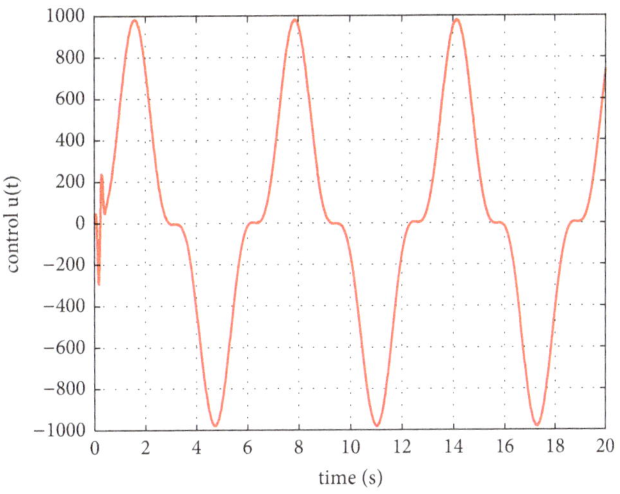

FIGURE 6: The time response of the control signal u(t).

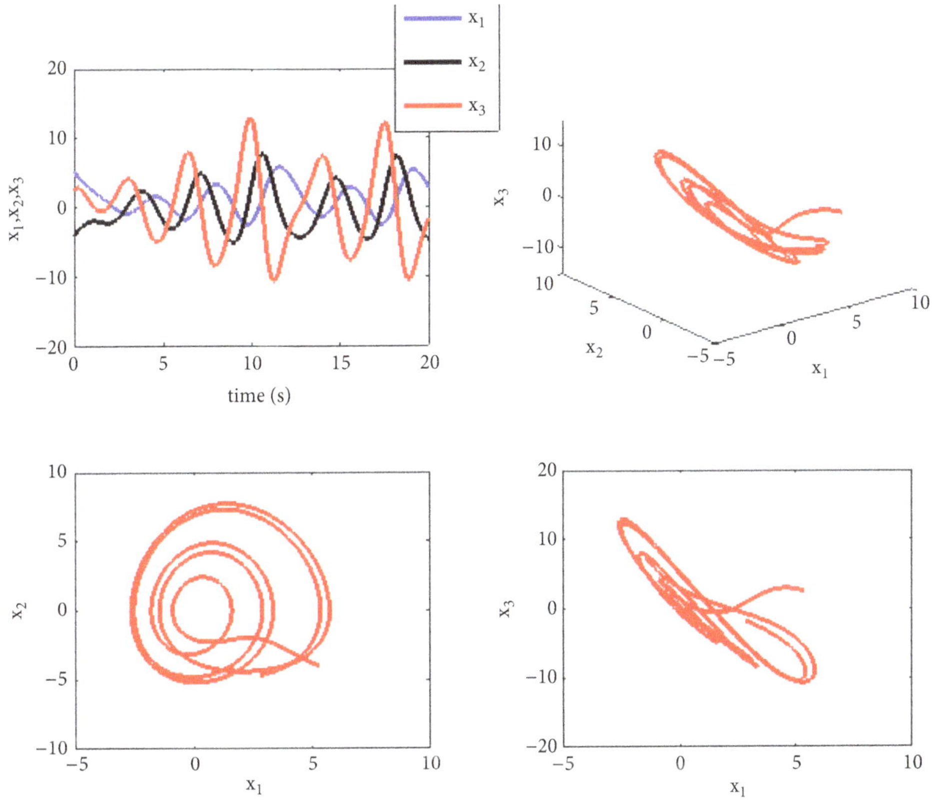

FIGURE 7: Chaotic trajectories of Genesio–Tesi system (GTS).

$$\dot{x}_3 = -ax_1 - bx_2 - cx^3 + mx_1^2 + \Delta f(x_1, x_2, x_3, t) + d(t) + u(t) \quad (21)$$

where x_1, x_2, and x_3 are the system states with $[x_1, x_2, x_3]^T$ the full-state vector, a, b, c, and m are unknown positive real constants satisfying $ab < c$, $\Delta f(x_1, x_2, x_3, t)$ is an unknown function depending on the full state and time, $u(t)$ is the control signal to be designed with control input $u(t) = 0$, $\Delta f(x_1, x_2, x_3, t) = 0$, and $d(t) = 0$. The chaotic response of the GTS (Figure 7) was established using the initial conditions $[x_{10} = 5.3, x_{20} = -4, x_{30} = 2.5]$ for the variables and a running time of 20 seconds.

To test the performance of the ADRC in the task of tracking, the GTS is controlled to follow the reference signal and its derivatives given by

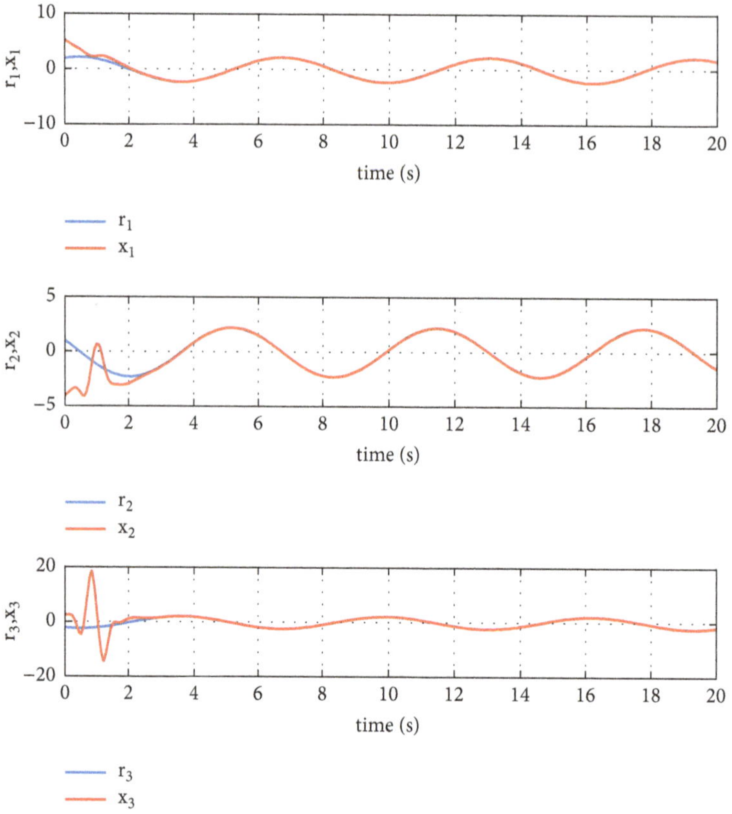

FIGURE 8: Time responses of the states.

$$r_1(t) = 2\cos(t) + \sin(t)$$
$$r_2(t) = \cos(t) - 2\sin(t) \tag{22}$$
$$r_3(t) = -2\cos(t) - \sin(t)$$

The uncertain term added is

$$\Delta f(x_1, x_2, x_3, t) = 0.05\sin(\pi x_1)\sin(2\pi x_2)\sin(3\pi x_3) \tag{23}$$

and the external disturbance is

$$d(t) = 0.002\cos(t) \tag{24}$$

To estimate the three states and the total uncertainty, the following fourth-order ESO is required:

$$\dot{\hat{x}}_1(t) = \hat{x}_2(t) + 4_1\omega_o(t)(x_1(t) - \hat{x}_1(t))$$
$$\dot{\hat{x}}_2(t) = \hat{x}_3(t) + 6\omega_o^2(t)(x_1(t) - \hat{x}_1(t))$$
$$\dot{\hat{x}}_3(t) = \hat{x}_4(t) + 4\omega_o^3(t)(x_1(t) - \hat{x}_1(t)) + u(t) \tag{25}$$
$$\dot{\hat{x}}_4(t) = \alpha_4\omega_o^4(t)(x_1(t) - \hat{x}_1(t))$$

The feedback controller is described by

$$u(t) = -\hat{x}_4 + \omega_c^3(t)\hat{e}_1 + 3\omega_c^2(t)\hat{e}_2 + 3\omega_c(t)\hat{e}_3 \tag{26}$$

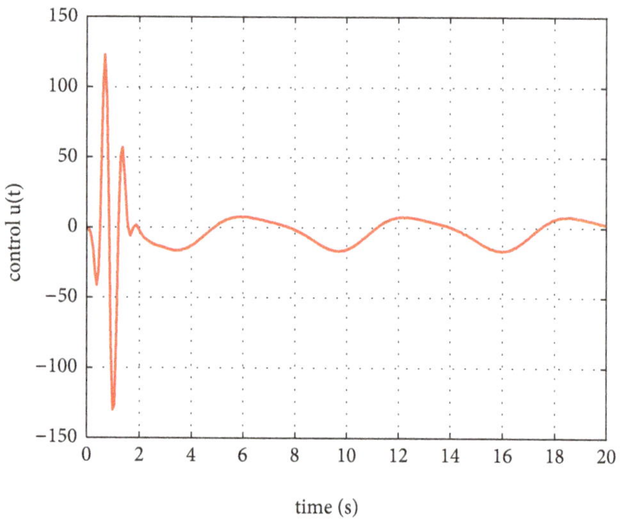

FIGURE 9: Time response of control input (u(t)).

Using GA optimization, the control parameter settings obtained are $A = 24.960$, $B = 0.535$, $C = 1.985$ using the same GA parameters and options adopted in the previous section. The simulation results are presented in Figures 8–12. The time-dependent responses of the states of the GTS (Figure 8) show again that the states exactly follow the reference signal and its derivative after a short time. Figure 9 shows the control

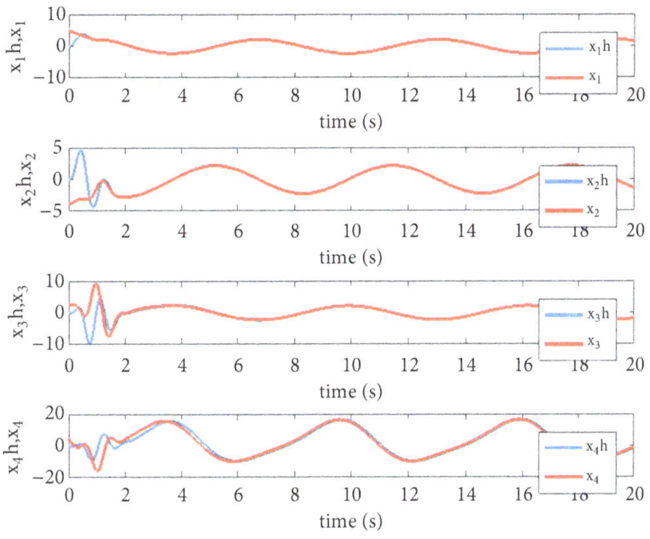

FIGURE 10: The performance of the ESO in estimating the states.

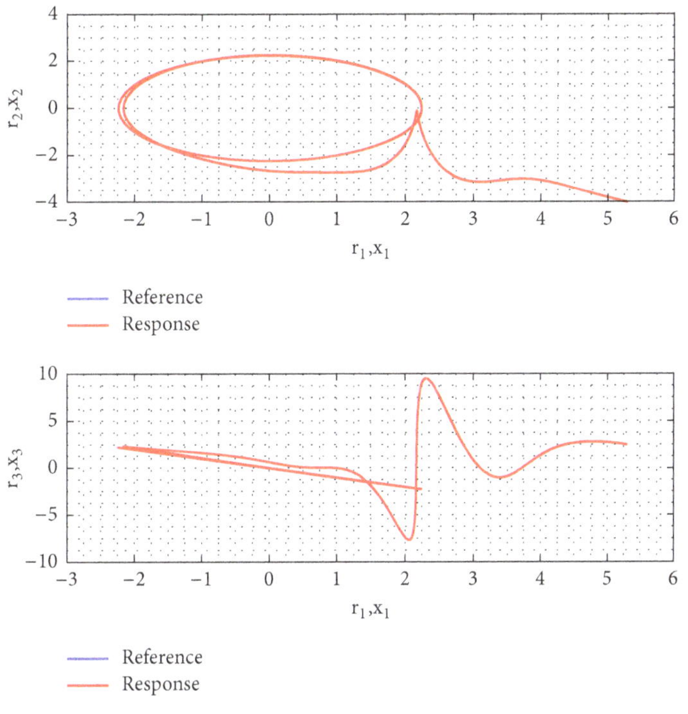

FIGURE 11: The controlled trajectories of the GT system.

effort required to follow the reference signal. The ESO also performs well in estimating the three states of the system and the extended state (Figure 10). Figure 11 shows the controlled trajectories of GTS in the x_1x_2 and x_1x_3 planes. Figure 12 shows the time-varying observer and controller bandwidths.

6. Conclusion

An ADR controller has been developed to control chaotic systems and applied in two well-known examples to demonstrate its capabilities. The merit of this approach is the use of time-varying parameters for the observer as well as for the feedback controller. The results obtained in simulations clearly show the controller performs well in controlling both chaotic systems with bounded uncertainty and disturbance to any arbitrarily desired reference signal with high accuracy. The results also show that the ADR method holds promise in providing solutions in controlling chaos. The ADR has been applied to control a second-order DS and a third-order GTS. A suitable selection of the control parameter settings was observed to influence the control effort and error. Hence, the genetic algorithm was applied to obtain optimal parameter

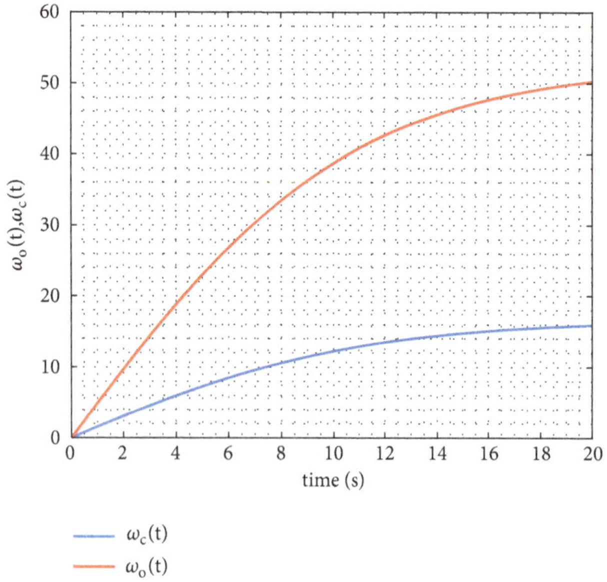

FIGURE 12: The time-varying observer and controller bandwidths.

settings of the ADR controller. In further study, the ADR controller is to be applied to the synchronization of chaos.

Conflicts of Interest

The author declares that there are no conflicts of interest.

References

[1] L. M. Pecora, T. L. Carroll, G. A. Johnson, D. J. Mar, and J. F. Heagy, "Fundamentals of synchronization in chaotic systems, concepts, and applications," *Chaos: An Interdisciplinary Journal of Nonlinear Science*, vol. 7, no. 4, pp. 520–543, 1997.

[2] X. Zhang, X. Liu, and Q. Zhu, "Adaptive chatter free sliding mode control for a class of uncertain chaotic systems," *Applied Mathematics and Computation*, vol. 232, pp. 431–435, 2014.

[3] B. Wang and S. Zhong, "Observer-based control on a chaotic system with unknowns and uncertainties," *Optik - International Journal for Light and Electron Optics*, vol. 137, pp. 167–174, 2017.

[4] R. Luo and H. Su, "Finite-time control and synchronization of a class of systems via the twisting controller," *Chinese Journal of Physics*, vol. 55, no. 6, pp. 2199–2207, 2017.

[5] S. Dadras, H. R. Momeni, and V. J. Majd, "Sliding mode control for uncertain new chaotic dynamical system," *Chaos, Solitons & Fractals*, vol. 41, no. 4, pp. 1857–1862, 2009.

[6] J. Sun, Y. Wang, Y. Wang, and Y. Shen, "Finite-time synchronization between two complex-variable chaotic systems with unknown parameters via nonsingular terminal sliding mode control," *Nonlinear Dynamics*, vol. 85, no. 2, pp. 1105–1117, 2016.

[7] X. Su, X. Liu, P. Shi, and R. Yang, "Sliding Mode Control of Discrete-Time Switched Systems with Repeated Scalar Nonlinearities," *IEEE Transactions on Automatic Control*, vol. 62, no. 9, pp. 4604–4610, 2017.

[8] J. Sun, Y. Wu, G. Cui, and Y. Wang, "Finite-time real combination synchronization of three complex-variable chaotic systems with unknown parameters via sliding mode control," *Nonlinear Dynamics*, vol. 88, no. 3, pp. 1677–1690, 2017.

[9] K.-M. Chang, "Adaptive control for a class of chaotic systems with nonlinear inputs and disturbances," *Chaos, Solitons & Fractals*, vol. 36, no. 2, pp. 460–468, 2008.

[10] J. Wen and C.-S. Jiang, "Adaptive fuzzy control for a class of chaotic systems with nonaffine inputs," *Communications in Nonlinear Science and Numerical Simulation*, vol. 16, no. 1, pp. 475–492, 2011.

[11] C. Hua, X. Guan, X. Li, and P. Shi, "Adaptive observer-based control for a class of chaotic systems," *Chaos, Solitons & Fractals*, vol. 22, no. 1, pp. 103–110, 2004.

[12] J. Zhou and M. J. Er, "Adaptive output control of a class of uncertain chaotic systems," *Systems & Control Letters*, vol. 56, no. 6, pp. 452–460, 2007.

[13] N. Jia and T. Wang, "Chaos control and hybrid projective synchronization for a class of new chaotic systems," *Computers & Mathematics with Applications. An International Journal*, vol. 62, no. 12, pp. 4783–4795, 2011.

[14] S. Dadras and H. R. Momeni, "Control uncertain Genesio-Tesi chaotic system: Adaptive sliding mode approach," *Chaos, Solitons & Fractals*, vol. 42, no. 5, pp. 3140–3146, 2009.

[15] D. Zhang, J. Mei, and P. Miao, "Global finite-time synchronization of different dimensional chaotic systems," *Applied Mathematical Modelling: Simulation and Computation for Engineering and Environmental Systems*, vol. 48, pp. 303–315, 2017.

[16] H. Handa and B. B. Sharma, "Novel adaptive feedback synchronization scheme for a class of chaotic systems with and without parametric uncertainty," *Chaos, Solitons & Fractals*, vol. 86, pp. 50–63, 2016.

[17] Jin Xing and Fangfang Zhang, "An Approach of Tracking Control for Chaotic Systems," *Journal of Control Science and Engineering*, vol. 2016, Article ID 9735264, 8 pages, 2016.

[18] Y.-C. Chang, "A robust tracking control for a class of uncertain chaotic systems," *Asian Journal of Control*, vol. 15, no. 6, pp. 1752–1763, 2013.

[19] J. Yu, J. Lei, and L. Wang, "Backstepping synchronization of chaotic system based on equivalent transfer function method," *Optik - International Journal for Light and Electron Optics*, vol. 130, pp. 900–913, 2017.

[20] J. Lei and L. Wang, "Backstepping synchronous control of chaotic system with reduced number of active inputs," *Optik - International Journal for Light and Electron Optics*, vol. 127, no. 23, pp. 11364–11373, 2016.

[21] J. Q. Han, "From PID to active disturbance rejection control," *IEEE Transactions on Industrial Electronics*, vol. 56, no. 3, pp. 900–906, 2009.

[22] Z. Gao, "Scaling and bandwidth-parameterization based controller tuning," in *Proceedings of the American Control Conference*, pp. 4989–4996, Denver, Colo, USA, June 2003.

[23] Q. Zheng and Z. Gao, "Predictive active disturbance rejection control for processes with time delay," *ISA Transactions®*, vol. 53, no. 4, pp. 873–881, 2014.

[24] J. Liu, *Advanced Sliding Mode Control for Mechanical Systems: Design, Analysis and MATLAB Simulation: with 165 Figures*, Tsinghua University Press, Beijing, China, 2012.

17

Dynamic Operation Management of a Renewable Microgrid including Battery Energy Storage

Shaozhen Jin,[1] **Zhizhong Mao**,[1] **Hongru Li**,[1] **and Wenhai Qi**[2]

[1]*School of Information Science & Engineering, Northeastern University, Shenyang 110819, China*
[2]*School of Engineering, Qufu Normal University, Rizhao 276800, China*

Correspondence should be addressed to Shaozhen Jin; jinshaozhen@163.com

Academic Editor: Gaetano Zizzo

In this paper, a novel dynamic programming technique is presented for optimal operation of a typical renewable microgrid including battery energy storage. The main idea is to use the scenarios analysis technique to proceed the uncertainties related to the available output power of wind and photovoltaic units and dynamic programming technique to obtain the optimal control strategy for a renewable microgrid system in a finite time period. First, to properly model the system, a mathematical model including power losses of the renewable microgrid is established, where the uncertainties due to the fluctuating generation from renewable energy sources are considered. Next, considering the dynamic power constraints of the battery, a new performance index function is established, where the Lagrange multipliers and interior point method will be presented for the equality and inequality operation constraints. Then, a feedback control scheme based on the dynamic programming is proposed to solve the model and obtain the optimal solution. Finally, simulation and comparison results are given to illustrate the performance of the presented method.

1. Introduction

Nowadays, renewable energy sources (RESs) such as wind or photovoltaics have become more wide spread due to needs for satisfying the environment concerns. On the other hand, distributed generators (DGs) like diesel engines, microturbines, and fuel cells can be used to enhance the resiliency of power system and yield other social economic benefits. Therefore, renewable microgrid is expected to play an important role in future power systems [1, 2]. As a key enabling element of renewable microgrids, battery energy storages make microgrid become a strong coupling system in the time domain. In this regard, the methodologies applied to operation management of a renewable microgrid are getting more complicated and challengeable; therefore there is a strong need for more reliable scheduling of energy sources in renewable microgrid including battery energy storage.

So far, many researchers have dealt with the optimal operation scheduling of energy sources in microgrids [3–7]. Previously conventional mathematical programming such as Lagrange relaxation [8, 9], lambda iteration [10, 11], Newton-Raphson [12], interior point method [13], weighted minimax [14], and quadratic programming [15] have been used to determine the least cost solution. However, the conventional mathematical programming methods have major disadvantages such that they can be trapped in local optimal, exhibited sensitivity to the initial starting points. And many of the methods cannot solve the nonsmooth, convex, and nonmonotonically increasing cost functions. Recently, computational intelligence [16, 17] and artificial intelligence based nonconventional methods [18] have been used to solve the optimal operation scheduling of energy sources in microgrids. Artificial intelligence based methods such as artificial neural network and computational intelligence methods such as genetic algorithm, particle swarm optimization, harmony search, simulated annealing, differential evolution, gravitational search algorithm, biogeography based optimization, bacterial foraging algorithm, ant colony optimization, cuckoo search, bat algorithm, artificial bee colony, firefly algorithm, and flower pollination algorithm have been used to solve the problem. These methods can enable us to solve the nonlinear and no-convex cost functions and can obtain nearly the global solutions. However, these methods have major disadvantages such as the evolutionary algorithms greatly depending on their parameters and having high computational time. Besides, hybrid methods which combine different algorithms have been used to solve the optimal operation scheduling of energy sources in microgrid. But, these methods usually have long computational time. Also, all previous researchers search the optimal solutions in each separated time periods. In other words, they are static optimal operation scheduling which do not consider the relationships among different time periods. Dynamic optimal operation scheduling problem is another fundamental part of the renewable microgrid operation to maintain the power balance [19–24]. Among these methods, Cheng et al. [23] have used an enhanced quorum sensing based particle swarm optimization to deal with the dynamic operation optimization problem. However, it is hard to solve the complex optimization problem with high dimension variables and multiple constraints by such metaheuristic methods. Liu et al. [19, 22] have used the dynamic programming algorithm for handing the dynamic economic dispatch problem. Generally, dynamic programming can be used to solve nonlinear dynamics when the problems can be discredited with time, state, and decision variable [25]. In other words, dynamic programming can be used to solve the problem by dividing the decision process into different stages. Meanwhile, these stages interact and interconnect with decision variables. And the target of the dynamic programming is used to find the decision set over the whole searching trajectory. However, such methods may have some deficiencies handling large scale problems. And a challenging aspect of this method is determining which of the inequality constraints are binding at the solution. Another relevant aspect in renewable microgrid operation management should be coping with uncertainty in the renewable energy sources, load demand, and market prices [24–26]. Similarly, power line losses should also be considered in the renewable microgrid system. There have been some researches including the power losses in a renewable microgrid [5, 27, 28]. For example, the output power losses of distributed generators are modeled in articles [5, 29]. The power line losses which are calculated using B-coefficients are considered in the economic dispatch problems [30, 31].

This paper presents several contributions as follows:

(1) The power losses of the renewable microgrid are included. It is noteworthy that, from an optimal operation planning point of view, such method can be optimized a renewable microgrid globally.

(2) Uncertainty in the renewable energy resources is considered. A methodology based on scenario analysis is used to proceed the uncertainties related to the available output power of wind and photovoltaic units.

(3) The dynamic performance of the battery storage system is considered. Therefore, this dynamic operation constraint makes renewable microgrid to become a strong coupling system in the time domain. Then, the Lagrange multipliers and interior point method will be presented for these operation constraints.

(4) A feedback control scheme based on the dynamic programming is proposed to solve the model and obtain the optimal solution. The advantage of the scheme is that the number of numerical operations is linear in the number of optimization parameters, which enables one to solve for a smooth input/control history. Meantime, this scheme enables one to initialize the optimization problem with a zero-initial guess. Finally, this scheme is a direct method to obtain the same results.

2. Description of the Renewable Microgrid

2.1. Renewable Microgrid. In this paper, the renewable microgrid system considers diesel generators, microturbines, fuel cells, wind turbines, photovoltaics, and a battery energy storage system in Figure 1. The renewable microgrid can operate in parallel to the main grid or as an island.

2.2. Active Power Balance of the Renewable Microgrid. In general, the active power balance between electrical energy production and consumption must be met at each time interval expressed as follows [28, 33]:

$$\begin{aligned}\sum_{i=1}^{N_w} p_w(t) + \sum_{i=1}^{N_s} p_s(t) + \sum_{i=1}^{N_d} p_d(t) + \sum_{i=1}^{N_m} p_m(t) + \sum_{i=1}^{N_f} p_f(t) \\ + \sum_{i=1}^{N_b} p_b(t) + p_g(t), \\ = \sum_{j=1}^{N_D} p_L(t) + L\left(p_{dg}(t)\right), \\ t = 1, \dots, T, \end{aligned} \tag{1}$$

where $p_w(t), p_s(t), p_d(t), p_m(t), p_f(t), p_b(t), p_g(t), p_{dg}(t), p_L(t)$ are the power set-point for the wind turbines, photovoltaics, diesel generators, microturbines, fuel cells, battery energy storage, power exchange between the main grid and renewable microgrid, controllable distribution generators, and load levels. $L(p_{dg}(t))$ is the power loss of the renewable microgrid. $N_w, N_s, N_d, N_m, N_f, N_b, N_D$ are the total number of wind turbines, photovoltaics, diesel generators, microturbines, fuel cells, battery energy storage, and load levels. In this paper, the controllable distribution generators (DGs) included diesel generators, microturbines, and fuel cells.

2.2.1. Estimation of Power Losses Coefficients. When the power losses of the renewable microgrid are considered, these power losses coefficients can be calculated as follows [28, 34]:

$$\begin{aligned} L\left(p_{dg}(t)\right) = \sum_{m=1}^{N_{dg}} \sum_{n=1}^{N_{dg}} p_{dg,m}(t) . B_{mn} p_{dg,n}(t) \\ + \sum_{n=1}^{N_{dg}} B_{m0} \cdot p_{dg,m} + B_{00}, \end{aligned} \tag{2}$$

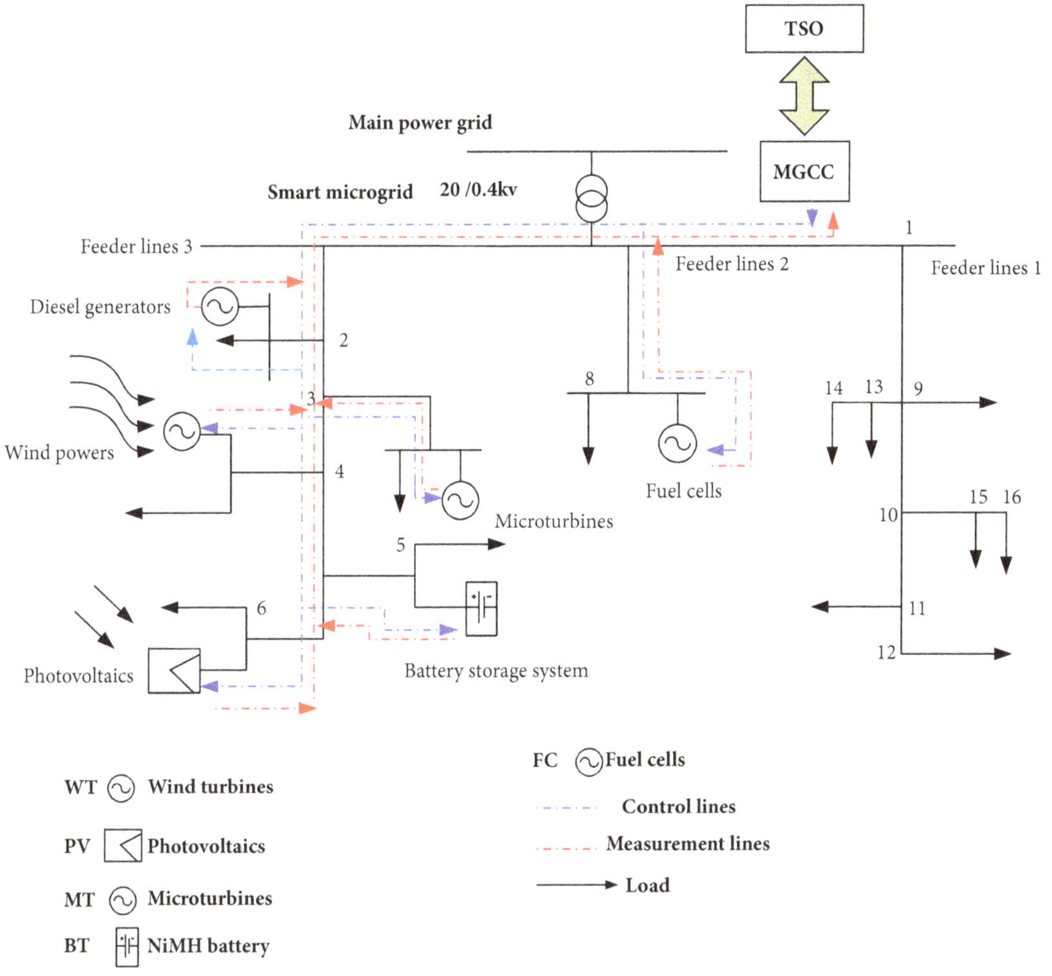

FIGURE 1: The smart microgrid structure diagram.

where $B = [B_{mn}]$ and B_{m0} and B_{00} are the power losses coefficients.

2.3. Subsystem Model of the Renewable Microgrid. This paper considers a renewable microgrid containing several types of devices, such as battery energy storages, distributed generators, wind turbines, and photovoltaics. A block diagram is given in Figure 1.

2.3.1. The Model of Battery Energy Storage System. Battery energy storage can store energy when charging. On the other hand, it can supply energy to the load demand when discharging.

The degradation cost of battery energy storage can be calculated as follows [33]:

$$C_{b,c}(t) = \sum_{j=1}^{N_b} \sum_{i=1}^{t} r_b \cdot (p_b(i)), \tag{3}$$

where r_b is the cost coefficient which can be determined by battery energy storage charging or discharging power.

$p_{b,c}(t)$ is the permitted rate of charge during a definite period of time and $p_{b,d}(t)$ is the permitted rate of discharge.

The following equation can be expressed for a battery energy storage

$$p_b(t) = u_c(t) \cdot p_{b,c}(t) + u_d(t) \cdot p_{b,d}(t), \tag{4}$$

where $u_c(t)$ or $u_d(t)$ indicate the working mode of the battery energy storage, $u_c(t), u_d(t) \in \{0, 1\}$. Define $u_c(t) = 1$ or $u_d(t) = 1$ if the battery energy storage is charge or discharge. These variables satisfy

$$u_c(t) + u_d(t) \leq 1, \quad t = 1, \ldots, T. \tag{5}$$

The permitted rate of charge and discharge are limited by the following constrains:

$$\begin{aligned} 0 \leq p_{b,c}(t) \leq p_{b,c}^{max}, \\ 0 \leq p_{b,d}(t) \leq p_{b,d}^{max}, \end{aligned} \tag{6}$$

where $p_{b,c}^{max}$ and $p_{b,d}^{max}$ are the maximum charging and discharging rates of battery energy storage.

After charge or discharge, the state of charge of the battery energy storage can be calculated as follows [33]:

$$SOC_b(t) = SOC_b(t-1) + \frac{(\eta_{b,c} \cdot p_{b,c}(t) - p_{b,d}(t)/\eta_{b,d})}{Q_{b,c}} . \Delta t, \quad (7)$$

where $\eta_{b,c}, \eta_{b,d}$ are the energy conversion coefficients. $Q_{b,c}$ is the energy capacity of the battery energy storage. On the other hand, the state of charge of the battery energy storage must be maintained in the following range. Δt is the operating time of the battery energy storage.

$$SOC_{b,min} \le SOC_b(t) \le SOC_{b,max}, \quad (8)$$

where $SOC_{b,min}, SOC_{b,max}$ are required limits of the state of charge.

2.3.2. The Model of Controllable Distributed Generators. In this paper, the distribution generators (DGs) are included diesel generators, microturbines, and fuel cells.

(A) Diesel Generators. In general, diesel generators serve as a backup power source. Meanwhile, the operating cost of diesel generators is formulated as follows:

$$C_d(t) = \sum_{j=1}^{N_d} \sum_{i=1}^{t} \left[C_{d,f}(i) + C_{d,om}(i) + R_d(i) . (1 - \delta_d(i-1)) \right] . \delta_d(i), \quad (9)$$

where $C_{d,f}$ is the fuel cost of diesel generator; $C_{d,om}$ is the operating and maintenance cost; R_d is the start-up cost of diesel generator.

The fuel cost of diesel generators is modeled as a function of their actual output power

$$C_{d,f}(t) = a . [p_d(t)]^2 + b . p_d(t) + c, \quad (10)$$

where $p_d(t)$ is the actual output power. a, b, c are the fuel consumption curve fitting coefficients.

The operating and maintenance cost can be expressed as

$$C_{d,om}(t) = k_{d,om} \cdot p_d(t) . \Delta t, \quad (11)$$

where $k_{d,om}$ is the operating and maintenance cost coefficient. Δt is the operating time of diesel generator.

The start-up cost of diesel generator is relative to its operating state

$$R_d(t) = \begin{cases} R_{d,h}, & T_{d,off} \le T_{d,d} + T_{d,c}, \\ R_{d,c}, & T_{d,off} > T_{d,d} + T_{d,c}, \end{cases} \quad (12)$$

where $R_{d,h}, R_{d,c}$ are the hot start-up cost and cold start-up cost. $T_{d,c}$ is the cooling time of diesel generator. $T_{d,off}, T_{d,d}$ are the off time and shut down time of diesel generator.

For a stable operation, the actual output power of diesel generator is limited by lower and upper bounds as follows:

$$p_{d,min} \le p_d(t) \le p_{d,max}, \quad (13)$$

where $p_{d,min}$ is the minimum active power of diesel generator. $p_{d,max}$ is the maximum power of diesel generator.

The following constraint ensures diesel generators do not exceed their ramp limits:

$$-r_d \cdot p_{d,max} \le p_d(t) - p_d(t-1) \le r_d \cdot p_{d,max}, \quad (14)$$

where r_d is the ramp constraint coefficient.

Due to the physical constraints that state shifting can only take place after a fixed time interval, the state variable of two adjacent times is as follows:

$$\left(T_{d,on}(t-1) - T_{d,u}\right) . \left(\delta_d(t-1) - \delta_d(t)\right) \ge 0, \quad (15)$$

$$\left(T_{d,off}(t-1) - T_{d,d}\right) . \left(\delta_d(t) - \delta_d(t-1)\right) \ge 0, \quad (16)$$

where $T_{d,on}, T_{d,u}$ are the cumulative uptime and the minimum turn on time.

(B) Fuel Cells. Fuel cell directly converts chemical energy to electrical energy by electrochemical reactions, which is one of the most promising energy conversion technologies. The operating cost of fuel cells are formulated as

$$C_f(t) = \sum_{j=1}^{N_f} \sum_{i=1}^{t} \left[C_{f,f}(i) + C_{f,om}(i) + R_f(i) . \left(1 - \delta_f(i-1)\right) \right] . \delta_f(i), \quad (17)$$

where $C_{f,f}$ is the fuel cost of fuel cell; $C_{f,om}$ is the operating and maintenance cost; R_f is the start-up cost of fuel cell.

The fuel cost of fuel cells is modeled as

$$C_{f,f}(t) = c_f . T_f . \sum_{i=1}^{t} \frac{p_f(i) + p_a}{\eta_f(i)}, \quad (18)$$

where c_f is the fuel price; T_f is the operating time of changing two adjacent states; p_a is the consumption with active power for the auxiliary equipment. η_f is the efficiency of fuel cells.

The operating and maintenance cost can be obtained by

$$C_{f,om}(t) = k_{f,om} \cdot T_f \cdot \sum_{i=1}^{t} p_f(i), \quad (19)$$

where $k_{f,om}$ is the operating and maintenance cost coefficient.

The start-up cost of fuel cells can be expressed as

$$R_f(t) = R_{f,h} + \beta_f \cdot \left(1 - e^{-T_f^{off}(t)/\tau_f}\right), \quad (20)$$

where $R_{f,h}$ is the hot start-up cost of fuel cells; $R_{f,h} + \beta_f$ is the cold start-up cost of fuel cells. T_f^{off} is the off time of fuel cells; τ_f is the cooling time of fuel cells.

The actual output power of fuel cell is limited by lower and upper bounds as

$$p_{f,min} \leq p_f(t) \leq p_{f,max}, \quad (21)$$

where $p_{f,min}, p_{f,max}$ are the minimum and maximum active power of fuel cells.

The following constraint ensures fuel cell does not exceed their ramp limits

$$-\Delta p_{f,D} \leq p_f(t) - p_f(t-1) \leq \Delta p_{f,U}, \quad (22)$$

where $\Delta p_{f,D}, \Delta p_{f,U}$ are the minimum and maximum ramp rate of fuel cells.

The minimum on or minimum down time constraints for fuel cells can be expressed as

$$\left(T_{f,on}(t-1) - T_{f,u}\right).\left(\delta_f(t-1) - \delta_f(t)\right) \geq 0, \quad (23)$$

$$\left(T_{f,off}(t-1) - T_{f,d}\right).\left(\delta_f(t) - \delta_f(t-1)\right) \geq 0, \quad (24)$$

where $T_{f,on}$ is the cumulative uptime. $T_{f,u}, T_{f,d}$ are the minimum turn on time and shut down time of fuel cells.

The number of turn on and turn off can be expressed as

$$n_{f,st-sp}(t) \leq N_{f,max}, \quad (25)$$

where $n_{f,st-sp}(t)$ is the number of turn on and turn off in time t; $N_{f,max}$ is the maximum number of turn on and turn off.

(C) Microturbine. Microturbines have the higher power density, produce less noise, and emit much less pollutants, especially *NOx*. So microturbines are effective devices of converting the fuel energy into electrical energy. Meanwhile, the operating cost of microturbines are formulated as follows:

$$C_m(t) = \sum_{j=1}^{N_m} \sum_{i=1}^{t} \left[C_{m,f}(i) + C_{m,om}(i) + R_m(i).\left(1 - \delta_m(i-1)\right).\delta_m(i)\right], \quad (26)$$

where $C_{m,f}$ is the fuel cost of microturbine; $C_{m,om}$ is the operating and maintenance cost; R_m is the start-up cost of microturbine.

The fuel cost of microturbines is modeled as

$$C_{m,f}(t) = c_m \frac{p_m(t)}{\eta_m(t)}, \quad (27)$$

where c_m is the fuel price of microturbines; η_m is the efficiency of microturbines.

The operating and maintenance cost of microturbines can be expressed as

$$C_{m,om}(t) = k_{m,om}.p_m(t).\Delta t, \quad (28)$$

where $k_{m,om}$ is the operating and maintenance cost coefficient of microturbines.

The start-up cost of microturbines can be expressed as

$$R_m(t) = R_{m,h} + \beta_m.\left(1 - e^{-T_m^{off}(t)/\tau_m}\right), \quad (29)$$

where $R_{m,h}$ is the hot start-up cost of microturbines; $R_{m,h}+\beta_m$ is the cold start-up cost of microturbines. T_m^{off} is the off time of microturbines; τ_m is the cooling time of microturbines.

The actual output power of microturbines is limited by lower and upper bounds as follows:

$$p_{m,min} \leq p_m(t) \leq p_{m,max}, \quad (30)$$

where $p_{m,min}, p_{m,max}$ are the minimum and maximum active power of microturbines.

The following constraint ensures microturbines do not exceed their ramp limits

$$-\Delta p_{m,D} \leq p_m(t) - p_m(t-1) \leq \Delta p_{m,U}, \quad (31)$$

where $\Delta p_{m,D}, \Delta p_{m,U}$ are the minimum and maximum ramp rate of microturbines.

The minimum on or minimum down time constraints for microturbines can be expressed as

$$\left(T_{m,on}(t-1) - T_{m,u}\right).\left(\delta_m(t-1) - \delta_m(t)\right) \geq 0, \quad (32)$$

$$\left(T_{m,off}(t-1) - T_{m,d}\right).\left(\delta_m(t) - \delta_m(t-1)\right) \geq 0, \quad (33)$$

where $T_{m,on}$ is the cumulative uptime; $T_{m,u}, T_{m,d}$ are the minimum turn on time and shut down time of microturbines.

2.3.3. The Probability Model of Wind Turbines. Wind powers as renewable energy sources (RESs) are dependent on numerous factors such as wind velocity and efficiency. The approximate relationship between the wind power and wind speed can be expressed as

$$p_w(v) = \begin{cases} 0, & v < v_i, v \geq v_0, \\ \dfrac{(v - v_i).p_r}{v_r - v_i}, & v_i \leq v < v_r, \\ p_r, & v_r \leq v \leq v_0, \end{cases} \quad (34)$$

where v_i, v_o, v_r are the cut-in wind speed, cut-out wind speed, and rated wind speed. p_r is the rated output power of wind turbines.

Prior research has shown that the wind speed profile follows the Weibull distribution over time, which is [29, 32]

$$F_v(v) = 1 - \exp\left[-\left(\frac{v}{c}\right)^k\right], \quad (35)$$

where positive variables c and k are the scale parameter and shape parameter, respectively.

Based on the characteristic of the wind power (34) and the probability distribution function (35), the probability distribution function of wind power is obtained by

$$F\left(P_w(v)\right)=\begin{cases}0, & p_w(v)<0,\\ 1-\exp\left\{-\left[\dfrac{v_i+(v_r-v_i)\left(p_w(v)/p_r\right)}{c}\right]\right\}, & \\ +\exp\left[-\left(\dfrac{v_0}{c}\right)^k\right], & 0\le p_w(v)\le p_r,\\ 1, & p_w(v)\ge p_r.\end{cases} \quad (36)$$

2.3.4. The Probability Model of Photovoltaics. The output power of photovoltaic can be calculated by its rated output power at the standard test condition and the operating ambient temperature [35]

$$p_s=A_c.\eta.I_\beta=A_c.\eta.\left(T.k_t-T'.k_t^2\right), \quad (37)$$

where A_c is the array surface area; η is the efficiency of the photovoltaic in realistic reporting conditions; I_β is the irradiance on a surface with inclination β to the horizontal plane; T, T' are the parameters that depend on inclination β, reflectance of the ground ρ, etc.

$$T=\left[\left(R_b+\rho.\frac{1-\cos\beta}{2}\right)+\left(\frac{1+\cos\beta}{2}-R_b\right).p\right]. \quad (38)$$

$$r_d.\frac{H_0}{3600},$$

$$T'=\left(\frac{1+\cos\beta}{2}-R_b\right).q.r_d.\frac{H_0}{3600}, \quad (39)$$

where R_b is the ratio of beam radiation on the tilted surface to that on a horizontal surface at any time and H_0 is the solar radiation, for that day, both referring to a horizontal surface; r_d is ratio, diffuse radiation in hour or diffuse in day; p, q are the parameters of the daily clearness index k_t.

Many researches have proved that cloudiness is the main factor affecting the difference between the values of solar radiation measured outside the atmosphere and on earthly surface. The daily clearness index can be obtained by [36]

$$k_t=\frac{I_t}{I_0}, \quad (40)$$

where I_t, I_0 are the ratio of the irradiance on horizontal plane and the extraterrestrial total solar irradiance.

The effect of clouds on terrestrial irradiance is the daily clearness index; k_t can not be predicted with complete confidence; it must be treated as random variable

$$f\left(k_t,\overline{k}_t\right)=\frac{C}{k_{tu}.\lambda.\gamma_1}\left[e^{\lambda.k_t}\left(1-\gamma_1.k_t\right)-1\right], \quad (41)$$

$$\overline{k}_t=\frac{\left(v_s.k_{tu}-1\right).\exp\left(v_s.k_{tu}\right)-\left(v_s.k_{tl}-1\right).\exp\left(v_s.k_{tl}\right)}{v_s.\left(\exp\left(v_s.k_{tu}\right)-\exp\left(v_s.k_{tl}\right)\right)}, \quad (42)$$

where k_{tl}, k_{tu} are the lower and upper bounds of the observed range for k_t; v_s is the parameter of daily clearness index.

$$C=\lambda^2.\frac{k_{tu}}{\left(e^{\lambda.k_{tu}}-1-\lambda.k_{tu}\right)}, \quad (43)$$

$$\gamma_1=\frac{\lambda}{\left(1+\lambda.k_{tu}\right)}, \quad (44)$$

$$\lambda=\frac{\left(2\Gamma-17.519.\exp(-1.3118.\Gamma)-1062.\exp(-5.0426.\Gamma)\right)}{k_{tu}}, \quad (45)$$

$$\Gamma=\frac{k_{tu}}{\left(k_{tu}-\overline{k}_t\right)}, \quad (46)$$

where C, γ_1, λ are the parameters of daily clearness index.

In particular, if $T>0, T'<0$, the probability density function $f(p_s)$ can be expressed as

$$f(p_s)=\begin{cases}\dfrac{C.\left(k_u-\left(\alpha+\alpha'\right)/2\right)}{-k_{tu}.A_c.\eta.T'.\alpha'}.e^{\lambda.(\alpha+\alpha')/2}, & p_s\in\left[0,p_s\left(k_{tu}\right)\right],\\ 0, & p_s\notin\left[0,p_s\left(k_{tu}\right)\right],\end{cases} \quad (47)$$

while if $T>0, T'>0$, the probability density function can be obtained by

$$f(p_s)=\begin{cases}\dfrac{C.\left(k_u-\left(\alpha-\alpha'\right)/2\right)}{-k_{tu}.A_c.\eta.T'.\alpha'}.e^{\lambda.(\alpha-\alpha')/2}, & p_s\in\left[0,p_s\left(k_{tu}\right)\right],\\ 0, & p_s\notin\left[0,p_s\left(k_{tu}\right)\right],\end{cases} \quad (48)$$

where $\alpha=T/T', \alpha'=\sqrt{\alpha^2-4.p_s/(\eta.T'.A_c)}$.

2.3.5. The Model of Interaction with External Main Grid. In this paper, the renewable microgrid is connected to the external main grid and can trade energy with the main grid. The transaction incurs the following cost to the renewable microgrid:

$$C_g(t)=\sum_{i=1}^{t}\left[\rho_g.p_g(i)\right]. \quad (49)$$

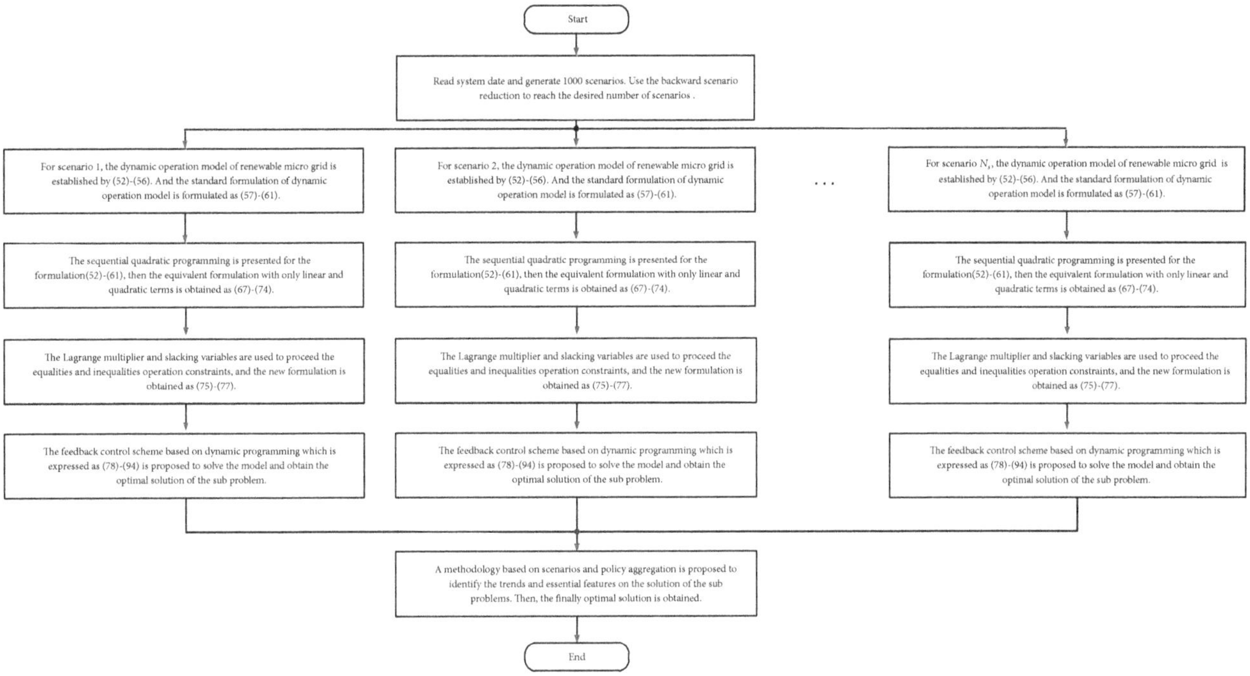

FIGURE 2: The flowchart of the proposed improved dynamic programming algorithm.

Let the amount of energy be bounded by

$$-p_{g,max} \le p_g(t) \le p_{g,max} \tag{50}$$

$$\rho_g = \begin{cases} \rho_s, & p_g \le 0. \\ \rho_b, & p_g > 0. \end{cases} \tag{51}$$

where $p_{g,max}$ is the maximum transaction limits; ρ_s, ρ_b are the price coefficients to purchase and sell energy.

2.4. Dynamic Operation Model of the Renewable Microgrid. Dynamic economic operation management of the renewable microgrid is to determine output power of distribution generators, in order to minimize total operation cost of the renewable microgrid and meet the dynamic operation constrains.

The total operating cost of the renewable microgrid can be defined as

$$\begin{aligned} \min F(t) &= \sum_{t=1}^{T} \left\{ C_{b,c}(t) + C_d(t) + C_f(t) + C_m(t) + C_g(t) \right\} \\ &= \min_{\Pi=\{u(1),\ldots,u(T)\}} E\left\{ \sum_{t=1}^{T} \left(C_t(x(t), u(t), w(t)) \right) \right\}, \end{aligned} \tag{52}$$

where $u(t)$ is the decision or control variables (specifically, action at state $x(t)$ in period t); $x(t)$ is the state variables of the renewable microgrid; $w(t)$ is the random variable such as the output power of wind turbine or photovoltaics.

$u(t)$, $x(t)$, $w(t)$ can be defined as follows:

$$u(t) = \left[\Delta p_b(t), \Delta p_d(t), \Delta p_m(t), \Delta p_f(t), \Delta p_g(t)\right], \tag{53}$$

$$x(t) = \left[SOC_b(t), p_b(t), p_d(t), p_m(t), p_f(t), p_g(t)\right], \tag{54}$$

$$w(t) = \left[p_w(t), p_s(t)\right]. \tag{55}$$

The state variables of renewable microgrid in $t+1$ period can be defined as follows:

$$\begin{aligned} x(t+1) &= f(x(t), u(t), w(t)) \\ &= \left[SOC_b(t+1), p_b(t) + \Delta p_b(t), \; p_d(t) \right. \\ &\quad + \Delta p_d(t), p_m(t) + \Delta p_m(t), p_f(t) \\ &\quad \left. + \Delta p_f(t), p_g(t) + \Delta p_g(t)\right]. \end{aligned} \tag{56}$$

The operation constraints of the renewable microgrid can be expressed as power balance (1)-(2), battery energy storage limits (5)-(8), distributed generations limits (13)-(16), (21)-(25), (30)-(33), and interaction with external main grid limits (50).

2.4.1. The Standard Formulation of the Dynamic Operation Model. According to dynamic programming formulation, the dynamic operation model of the renewable microgrid can be formulated as

$$J_t(x(t)) = \min_{u(t)\in U(t)} \{C_t(x(t), u(t), w(t)) + E[J_{t+1}(x(t+1))]\}. \quad (57)$$

The operation constraints can be expressed as

$$x(t+1) = f(x(t), u(t), w(t)), \quad (58)$$

$$c_{tE}(x(t), u(t), w(t)) = 0, \quad c_{TE}(x(T)) = 0, \quad (59)$$

$$c_{tI}(x(t), u(t), w(t)) \le 0, \quad c_{TI}(x(T)) \le 0, \quad (60)$$

where $c_{tE}, c_{TE}, c_{tI}, c_{TI}$ are, respectively, the equality and inequality constraints.

Meanwhile, the constraint of condition should be satisfied as follows:

$$J_{T+1}(x(T+1)) \equiv 0. \quad (61)$$

3. Solution Methodology

Based on Section 2, dynamic operation management of the renewable microgrid can be regarded as a discrete time system under uncertainty. In order to solve the proposed problem, the random variables can be realized by the scenario analysis technique. Then, a feedback control scheme based on the dynamic programming is proposed to solve the model and obtain the optimal solution. The flowchart of the whole process is given in Figure 2.

3.1. The Scenario Analysis Technique. In this paper, a discrete set of scenarios can be used to represent the probability realization of the output power of wind turbines or photovoltaics. On the other hand, these scenarios are generated using the Roulette wheel mechanism and Monte Carlo simulation method [19]. And the probability distribution function of the random variables can be obtained by (36) and (47)-(48).

The Lattice Monte Carlo simulation(LMCS) can be used to generate the random numbers [29]

$$\sum_{j=1}^{r} \left(\frac{k_j}{n_j} \sum_{i=1}^{N_L} v_i \right) \quad mod 1, \; k_j = 1, \ldots, N_L; \; j = 1, \ldots, r, \quad (62)$$

where r is the number of random variables; N_L is the number of random sampling; v_i is vector with dimension d.

According to the desired preciseness, the probability distribution functions (36) and (47)-(48) are divided into n_l class intervals. Each class interval determines mean value $Int_i, i = 1, \ldots, n_l$; and each interval is associated with a probability denoted by $\beta_i = 1, \ldots, n_l$. Meanwhile, the probabilities of different intervals are normalized in which their summations become equal to unity.

Therefore, each scenario comprises a vector identifying the output power of wind turbine or photovoltaics:

$$S_{t,sc} = \{W^w_{1,t,sc}, \ldots, W^w_{n_l,t,sc}, W^s_{1,t,sc}, \ldots, W^s_{n_l,t,sc}\}, \quad t = 1, \ldots, T, \; sc = 1, \ldots, N_{sc}, \quad (63)$$

where $W^w_{n_l,t,sc}, W^s_{n_l,t,sc}$ are binary parameters indicating where n_l wind power interval or photovoltaic power output whether are selected in scenarios. On the other hand, comparing the random number which follows the LMCS strategy and the probability β_i, the binary parameter can be selected.

Thus, the output power of wind turbine or photovoltaics for each scenario can be obtained by

$$p_w(t) = p_{w,sc}(t) = \sum_{uw=1}^{n_l} \left(W^w_{uw,t,sc} . Int^w_{uw,t} \right), \quad t = 1, \ldots, T, \; sc = 1, \ldots, N_{sc}, \quad (64)$$

$$p_s(t) = p_{s,sc}(t) = \sum_{us=1}^{n_l} \left(W^s_{us,t,sc} . Int^s_{us,t} \right), \quad t = 1, \ldots, T, \; sc = 1, \ldots, N_{sc}. \quad (65)$$

The normalized probability of each scenario can be expressed as follows:

$$\pi_{sc} = \frac{\prod_{t=1}^{T} \left(\left(\sum_{uw=1}^{n_l} \left(W^w_{uw,t,sc} \beta_{uw,t} \right) \right) \left(\sum_{us=1}^{n_l} \left(W^s_{us,t,sc} \beta_{us,t} \right) \right) \right)}{\sum_{sc=1}^{N_s} \prod_{t=1}^{T} \left(\left(\sum_{uw=1}^{n_l} \left(W^w_{uw,t,sc} \beta_{uw,t} \right) \right) \left(\sum_{us=1}^{n_l} \left(W^w_{us,t,sc} \beta_{us,t} \right) \right) \right)}, \quad t = 1, \ldots, T, \; sc = 1, \ldots, N_{sc}, \quad (66)$$

where N_{sc} is the number of scenarios.

3.2. The Feedback Control Scheme Based on the Dynamic Programming. In the following section, we formulate the dynamic operation management of renewable microgrid which satisfies dynamic operations constraints. Firstly, the Lagrange multipliers and interior point method will be presented for the equality and inequality operation constraint. Then, a feedback control scheme based on the dynamic programming is proposed to solve the model and obtain the optimal solution.

3.2.1. The Sequential Quadratic Programming Subproblem Formulation. In this paper, the sequential quadratic programming will be presented for the formulation (52)-(61). The purpose of sequential quadratic programming is to leave the model with only linear and quadratic terms. The equivalent formulation can be obtained as follows:

$$\min \Gamma_q = \sum_{t=1}^{T}\left[\Gamma_t + \overline{x}(t)^T.y(t) + \overline{u}(t)^T.z(t) + \frac{\left(\overline{x}(t)^T.Q(t).\overline{x}(t) + 2.\overline{x}(t)^T.R(t).\overline{u}(t) + \overline{u}(t)^T.S(t).\overline{u}(t)\right)}{2}\right] + \Gamma_T$$
$$+ \overline{x}(T)^T.y(T) + \overline{x}(T)^T.Q(T).\frac{\overline{x}(T)}{2}, \tag{67}$$

where $v(t)$ is a constant. $\overline{u}(t), \overline{x}(t)$ are the perturbation variables. $\hat{u}(t), \hat{x}(t)$ are the nominal values of $x(t), u(t)$. And these variables can be introduced and defined as

$$\Gamma_t = x(t)^T.y(t) + x(t)^T.Q(t).x(t) + v(t), \tag{68}$$

$$\Gamma_T = x(T)^T.y(T) + x(T)^T.Q(T).x(T) + v(T), \tag{69}$$

$$\overline{u}(t) = u(t) - \hat{u}(t), \tag{70}$$

$$\overline{x}(t) = x(t) - \hat{x}(t). \tag{71}$$

The operation constraints can be expressed as

$$\overline{x}(t+1) = A(t).\overline{x}(t) + B(t).\overline{u}(t), \tag{72}$$

$$A_E(t).\overline{x}(t) + B_E(t).\overline{u}(t) + \hat{c}_E(t) = 0,$$
$$A_E(T).\overline{x}(T) + \hat{c}_E(T) = 0, \tag{73}$$

$$A_I(t).\overline{x}(t) + B_I(t).\overline{u}(t) + \hat{c}_I(t) \le 0,$$
$$A_I(T).x(T) + \hat{c}_I(T) \le 0, \tag{74}$$

where $y(t), z(t), Q(t), R(t), S(t)$ and $A(t), B(t), A_E(t), B_E(t), \hat{c}_E(t), A_I(t), B_I(t), \hat{c}_I(t)$ can be obtained in Appendix A.

3.2.2. Interior Point Methods. According to the sequential quadratic programming formulations (67), by applying the Lagrange multipliers and slacking variables to proceed the equalities and inequalities operation constraints, a new formulation is shown as

$$L_q^k = \sum_{t=1}^{T}\left[\overline{v}^k(t + \tilde{x}(t^T.\overline{y}(t) + \tilde{u}(t)^T.\overline{z}(t)\right.$$
$$+ \left(\tilde{x}(t)^T.\overline{Q}(t).\tilde{x}(t) + 2.\tilde{x}(t)^T.\overline{R}(t).\tilde{u}(t)\right.$$
$$\left. + \tilde{u}(t)^T.\overline{S}(t).\frac{\tilde{u}(t)}{2}\right] + \left[\left(\overline{A}_E(t).\tilde{x}(t)\right.\right.$$
$$+ \overline{B}_E(t).\tilde{u}(t) + \overline{c}_E(t)\Big)^T.\tilde{\lambda}_E(t) + \left(\overline{A}_I(t).\tilde{x}(t)\right.$$
$$\left. + \overline{B}_I(t).\tilde{u}(t) + \overline{c}_I(t)\right)^T.\tilde{\lambda}_I(t) - \tilde{\lambda}_I(t)^T.\overline{G}_t.\frac{\tilde{\lambda}_I(t)}{2}\Bigg]$$
$$+ \tilde{x}(T)^T.\overline{y}(T) + \tilde{x}(T)^T.Q(T).\frac{\tilde{x}(T)}{2} + \left(\overline{A}_E(T).\right.$$
$$\tilde{x}(T) + \overline{c}_E(T)\Big)^T.\tilde{\lambda}_E(T) + \left(\overline{A}_I(T).\tilde{x}(T) + \overline{c}_I(T)\right)^T.$$
$$\tilde{\lambda}_I(T) - \tilde{\lambda}_I(T)^T.\overline{G}_T.\frac{\tilde{\lambda}_I(T)}{2}, \tag{75}$$

where k is considered as the kth iteration. $\overline{v}^k(t)$ is a constant.

Meanwhile, the parameters can be defined as follows:

$$\tilde{z} = \left[\tilde{u}(1) \;\ldots\; \tilde{u}(T-1) \;\; \tilde{x}(1)\right]^T,$$
$$\tilde{\xi} = \left[\tilde{\lambda}_E(1) \;\ldots\; \tilde{\lambda}_E(t) \;\; \tilde{\lambda}_E(T)\right]^T, \tag{76}$$

$$\tilde{\lambda} = \left[\tilde{\lambda}_I(1) \;\ldots\; \tilde{\lambda}_I(t) \;\; \tilde{\lambda}_I(T)\right]^T,$$
$$\tilde{s} = \left[\tilde{s}_I(1) \;\ldots\; \tilde{s}_I(t) \;\; \tilde{s}_I(T)\right]^T. \tag{77}$$

The approach in solving $(\tilde{z}, \tilde{\xi}, \tilde{\lambda}, \tilde{s})$ is given in Appendix B.

3.2.3. Total Algorithm Procedures. In this paper, an algorithm is presented based on an adaptation of Mehrotra's predictor-corrector algorithm [37]. Given the initial parameters $\delta > 0, k = 0$ and $(z^0, \xi^0, \lambda^0, s^0)$ the residual terms are defined as

$$\left(r_1^k\right)^2 = \sum_{t=1}^{T}\left(\frac{\partial L_q^k}{\partial \tilde{u}(t)}\right)^T\left(\frac{\partial L_q^k}{\partial \tilde{u}(t)}\right), \tag{78}$$

$$\left(r_2^k\right)^2 = \sum_{t=1}^{T}\left(A_E(t).x(t)^k + B_E(t).u(t)^k + \hat{c}_E(t)\right)^T.$$
$$\left(A_E(t).x(t)^k + B_E(t).u(t)^k + \hat{c}_E(t)\right), \tag{79}$$

$$\left(r_3^k\right)^2 = \sum_{t=1}^{T}\left(A_I(t).x(t)^k + B_I(t).u(t)^k + \hat{c}_I(t)\right.$$
$$\left. + s_I^k(t)\right)^T.\left(A_I(t).x(t)^k + B_I(t).u(t)^k + \hat{c}_I(t)\right.$$
$$\left. + s_I^k(t)\right), \tag{80}$$

$$\left(r_4^k\right)^2 = \sum_{t=1}^{T}\left[\min\left(\lambda_I^k(t), s_I^k(t)\right)\right]^T.$$
$$\left[\min\left(\lambda_I^k(t), s_I^k(t)\right)\right]. \tag{81}$$

As the solution set approaches optimum point, the summation of all residual term should approach zero.

The total algorithm can be expressed as follows.

Step 1. Calculate the residual term as

$$r^k = \sqrt{\left(r_1^k\right)^2 + \left(r_2^k\right)^2 + \left(r_3^k\right)^2 + \left(r_4^k\right)^2}, \tag{82}$$

where the expressions for $r_1^k, r_2^k, r_3^k, r_4^k$ are given in (78)-(81). If $r^k < \delta$, then go to Step 5.

Step 2. Calculate the affine-scaling direction $\{\tilde{z}_a, \tilde{\xi}_a, \tilde{\lambda}_a, \tilde{s}_a\}$ by using the method introduced in Appendix B.

$d_k, c_I^k(t)$ can be defined as

$$d_k = -\text{diag}\left(\lambda^k\right).\text{diag}\left(s^k\right).e, \quad (83)$$

$$c_I^k(t) = A_I(t).x^k(t) + B_I(t).u^k(t) + \tilde{c}_I(t). \quad (84)$$

The scalars α_a, μ_a, and σ can be calculated as follows:

$$\mu_k = \left(\lambda^k\right)^T.\frac{s^k}{n_2}, \quad (85)$$

where n_2 is the dimension of s^k.

$$\alpha_a = \arg\max\left\{\alpha \in [0,1] \mid \lambda^k + \alpha.\tilde{\lambda}_a \geq 0, s^k + \alpha\tilde{s}_a \geq 0\right\}, \quad (86)$$

$$\mu_a = \left(\lambda^k + \alpha.\tilde{\lambda}_a\right)^T.\frac{1}{n_2}.\left(s^k + \alpha_a.\tilde{s}_a\right). \quad (87)$$

Meanwhile, the centering parameter is defined as follows:

$$\sigma = \left(\frac{\mu_a}{\mu_k}\right)^3. \quad (88)$$

Step 3. Calculate the combined predictor-centering-corrector direction $\{\tilde{z}_p, \tilde{\xi}_p, \tilde{\lambda}_p, \tilde{s}_p\}$ and the scalar α_{max} as the method introduced in Appendix B.

Let

$$d_k = -\text{diag}\left(\lambda^k\right).\text{diag}\left(s^k\right).e - \text{diag}\left(\tilde{\lambda}_a\right).\text{diag}\left(\tilde{s}_a\right).e + \sigma.\mu_k.e, \quad (89)$$

$$c_I^k(t) = A_I(t).x^k(t) + B_I(t).u^k(t) + \hat{c}_I(t) + \text{diag}\left(\lambda_I^k(t)\right)^{-1}.\left[\sigma.\mu_k.e - \text{diag}\left(\tilde{\lambda}_a(t)\right).\text{diag}\left(\tilde{s}_a(t)\right).e\right], \quad (90)$$

where $e = [1 \ \ldots \ 1]$.

Calculate the scalar as follows:

$$\alpha_{max} = \arg\max\left\{\alpha \in [0,1] \mid \lambda^k + \alpha.\tilde{\lambda}_p \geq 0, s^k + \alpha.\tilde{s}_p \geq 0\right\}. \quad (91)$$

Step 4. Calculate $(z^{k+1}, \xi^{k+1}, \lambda^{k+1}, s^{k+1})$ from the equation

$$\left(z^{k+1}, \xi^{k+1}, \lambda^{k+1}, s^{k+1}\right) = \left(z^k, \xi^k, \lambda^k, s^k\right) + \alpha\left(\tilde{z}_p, \tilde{\xi}_p, \tilde{\lambda}_p, \tilde{s}_p\right), \quad (92)$$

where $\alpha = \min(0.995.\alpha_{max}, 1)$. Then set $k \longleftarrow k+1$ and return to Step 1.

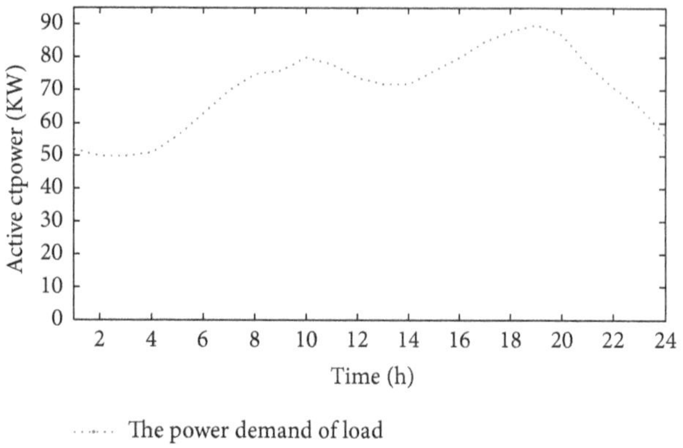

FIGURE 3: Load profiles for a period of one day.

Step 5. Calculate the search direction $\tilde{u}(t), \tilde{x}(1), \tilde{\lambda}_E(t), \tilde{\lambda}_I(t), \tilde{s}_I(t)$ and the optimal solution

$$u^*(t) = u^k(t) + \tilde{u}(t), \quad x^*(1) = x^k(1) + \tilde{x}(1). \quad (93)$$

Calculate all state variables by using the equation

$$x^*(t+1) = A(t).x^*(t) + B(t).u^*(t), \quad t = 1, \ldots, T. \quad (94)$$

4. Simulation

In this section, the renewable microgrid in the simulation is shown in Figure 1; it operates in parallel to the main power grid or as an island and comprises wind turbines with maximum power of 13kw, PV panels with maximum power of 15kw, and distributed generators such as microturbines, diesel generators, and full cells. A battery energy storage is included, bounded between 20 and 60 kWh and with maximal charge and discharge rates, respectively, 30 and -30kw. The charge and discharging efficiencies are both equal to 0.85. Tables 1–4 describe the distributed generators units parameters, based on data provided in [29, 32]. The simulations are carried out in the MATLAB environment on an Intel Core 2 Duo 3.00GHz running Windows 7. In the simulations study, we chose the sampling time of 1 hour. And simulations are performed over a horizon of 24 hours.

The electricity usage, wind power, and solar generation data in the simulations have been provided by National Renewable Energy Laboratory. The electricity usage data is from the utility operator customers with peak usage over 90kw; the maximum power demand of load is 90.1kw. The daily electricity usage (from Miami, FI, on a certain day) is shown in Figure 3.

4.1. Scenario Generation and Reduction Results. The approach used for generation of scenarios may have an impact on the determination of operational cost of renewable microgrid. In the study case, the dates from wind power profiles of the 24 hours are used as scenarios in the dynamic operation

Table 1: The capacities parameters of microgenerators (MGs) [29, 32].

The capacities of MGs	Wind power	photovoltaics	Battery	Diesel generator	microturbine	full cell
Unite(kw)	13	15	60	30	30	50

Table 2: The technologies parameters of distributed generators [29, 32].

Parameters	Parameters a	Parameters b	Parameters c
Diesel generator	0.4333	0.2333	0.0074
Parameters	Fuel price parameter	Efficiency parameter	Auxiliary equipment
Full cell	0.0164	0.6	1.228
Microturbine	0.0164	0.56	-

Table 3: The technologies parameters of battery storage system [29, 32].

Upper energy limits	Lower energy limits	Charging efficiency	Discharging efficiency
20 (kw)	60(kw)	0.85	0.85

Table 4: The electricity interaction parameters with dispatching networks [10, 19].

The upper energy limits	The lower energy limits
40 (kw)	-40(kw)

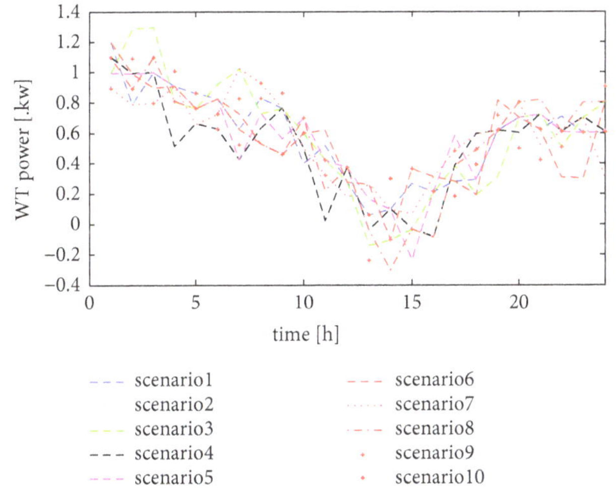

Figure 4: Renewable energy (wind power) scenarios for a period of one day.

management of renewable microgrid. The scenarios of wind power and photostatic chosen for this study are shown in Figures 4, 5, and 6.

4.2. The Base Case of the Dynamic Operation Management of Renewable Microgrid. In this part, its assumed that the renewable microgrid is disconnected from the main power grid. The active power of load is supplied by the microgenerators such as microturbines, diesel generators, fuel cells, wind turbines and photovoltaics, and a battery storage system. The optimal results of the dynamic operation management of renewable microgrid are given by Figures 7 and 8.

Table 5: The comparison of the results obtained by different modes for smart microgrid.

Modes	The base case	The second case
Cost ($)	**1.8636e+03**	**1.83677e+03**

Figures 7 and 8 show that microturbine and fuel cell have to generate more electricity power than the diesel generators. Because the diesel generators are expensive, they are restricted to their minimum value during most hours.

4.3. The Case of the Renewable Microgrid Operates in Parallel to the Main Power Grid. In the second case, its assumed that the renewable microgrid can operate in parallel to the main power grid. When the renewable microgrid is connected to the main power grid, its loads receive power from both the main power gird and the local distributed energy resources such as solar and wind power, distributed energy generators, and a battery storage system. The optimal results of the dynamic operation management of renewable microgrid are shown in Figures 9 and 10.

It should be noted that, in Figures 9 and 10, the main power grid has to generate more electricity power. Fuel cell and diesel generators are more expensive.

In Table 5, it is worthwhile to note that the total operation cost of second case is the lowest. In other word, the renewable microgrid which operates in parallel to the main power grid decreases the operation cost because of its loads receiving power from both the main power gird and the local distributed energy resources.

4.4. The Case of Dynamic Operation Management of Renewable Microgrid under Uncertainties. In this case, a methodology based on scenarios analysis is used to proceed the

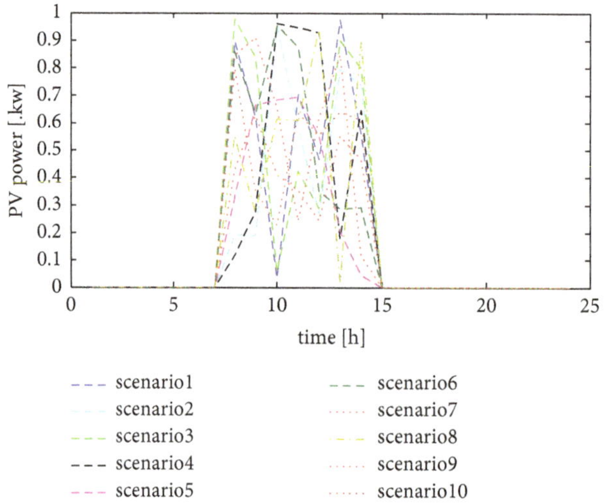

FIGURE 5: Renewable energy (photovoltaics) scenarios for a period of one day.

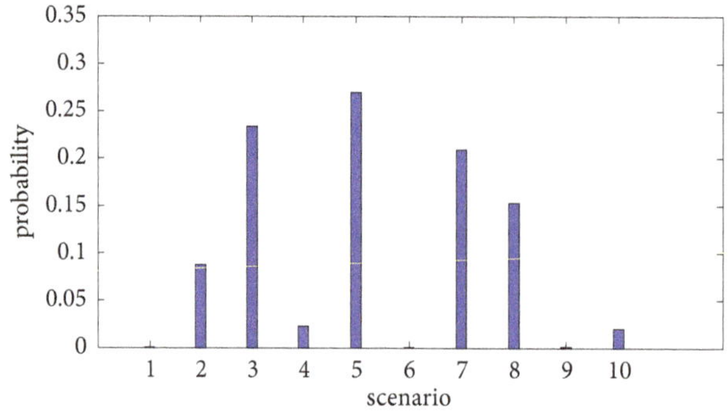

FIGURE 6: The probability of renewable energy scenarios for a period of one day.

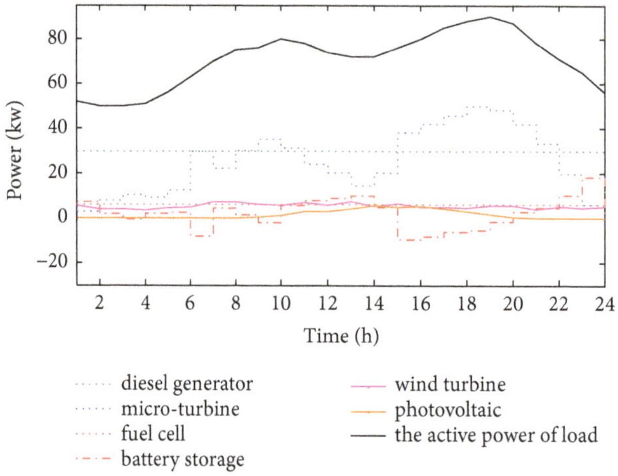

FIGURE 7: Optimal results of economic operation optimization of renewable microgrid.

uncertainty of the power output of distributed renewable generators, such as wind power and photovoltaics power and it is assumed that the renewable microgrid is disconnected from the main grid. The optimal results of the dynamic economic operation optimization of renewable microgrid are shown in Figures 11 and 12.

In Table 6, it is worthwhile to note that this scheme is a direct method to obtain the optimal solution without

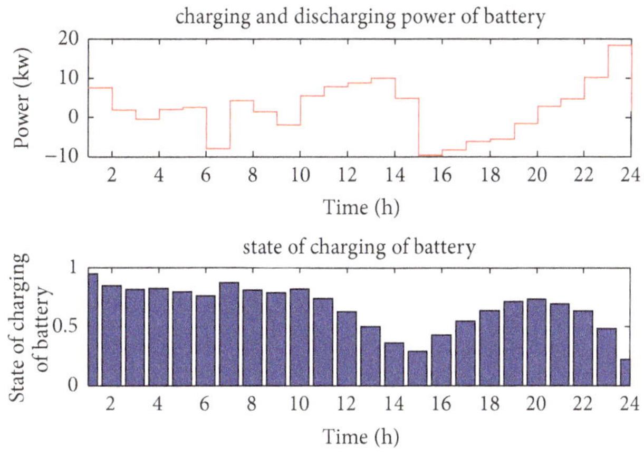

FIGURE 8: Optimal results of battery of renewable microgrid.

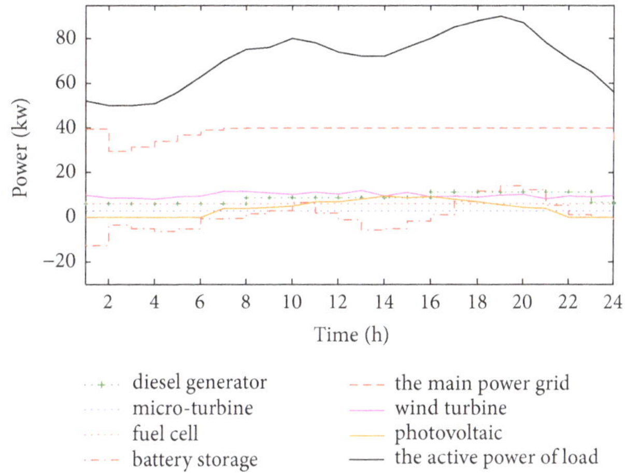

FIGURE 9: Optimal results of economic operation optimization of renewable microgrid (the main power grid).

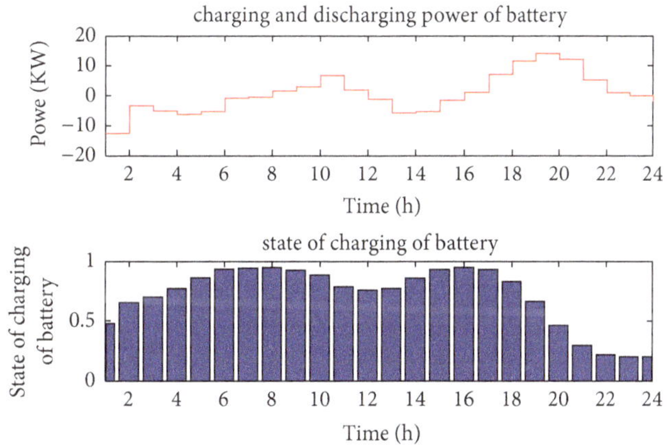

FIGURE 10: Optimal results of battery of renewable microgrid (the main power grid).

probability. And the total operation cost of the method of proposed by this paper is the lowest. In other words, modeling the renewable microgrid system under uncertainty decreases the operation cost. This is because of considering different scenarios in the stochastic model instead of single scenario in the deterministic scheme.

5. Conclusions

In this paper, we introduce a novel dynamic operation management of renewable microgrid. First, to properly model the system, a mathematical model including power losses of the renewable microgrid is established. Then,

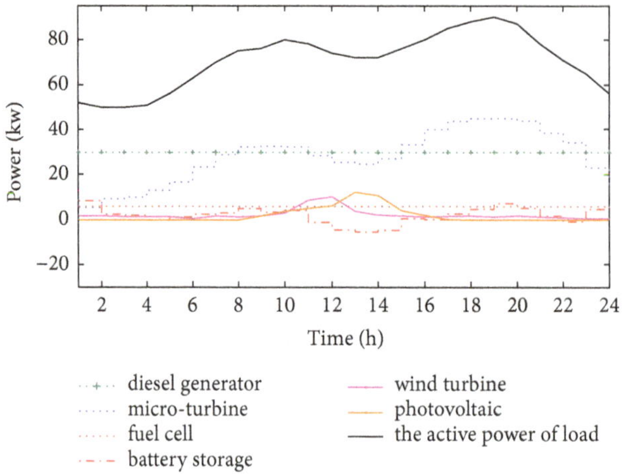

FIGURE 11: Optimal results of economic operation optimization of renewable microgrid under uncertainty.

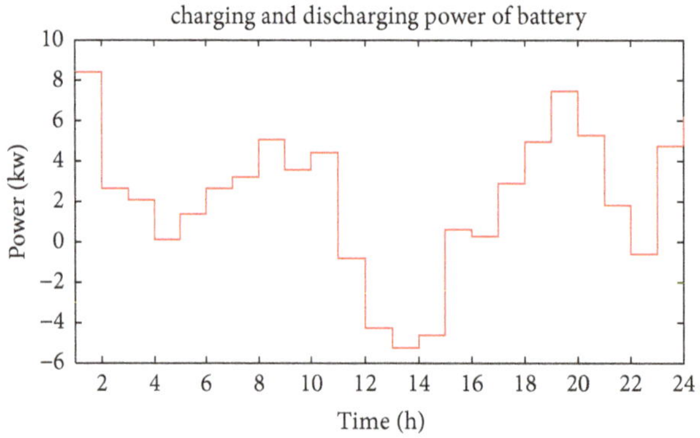

FIGURE 12: Optimal results of battery of renewable microgrid under uncertainty.

TABLE 6: The comparison of the results obtained by dynamic power dispatching under uncertainty.

Dynamic power dispatching strategies under uncertainty	The probability of scenario	The cost of scenario
The cost of scenario 1	0.003	1866.060
The cost of scenario 2	3.542121e-05	**1866.166**
The cost of scenario 3	**3.517950e-09**	1866.154
The cost of scenario 4	9.450705e-04	1866.152
The cost of scenario 5	0.232	1866.129
The cost of scenario 6	0.103	1866.166
The cost of scenario 7	0.096	1866.143
The cost of scenario 8	9.103001e-07	1866.140
The cost of scenario 9	**0.437**	1866.152
The cost of scenario 10	0.127	1866.165
The method of proposed by this paper	-	**1733.055**

the uncertainty in the distributed renewable generators, such as wind power and photovoltaics power, are modeled by choosing a set of possible scenarios. And each scenario is assigned a probability that reflects the scenario to be occurred. Finally, the problem of economic operation of renewable microgrid is solved by a feedback control scheme based on the dynamic programming. To solved the deterministic model efficiently, we use the Lagrange multipliers and the interior point method to solve the equality and inequality operation constraints. The numerical results indicate the validity of this optimization methodology.

Appendix

A. A Dynamic Programming Method for Constrained Optimal Control

A.1. Problem Formulation. In this paper, the discrete time, optimal control problem is presented as formulation (52)-(61). Find the controls $u(1), \ldots, u(T)$ and the initial state $x(1)$ which minimize the function

$$\Gamma = \sum_{t=1}^{T} \Gamma_t (x(t), u(t)) + \Gamma_T (x(T)). \tag{A.1}$$

The constraints are as follows:

$$x(t+1) = g_t (x(t), u(t)), \tag{A.2}$$

$$\begin{aligned} c_{tE} (x(t), u(t)) &= 0, \\ c_{TE} (x(T)) &= 0, \end{aligned} \tag{A.3}$$

$$\begin{aligned} c_{tI} (x(t), u(t)) &\leq 0, \\ c_{TI} (x(T)) &\leq 0, \end{aligned} \tag{A.4}$$

where the functions $g_t, \Gamma_t, c_{tE}, c_{tI}$ are assumed as twice differentiable.

A.2. Sequential Quadratic Programming Subproblem Formulations. The discrete time, optimal control problem can be approximated by a series of quadratic subproblems. And consider the Lagrange function associated with ((A.1)–(A.4)) and expressed as follows:

$$\begin{aligned} L = \sum_{i=1}^{T} \Gamma_t (x(t), u(t)) + c_{tE} (x(t), u(t))^T . \lambda_E (t) \\ + c_{tI} (x(t), u(t)) . \lambda_I (t) + \Gamma_T (x(T)) \\ + c_{TE} (x(T))^T . \lambda_{TE} + c_{TI} (x(T))^T . \lambda_I (T), \end{aligned} \tag{A.5}$$

where $\lambda_E(t), \lambda_I(t)$ are the Lagrange multipliers.

Expanding (A.2) and (A.5) as Taylor series about $\widehat{u}(t), \widehat{x}(1), \widehat{\lambda}_E(t), \widehat{\lambda}_I(t)$ yields the following quadratic approximation of Lagrange function:

$$\begin{aligned} L_q = \sum_{t=1}^{T} \Bigg[v(t) + \overline{x}(t)^T . y(t) + \overline{u}(t)^T . z(t) + \frac{\left(\overline{x}(t)^T . Q(t) . \overline{x}(t) + 2 . \overline{x}(t)^T . R(t) . \overline{u}(t)^T + \overline{u}(t)^T . S(t) . \overline{u}(t)\right)}{2} + (A_E (t) . \\ \overline{x}(t) + B_E (t) . \overline{u}(t) + \widehat{c}_E (t))^T . \lambda_E (t) + (A_I (t) . \overline{x}(t) + B_I (t) . \overline{u}(t) + \widehat{c}_I (t))^T . \lambda_I (t) \Bigg] + v(T) + \overline{x}(T)^T . y(T) + \overline{x}(T)^T . \\ Q(T) . \frac{\overline{x}(T)}{2} + (A_E (T) . \overline{x}(T) + \widehat{c}_E (T))^T . \lambda_E (T) + (A_I (T) . \overline{x}(T) + \widehat{c}_I (T))^T . \lambda_I (T). \end{aligned} \tag{A.6}$$

And the constraints are as follows:

$$\overline{x}(t+1) = A(t) . \overline{x}(t) + B(t) . \overline{u}(t), \tag{A.7}$$

where $y(t), z(t), Q(t), R(t), S(t)$ and $A(t), B(t), A_E(t), B_E(t), \widehat{c}_E(t), A_I(t), B_I(t), \widehat{c}_I(t)$ can be obtained by

$$\widehat{c}_E (t) = c_E (\widehat{x}(t), \widehat{u}(t)),$$

$$\widehat{c}_I (t) = c_I (\widehat{x}(t), \widehat{u}(t)),$$

$$y^\alpha (t) = \frac{\partial \Gamma_t}{\partial x(t)^\alpha},$$

$$z^\alpha (t) = \frac{\partial \Gamma_t}{\partial u(t)^\alpha},$$

$$a^{\alpha\beta} (t) = \frac{\partial g(t)^\alpha}{\partial x(t)^\beta},$$

$$b^{\alpha\beta} (t) = \frac{\partial g(t)^\alpha}{\partial u(t)^\beta},$$

$$a_E^{\alpha\beta} (t) = \frac{\partial c_E^\alpha (t)}{\partial x^\beta (t)},$$

$$b_E^{\alpha\beta} (t) = \frac{\partial c_E^\alpha (t)}{\partial u^\beta (t)},$$

$$a_I^{\alpha\beta} (t) = \frac{\partial c_I^\alpha (t)}{\partial x^\beta (t)},$$

$$b_I^{\alpha\beta} (t) = \frac{\partial c_I^\alpha (t)}{\partial u^\beta (t)},$$

$$\begin{aligned} q^{\alpha\beta} (t) = & \frac{\partial^2 \Gamma_t}{\partial x^\alpha (t)} . \partial x^\beta (t) \\ & + \sum_{\gamma=1}^{n_E(t)} \left(\frac{\partial^2 c_E^\gamma (t)}{\partial x^\alpha (t)} . \partial x^\beta (t) \right) . \widehat{\lambda}_E^\gamma (t) \\ & + \sum_{\gamma=1}^{n_I(t)} \left(\frac{\partial^2 c_I^\gamma (t)}{\partial x^\alpha (t)} . \partial x^\beta (t) \right) . \widehat{\lambda}_I^\gamma (t) \end{aligned}$$

$$+\sum_{\gamma=1}^{n}\left(\frac{\partial^2 g^\gamma(t)}{\partial x^\alpha(t)}.\partial x^\beta(t)\right).p^\gamma(t+1),$$

$$r^{\alpha\beta}(t)=\frac{\partial^2\Gamma_t}{\partial x^\alpha(t)}.\partial u^\beta(t)$$

$$+\sum_{\gamma=1}^{n_E(t)}\left(\frac{\partial^2 c_E^\gamma(t)}{\partial x^\alpha(t)}.\partial u^\beta(t)\right).\hat{\lambda}_E^\gamma(t)$$

$$+\sum_{\gamma=1}^{n_I(t)}\left(\frac{\partial^2 c_I^\gamma(t)}{\partial x^\alpha(t)}.\partial u^\beta(t)\right).\hat{\lambda}_I^\gamma(t)$$

$$+\sum_{\gamma=1}^{n}\left(\frac{\partial^2 g^\gamma(t)}{\partial x^\alpha(t)}.\partial u^\beta(t)\right).p^\gamma(t+1),$$

$$s^{\alpha\beta}(t)=\frac{\partial^2\Gamma_t}{\partial u^\alpha(t)}.\partial u^\beta(t)$$

$$+\sum_{\gamma=1}^{n_E(t)}\left(\frac{\partial^2 c_E^\gamma(t)}{\partial u^\alpha(t)}.\partial u^\beta(t)\right).\hat{\lambda}_E^\gamma(t)$$

$$+\sum_{\gamma=1}^{n_I(t)}\left(\frac{\partial^2 c_I^\gamma(t)}{\partial u^\alpha(t)}.\partial u^\beta(t)\right).\hat{\lambda}_I^\gamma(t)$$

$$+\sum_{\gamma=1}^{n}\left(\frac{\partial^2 g^\gamma(t)}{\partial u^\alpha(t)}.\partial u^\beta(t)\right).p^\gamma(t+1). \tag{A.8}$$

$p(t), p(T)$ are obtained from the recursive equation

$$p(t)=y(t)+A_E^T(t).\hat{\lambda}_E(t)+A_I^T(t).\hat{\lambda}_I(t)+A^T(t).p(t+1), \tag{A.9}$$

$$p(T)=y(T)+A_E^T(T).\hat{\lambda}_E(T)+A_I^T(T).\hat{\lambda}_I(T), \tag{A.10}$$

where $a^{\alpha\beta}$ refers to the element in row α and column β of the matrix A. Similarly, x^α refers to the element in row α of the vector x.

B. A Modification Dynamic Programming for Equivalent Problem

Expanding (A.2) and (A.5) as Taylor series about $\hat{u}(t), \hat{x}(1), \hat{\lambda}_E(t), \hat{\lambda}_I(t)$ yields the following quadratic approximation of Lagrange function:

$$L_q^k$$
$$=\sum_{t=1}^{T}\left[v^k(t)+\tilde{x}(t)^T.y^k(t)+\tilde{u}(t)^T.z^k(t)+\frac{\left(\tilde{x}(t)^T.Q(t).\tilde{x}(t)+2.\tilde{x}(t)^T.R(t).\tilde{u}(t)+\tilde{u}(t)^T.S(t).\tilde{u}(t)\right)}{2}\right]$$
$$+\left[\left(A_E(t).\tilde{x}(t)+B_E(t).\tilde{u}(t)+c_E^k(t)\right)^T.\tilde{\lambda}_E(t)+\left(A_I(t).\tilde{x}(t)+B_I(t).\tilde{u}(t)+c_I^k(t)\right)^T.\tilde{\lambda}_I(t)-\tilde{\lambda}_I(t)^T.G_t.\frac{\tilde{\lambda}_I(t)}{2}\right] \tag{B.1}$$
$$+\tilde{x}(T)^T.y(T)+\tilde{x}(T)^T.Q(T).\frac{\tilde{x}(T)}{2}\left(A_E(T).\tilde{x}(T)+c_E^k(T)\right)^T.\tilde{\lambda}_E(T)+\left(\overline{A}_I(T).\tilde{x}(T)+c_I^k(T)\right)^T.\tilde{\lambda}_I(T)$$
$$-\tilde{\lambda}_I(T)^T.G_T^k.\frac{\tilde{\lambda}_I(T)}{2}$$

and

$$\tilde{x}(t+1)=A(t).\tilde{x}(t)+B(t).\tilde{u}(t), \tag{B.2}$$

$$x^k(t+1)=A(t).x^k(t)+B(t).u^k(t), \tag{B.3}$$

where $v^k(t)$ is a constant and

$$y^k(t)=y(t)+Q(t).x^k(t)+R(t).u^k(t)+A_E^T(t).\lambda_E^k(t)+A_I^T(t).\lambda_I^k(t), \tag{B.4}$$

$$z^k(t)=z(t)+R^T(t).x^k(t)+S(t).u^k(t)+B_E^T(t).\lambda_E^k(t)+B_I^T(t).\lambda_I^k(t), \tag{B.5}$$

$$c_E^k(t)=A_E(t).x^k(t)+B_E(t).u^k(t)+\hat{c}_E(t), \tag{B.6}$$

$$c_I^k(t)=A_I(t).x^k(t)+B_I(t).u^k(t)+\hat{c}_I(t)+s_I^k(t)+\operatorname{diag}\left(\lambda_I^k(t)\right)^{-1}.d_k(t), \tag{B.7}$$

$$G^k(t)=\operatorname{diag}\left(\lambda_I^k(t)\right)^{-1}.\operatorname{diag}\left(s_I^k(t)\right). \tag{B.8}$$

The matrix $G^k(t)$ can be defined as follows:

$$G^k(t)=\left[P_b(t)\;\;P_f(t)\right].\begin{bmatrix}G_b(t) & 0\\ 0 & G_f(t)\end{bmatrix}.\begin{bmatrix}P_b^T(t)\\ P_f^T(t)\end{bmatrix}, \tag{B.9}$$

where $\left[P_b(t)\;\;P_f(t)\right]$ is a permutation matrix. For purposes of numerical stability, the number of columns in $P_b(t)$ is set equal to the number of diagonal elements in $G^k(t)$ which are less than unity.

So, the parameters of (75) can be expressed as follows:

$$\overline{y}(t) = y^k(t) + A_I^T(t).\overline{P}(t).c_I^k(t), \quad \overline{z}(t) = z^k(t) + B_I^T(t).\overline{P}(t).c_I^k(t), \tag{B.10}$$

$$\overline{P}(t) = P_f(t).G_f^{-1}(t).P_f^T(t), \tag{B.11}$$

$$\overline{Q}(t) = Q(t) + A_I^T(t).\overline{P}(t).A_I(t), \quad \overline{R}(t) = R(t) + A_I^T(t).\overline{P}(t).B_I(t), \tag{B.12}$$

$$\overline{S}(t) = S(t) + B_I^T(t).\overline{P}(t).B_I(t), \tag{B.13}$$

$$\overline{A}(t) = \begin{pmatrix} A_E(t) \\ P_b^T(t).A_I(t) \end{pmatrix}, \quad \overline{B}(t) = \begin{pmatrix} B_E(t) \\ P_b^T(t).B_I(t) \end{pmatrix}, \quad \overline{c}(t) = \begin{pmatrix} c_E^k(t) \\ P_b^T(t).c_I^k(t) \end{pmatrix}, \tag{B.14}$$

$$\widetilde{\lambda}(t) = \begin{pmatrix} \widetilde{\lambda}_E(t) \\ \widetilde{\lambda}_b(t) \end{pmatrix}, \quad \overline{G}(t) = \begin{bmatrix} 0 & 0 \\ 0 & G_b(t) \end{bmatrix}. \tag{B.15}$$

And then, a modification of the dynamic programming algorithm is presented to solve the equivalent problem (75). We can obtain

$$H_1(t) = \overline{Q}(t) + A^T(t).W(t+1).A(t), \quad H_2(t) = \overline{R}(t) + A^T(t).W(t+1).B(t), \tag{B.16}$$

$$H_3(t) = \overline{S}(t) + B^T(t).W(t+1).B(t), \quad H_4(t) = \overline{y}(t) + A^T(t).v(t+1), \tag{B.17}$$

$$H_5(t) = \overline{z}(t) + B^T(t).v(t+1), \tag{B.18}$$

$$F_1(t) = \begin{bmatrix} H_3(t) & \overline{B}(t)^T & B(t)^T.A_n(t+1)^T \\ \overline{B}(t) & -\overline{G}(t) & 0 \\ A_n(t+1).B(t) & 0 & B_n(t+1) \end{bmatrix}, \tag{B.19}$$

$$F_2(t) = \begin{bmatrix} H_2^T(t) \\ \overline{A}(t) \\ A_n(t+1).A(t) \end{bmatrix}, \tag{B.20}$$

$$f_3(t) = \begin{bmatrix} h_5(t) \\ \overline{c}(t) \\ c_n(t+1) \end{bmatrix}, \tag{B.21}$$

where $A_n(t), B_n(t), c_n(t), v(t), W(t)$ are calculated from recursive equations.

The matrix $F_1(t)$ is factored as follows:

$$F_1(t) = \Phi(t).\Sigma(t).\Phi^T(t), \tag{B.22}$$

where $\Phi(t)$ is an orthogonal matrix of eigenvectors.

$$\Sigma(t) = \operatorname{diag}\left(\sigma(t), \ldots, \sigma_{n_f(t)}(t)\right). \tag{B.23}$$

For convenience, the eigenvalues $\sigma(t)$ can be obtained by

$$|\sigma(t)| \geq |\sigma(j+1)|, \quad j = 1, \ldots, n_f(t) - 1. \tag{B.24}$$

Define

$$n_r(t) = \max\left(\max\left\{j \mid \left|\frac{\sigma(t)}{\sigma(1)}\right| \geq \frac{1}{c_{max}}\right\}, n_f(t) - n\right). \tag{B.25}$$

$n_n(t) = n_f(t) - n_r(t)$, c_{max} is a positive scalar.

$$\sum_{r(t)} = \operatorname{diag}\left(\sigma(1), \ldots, \sigma_{n_r}(t), \sum_{n(t)} = \operatorname{diag}\left(\sigma_{n_r(t)+1}, \ldots, \sigma_{n_f(t)}\right)\right) \tag{B.26}$$

$$\Phi(t) = \left[\Phi_{r(t)}, \Phi_{n(t)}\right], \tag{B.27}$$

where $\Phi_{r(t)} \in R^{n_f(t).n_r(t)}$, $\Phi_{n(t)} \in R^{n_f(t).n_n(t)}$.

The modification of the dynamic programming algorithm to solve the problem can be obtained by the following.

Step 1. Calculate the parameters as follows:

$$A_n(t) = \Phi_n^T(t).F_2(t), \quad B_n(t) = \sum_{n(t)}, \quad c_{n(t)} = \Phi_{n(t)}^T.F_3(t), \tag{B.28}$$

$$W(t) = H_1(t) - F_2^T(t).\Phi_r(t).E_1(t), \quad v(t) = H_4(t) - F_2^T(t).\Phi_r(t).e_2(t), \tag{B.29}$$

where the parameters $E_1(t)$ can be obtained as

$$E_1(t) = \sum_{r(t)}^{-1} \Phi_{r(t)}^T.F_2(t), \quad e_2(t) = \sum_{r(t)}^{-1} \Phi_{r(t)}^T.F_3(t), \tag{B.30}$$

for $t = T, \ldots, 1$, storing the terms $E_1(t), e_2(t), \Phi(t)$.

Step 3. Solve the linear equation as follows:

$$\begin{bmatrix} W(1) & A_n^T(1) \\ A_n(1) & B_n(1) \end{bmatrix} \cdot \begin{bmatrix} \tilde{x}(1) \\ \tilde{a}_n(1) \end{bmatrix} = - \begin{bmatrix} v(1) \\ c_n(1) \end{bmatrix}. \quad \text{(B.31)}$$

Step 4. Calculate

$$\tilde{a}_{r(t)} = -\left(E_1(t).\tilde{x}(t) + e_2(t)\right),$$
$$\tilde{a}_t(t) = \Phi(t).\left[\tilde{a}_r^T(t) \;\; \tilde{a}_n^T(t)\right], \quad \text{(B.32)}$$

$$\left[\tilde{u}^T(t),\; \tilde{\lambda}^T(t),\; \tilde{a}_n^T(t+1)\right] = \tilde{a}_t^T(t), \tilde{x}(t+1)$$
$$= A(t).\tilde{x}(t) + B(t)\tilde{u}(t), \quad \text{(B.33)}$$

and set $\tilde{\lambda}^T(t) = \tilde{a}_n^T(t)$.

This algorithm is used to determine the controls $\tilde{u}(1), \ldots, \tilde{u}(T)$, initial state $\tilde{x}(1)$, and the Lagrange multipliers $\tilde{\lambda}(1), \ldots, \tilde{\lambda}(T)$. Equations (B.3) and (B.15) are then used to calculate the states $\tilde{x}(2), \ldots, \tilde{x}(T)$ and the Lagrange multipliers $\tilde{\lambda}_E(t), \tilde{\lambda}_I(t)$ for $t = 1, \ldots, T$. The slack variable $\tilde{s}_I(t)$ is given by

$$\tilde{s}_I(t) = -\left[A_I(t).x^k(t) + \tilde{x}(t) + B_I(t).u(t)^k + \tilde{u}(t) + \tilde{c}_I(t) + s^k(t)\right] \quad \text{(B.34)}$$

Conflicts of Interest

The authors declare that they have no conflicts of interest.

Acknowledgments

This work is supported by Key Program of National Natural Science Foundation of China (61164015, 61305132, and 61703231).

References

[1] A. G. Tsikalakis and N. D. Hatziargyriou, "Centralized control for optimizing microgrids operation," *IEEE Transactions on Energy Conversion*, vol. 23, no. 1, pp. 241–248, 2008.

[2] R. H. Lasseter, "MicroGrids," in *Proceedings of the IEEE Power Engineering Society Winter Meeting (PESWM '02)*, pp. 305–308, New York, NY, USA, January 2002.

[3] C. A. Hernandez-Aramburo, T. C. Green, and N. Mugniot, "Fuel consumption minimization of a microgrid," *IEEE Transactions on Industry Applications*, vol. 41, no. 3, pp. 673–681, 2005.

[4] W. Su and J. Wang, "Energy Management Systems in Microgrid Operations," *The Electricity Journal*, vol. 25, no. 8, pp. 45–60, 2012.

[5] D. E. Olivares, C. A. Canizares, and M. Kazerani, "A centralized energy management system for isolated microgrids," *IEEE Transactions on Smart Grid*, vol. 5, no. 4, pp. 1864–1875, 2014.

[6] C. M. Colson, M. H. Nehrir, and S. A. Pourmousavi, "Towards real-time microgrid power management using computational intelligence methods," in *Proceedings of the IEEE Power and Energy Society General Meeting (PES '10)*, pp. 1–8, IEEE, Minneapolis, Minn, USA, July 2010.

[7] A. Parisio, E. Rikos, and L. Glielmo, "A model predictive control approach to microgrid operation optimization," *IEEE Transactions on Control Systems Technology*, vol. 22, no. 5, pp. 1813–1827, 2014.

[8] S. Shalini and K. Lakshmi, "Solution to Economic Emission Dispatch problem using Lagrangian relaxation method," in *Proceedings of the 2014 International Conference on Green Computing Communication and Electrical Engineering (ICGCCEE)*, pp. 1–6, Coimbatore, India, March 2014.

[9] S. Krishnamurthy, "Comparison of the Lagrange's and particle swarm optimization solutions of an economic emission dispatch problem with transmission constraints," *Energy*, vol. 36, pp. 6490–6507, 2012.

[10] P. K. Singhal, R. Naresh, V. Sharma, and . Goutham Kumar N, "Enhanced lambda iteration algorithm for the solution of large scale economic dispatch problem," in *Proceedings of the 2014 Recent Advances and Innovations in Engineering (ICRAIE)*, pp. 1–6, Jaipur, India, May 2014.

[11] J. P. Zhan, "Fast lambda-iteration method for economic dispatch," *IEEE Transaction on Power System*, vol. 29, pp. 990-991, 2014.

[12] S.-D. Chen and J.-F. Chen, "A direct Newton-Raphson economic emission dispatch," *International Journal of Electrical Power & Energy System*, vol. 25, no. 5, pp. 411–417, 2003.

[13] H. M. Bishe, "A primal dual interior point method for solving environ-mental/economic power dispatch problem," *Int Rev Electr Eng-IREE*, vol. 6, pp. 1463–1473, 2011.

[14] J. S. Dhillon, S. C. Parti, and D. P. Kothari, "Stochastic economic emission load dispatch," *Electric Power Systems Research*, vol. 26, no. 3, pp. 179–186, 1993.

[15] J. Y. Fan and L. Zhang, "Real-time economic dispatch with line flow and emission constraints using quadratic programming," *IEEE Transactions on Power Systems*, vol. 13, no. 2, pp. 320–325, 1998.

[16] A. A. Moghaddam, A. Seifi, T. Niknam, and M. R. Alizadeh Pahlavani, "Multi-objective operation management of a renewable MG (micro-grid) with back-up micro-turbine/fuel cell/battery hybrid power source," *Energy*, vol. 36, no. 11, pp. 6490–6507, 2011.

[17] A. A. Moghaddam, A. Seifi, and T. Niknam, "Multi-operation management of a typical micro-grids using Particle Swarm Optimization: A comparative study," *Renewable & Sustainable Energy Reviews*, vol. 16, no. 2, pp. 1268–1281, 2012.

[18] A. Chaouachi, R. M. Kamel, R. Andoulsi, and K. Nagasaka, "Multiobjective intelligent energy management for a microgrid," *IEEE Transactions on Industrial Electronics*, vol. 60, no. 4, pp. 1688–1699, 2013.

[19] X. P. Liu, "Dynamic economic dispatch for microgrids including battery energy storage," *Journal of energy and power engineering*, vol. 5, pp. 461–465, 2011.

[20] G. K. Venayagamoorthy, R. K. Sharma, P. K. Gautam, and A. Ahmadi, "Dynamic energy management system for a smart microgrid," *IEEE Transactions on Neural Networks and Learning Systems*, vol. 27, no. 8, pp. 1643–1656, 2016.

[21] C. Chen, S. Duan, T. Cai, B. Liu, and G. Hu, "Smart energy management system for optimal microgrid economic operation," *IET Renewable Power Generation*, vol. 5, no. 3, pp. 258–267, 2011.

[22] L. Xiaoping, D. Ming, H. Jianghong, H. Pingping, and P. Yali, "Dynamic economic dispatch for microgrids including battery energy storage," in *Proceedings of the 2010 2nd IEEE*

International Symposium on Power Electronics for Distributed Generation Systems (PEDG), pp. 914–917, Hefei, China, June 2010.

[23] S. Cheng, G. C. Su, L. L. Zhao, and T. L. Huang, "Dynamic dispatch optimization of microgrid based on a QS-PSO algorithm," *Journal of Renewable and Sustainable Energy*, vol. 9, no. 4, 2017.

[24] X. Li and Y. Fang, "Dynamic Environmental/Economic Scheduling for Microgrid Using Improved MOEA/D-M2M," *Mathematical Problems in Engineering*, vol. 2016, Article ID 2167153, 14 pages, 2016.

[25] X. P. Chen, N. Hewitt, Z. T. Li, Q. M. Wu, X. Yuan, and T. Roskilly, "Dynamic programming for optimal operation of a biofuel micro CHP-HES system," *Applied Energy*, vol. 208, pp. 132–141, 2017.

[26] Y. Zhang, N. Gatsis, and G. B. Giannakis, "Robust energy management for microgrids with high-penetration renewables," *IEEE Transactions on Sustainable Energy*, vol. 4, no. 4, pp. 944–953, 2013.

[27] M. Ross, C. Abbey, F. Bouffard, and G. Joos, "Microgrid Economic Dispatch With Energy Storage Systems," *IEEE Transactions on Smart Grid*, vol. 9, no. 4, pp. 3039–3047, 2018.

[28] E. Mojica-Nava, S. Rivera, and N. Quijano, "Game-theoretic dispatch control in microgrids considering network losses and renewable distributed energy resources integration," *IET Generation, Transmission & Distribution*, vol. 11, no. 6, pp. 1583–1590, 2017.

[29] T. Niknam, R. Azizipanah-Abarghooee, and M. R. Narimani, "An efficient scenario-based stochastic programming framework for multi-objective optimal micro-grid operation," *Applied Energy*, vol. 99, pp. 455–470, 2012.

[30] B. R. Adarsh, T. Raghunathan, T. Jayabarathi, and X.-S. Yang, "Economic dispatch using chaotic bat algorithm," *Energy*, vol. 96, pp. 666–675, 2016.

[31] T. Jayabarathi, T. Raghunathan, B. R. Adarsh, and P. N. Suganthan, "Economic dispatch using hybrid grey wolf optimizer," *Energy*, vol. 111, pp. 630–641, 2016.

[32] T. Niknam, F. Golestaneh, and A. Malekpour, "Probabilistic energy and operation management of a microgrid containing wind/photovoltaic/fuel cell generation and energy storage devices based on point estimate method and self-adaptive gravitational search algorithm," *Energy*, vol. 43, no. 1, pp. 427–437, 2012.

[33] Z. Yang, R. Wu, J. Yang, K. Long, and P. You, "Economical Operation of Microgrid with Various Devices Via Distributed Optimization," *IEEE Transactions on Smart Grid*, vol. 7, no. 2, pp. 857–867, 2016.

[34] E. F. Hill and W. D. Stevenson, "A New Method of Determining Loss Coefficients," *IEEE Transactions on Power Apparatus and Systems*, vol. 87, no. 7, pp. 1548–1553, 1968.

[35] G. Tina, S. Gagliano, and S. Raiti, "Hybrid solar/wind power system probabilistic modelling for long-term performance assessment," *Solar Energy*, vol. 80, no. 5, pp. 578–588, 2006.

[36] K. G. T. Hollands and R. G. Huget, "A probability density function for the clearness index, with applications," *Solar Energy*, vol. 30, no. 3, pp. 195–209, 1983.

[37] R. D. Robinett, *Applied Dynamic Programming for Optimization of Dynamic System*, Society for industrial and applied mathematics, Philadelphia, 2005.

18

Analytical and Numerical Study of Soret and Dufour Effects on Double Diffusive Convection in a Shallow Horizontal Binary Fluid Layer Submitted to Uniform Fluxes of Heat and Mass

A. Lagra,[1] M. Bourich,[2] M. Hasnaoui,[1] A. Amahmid,[1] and M. Er-Raki[3]

[1] *Faculty of Sciences Semlalia, Department of Physics, LMFE, BP 2390, Marrakesh, Morocco*
[2] *National School of Applied Sciences, Physics Department, LMFE, BP 575, Marrakesh, Morocco*
[3] *High School of Technology, Cadi Ayyad University, LMFE, Essaouira, Morocco*

Correspondence should be addressed to M. Hasnaoui; hasnaoui@uca.ac.ma

Academic Editor: Filippo de Monte

Combined Soret and Dufour effects on thermosolutal convection induced in a horizontal layer filled with a binary fluid and subject to constant heat and mass fluxes are investigated analytically and numerically. The thresholds marking the onset of supercritical and subcritical convection are predicted analytically and explicitly versus the governing parameters. The present investigation shows that different regions exist in the N-Du plane corresponding to different parallel flow regimes. The number, the extent, and the locations of these regions depend on whether $\mathrm{SrDu} > -(1+\mathrm{Le}^2)/2\mathrm{Le}^2 = f(\mathrm{Le})$ or $\mathrm{SrDu} < -(1+\mathrm{Le}^2)/2\mathrm{Le}^2$. Conjugate effects of cross-phenomena on thresholds of fluid flow and heat and mass transfer characteristics are illustrated and discussed.

1. Introduction

Great interest in the study of thermosolutal convection in fluid and porous media has been motivated by its presence in many engineering applications, such as in hydrology, petrology, geophysics, and material processing technology where melting and solidification of binary alloys are involved [1]. More specifically, such flows are encountered in nature (lakes, solar ponds, and atmosphere). In the industrial field, examples include food processing, chemical processes, crystal growth, energy storage, material processing, and many other examples. Important experimental, analytical, and numerical results on convective heat and mass transfer are documented in earlier books of Nield and Bejan [2] and De Groot and Mazur [3]. Most studies on this topic are concerned with double diffusive convection in vertical/horizontal cavities for which the flows induced by the buoyancy forces result from the imposition of both thermal and solutal boundary conditions on the vertical/horizontal walls [4–6].

The diffusion of mass due to temperature gradient is called Soret or thermodiffusion effect that often counts among the main drivers of various convective phenomena occurring within thermal stratified media. Recently, Rahman and Saghir [7] proposed a detailed historical review of works, focusing on different aspects of Soret effect. Examples of interesting phenomena resulting from the coupling between double diffusive convection and Soret effect are available in the paper by Bourich et al. [8] who investigated analytically and numerically Soret-driven thermosolutal convection within a shallow porous or fluid layer subject to a vertical gradient of temperature, using a Brinkman-Hazen-Darcy model in its transient form. The critical Rayleigh numbers for the onset of subcritical, oscillatory, and stationary convection were determined explicitly as functions of the governing parameters for infinite and finite layers.

Generally, Soret and Dufour effects are assumed negligible in problems related to double diffusive convection. However, such effects could be of significant effect when density differences exist in the flow regime. In fact, energy flux can be generated by composition of gradients (Dufour or diffusion-thermoeffect). Similarly, mass fluxes can be created by temperature gradients (Soret or thermodiffusion effect). In an earlier study, Malashetty [9] investigated the effect of anisotropic thermoconvective currents, in the presence of Soret and Dufour effects, on the critical Rayleigh number for both marginal and overstable motions. Mortimer and Eyring [10] used the elementary transition state approach to obtain a simple model theory for the Soret and Dufour effects. They found that the results of the theory conform to the Osanger reciprocal relationship. Gaikwad et al. [11] studied the onset of double diffusive convection in a two-component couple stress fluid layer with Soret and Dufour effects using both linear and nonlinear stability analysis. The effects of Soret and Dufour parameters together with the couple stress parameter on the stationary and oscillatory convection are graphically illustrated and discussed. In the presence of Soret and Dufour effects, Nithyadevi and Yang [12] presented numerical results on natural convection in a square enclosure filled with water, partially heated from one vertical wall, and totally cooled from the opposite vertical wall. The study was conducted around the maximum density for three different combinations of the heating element location. Makinde et al. [13] described a theoretical study used to analyze the hydromagnetic flow and mass diffusion of chemical reactive species with first- and higher-order reactions of an electrically conducting fluid over a moving vertical plate. The study was conducted in the presence of Soret and Dufour effects with convective heat exchange at the plate surface. Pal and Mondal [14] considered the problem of steady laminar, hydromagnetic two-dimensional mixed convection flow due to stretching sheet in the presence of Soret and Dufour effects. Cheng [15] examined the Dufour and Soret effects on the steady boundary layer flow due to natural convection heat and mass transfer over a vertical cone embedded in a porous medium with constant wall temperature and concentration. The results presented show that the effects of Dufour and Soret parameters on the local surface temperature are increased by increasing the Lewis number. Tsai and Huang [16] investigated heat and mass transfer from natural convection flow along a vertical surface with variable heat fluxes embedded in a porous medium due to Soret and Dufour effects. They concluded that Soret and Dufour effects could play a significant role. Soret and Dufour effects have been also considered by Hayat et al. [17] who studied mixed convection boundary layer flow about a linearly stretching vertical surface in a porous medium filled with a viscoelastic fluid and, more recently, by Wang et al. [18] who studied the onset of double diffusive convection in a horizontal cavity.

The main purpose of the present investigation is to study analytically and numerically the combined effects of Soret and Dufour parameters on double diffusive convection developed in a horizontal layer filled with a binary fluid. This paper is an extended version of preliminary results presented in a conference [19]. Analytical predictions are developed and validated numerically for shallow enclosures. The Dufour parameter effects on thresholds of stationary convection, subcritical convection, flow structure, and heat and mass transfer are also discussed.

2. Mathematical Formulation

The system under study is a two-dimensional shallow cavity of length L' and height H', filled with a binary fluid. The vertical end-walls of the layer are adiabatic and impermeable to mass transfer while its horizontal walls are subject to uniform fluxes of heat, q', and mass, j'. The flow is assumed to obey the Boussinesq approximation. Using the vorticity and the stream function formulation and taking into account the cross-phenomena (Soret and Dufour effects), the dimensionless governing equations are obtained as follows:

$$\frac{1}{\mathrm{Pr}}\left[\frac{\partial(\nabla^2\psi)}{\partial t}+u\frac{\partial(\nabla^2\psi)}{\partial x}+v\frac{\partial(\nabla^2\psi)}{\partial y}\right] = \nabla^4\psi - R_T\left(\frac{\partial T}{\partial x}+N\frac{\partial S}{\partial x}\right), \tag{1}$$

$$\frac{\partial T}{\partial t}+u\frac{\partial T}{\partial x}+v\frac{\partial T}{\partial y} = \nabla^2 T + D_u N\nabla^2 S, \tag{2}$$

$$\frac{\partial S}{\partial t}+u\frac{\partial S}{\partial x}+v\frac{\partial S}{\partial y} = \frac{\mathrm{Sr}}{N}\nabla^2 T + \frac{1}{Le}\nabla^2 S, \tag{3}$$

$$\nabla^2\psi = -\zeta. \tag{4}$$

The associated hydrodynamic, thermal, and solutal boundary conditions are

$$x = \pm\frac{A_r}{2}:\quad \psi = 0,\quad \frac{\partial T}{\partial x} = 0,\quad \frac{\partial S}{\partial x} = 0, \tag{5a}$$

$$y = \pm\frac{1}{2}:\quad \psi = 0,\quad \frac{\partial T}{\partial y} = -\frac{1-N\mathrm{Du}}{1-\mathrm{LeSrDu}} = -\varphi_T,\quad \frac{\partial S}{\partial y} = -\frac{1-\mathrm{LeSr}/N}{1-\mathrm{LeSrDu}} = -\varphi_S. \tag{5b}$$

The parameters governing the problem are the thermal Darcy-Rayleigh number, $R_T = g\beta_T\Delta T' H'^3/(\kappa_{11}\nu)$, the Prandlt number, $\mathrm{Pr} = \nu/\kappa_{11}$, the Lewis number, $\mathrm{Le} = \kappa_{11}/\kappa_{22}$, the buoyancy ratio, $N = \beta_S\Delta S'/(\beta_T\Delta T')$, and the aspect ratio of the cavity, $A_r = L'/H'$. The parameters Sr and Du are, respectively, the Soret and Dufour parameters expressed as $\mathrm{Sr} = \beta_S\kappa_{21}/(\beta_T\kappa_{11})$ and $\mathrm{Du} = \beta_T\kappa_{12}/(\beta_S\kappa_{11})$, with κ_{11} being the thermal diffusivity, κ_{12} being cross-diffusion due to solute concentration component, κ_{21} being cross-diffusion due to temperature component, and κ_{22} being solute diffusivity.

In the presence of the cross-diffusion phenomena, the Nusselt and Sherwood numbers are defined as follows:

$$\text{Nu} = \frac{1}{\Delta T + N\text{Du}\Delta S},$$
$$\text{Sh} = \frac{1}{\Delta S + ((\text{Sr}\cdot\text{Le})/N)\,\Delta T}, \tag{6}$$

where $\Delta T = T(0,-1/2) - T(0,1/2)$ and $\Delta S = S(0,-1/2) - S(0,1/2)$ are the temperature and concentration differences, evaluated at $x = 0$ with the origin of the coordinate system being taken at the center of the cavity.

3. Methods

3.1. Numerical Solution. The numerical solution of (1) to (3) was obtained using a finite-difference method, described in detail by Bourich et al. [8]. A second-order scheme was used for the discretization of the spatial derivatives. Equations (2)-(3) were marched in time using the Alternating Direction Implicit (ADI) method. The stream function equation (4) was solved at each time step with the Point Successive Overrelaxation (PSOR) method with an optimum overrelaxation coefficient calculated for the used grid. In addition, a convergence criterion was adopted for the stream function to satisfy a variation by less than 10^{-5} for each time step. A second criterion $\sum\sum|\Gamma_{i,j}^{n+1} - \Gamma_{i,j}^{n}|/\sum\sum|\Gamma_{i,j}^{n+1}| \leq 10^{-6}$ was used to check the convergence of the numerical code. Here, Γ stands for any of the variables T, S, or ψ. The superscripts n and $(n+1)$ indicate the iterations numbers, and the subscripts i and j indicate locations in the grid system. For large aspect ratio enclosures, nonuniform grid was used in the x-direction near the short walls to capture the flow details near the enclosures end-walls and also in the y-direction to obtain a finer grid in the close vicinity of the horizontal walls.

3.2. Analytical Solution. For a shallow enclosure with constant heat and mass flux boundary conditions, an approximate analytical solution based on the parallel flow concept is possible, which renders the problem amenable to a parametric study while retaining the essential physics of the problem. The analytical solution is developed for steady-state flows using the parallel flow approximation (see, e.g., Bourich et al. [8]), which leads to the following simplifications (justified by examining the streamlines, isotherms, and iso-concentration lines obtained numerically). These simplifications lead to $\psi(x,y) = \psi(y)$, $T(x,y) = C_T x + \theta_T(y)$, and $S(x,y) = C_S x + \theta_S(y)$, where C_T and C_S are, respectively, unknown constant temperature and concentration gradients in the x-direction (the direction of the long sides of the cavity). Using these approximations together with the boundary conditions (5a) and (5b), (1)–(3) reduceto a set of ordinary differential equations for which the solution is obtained as follows:

$$\begin{aligned}
&\psi(y) = \psi_0\left(4y^2-1\right)^2,\\
&T(x,y)\\
&\quad = C_T x + \frac{C_T - C_S\text{Du}N\text{Le}}{1-\text{LeSrDu}}\frac{\psi_0}{15}y\left[48y^4 - 40y^2 + 15\right]\\
&\quad - \varphi_T y,\\
&S(x,y)\\
&\quad = C_S x + \frac{C_S - C_T\text{Sr}/N}{1-\text{LeSrDu}}\cdot\text{Le}\frac{\psi_0}{15}\\
&\quad \cdot y\left[8y^4 - 40y^2 + 15\right] - \varphi_S y,
\end{aligned} \tag{7}$$

where ψ_0 is the stream function value at the midheight of the layer, given by

$$\psi_0 = \frac{R_T}{384}\left(C_T + NC_S\right). \tag{8}$$

The analytical expressions of C_T and C_S were determined by using thermal and solutal balances in the layer, which leads to the following expressions:

$$C_T = -\frac{8}{15}\psi_0 \cdot \frac{2a\psi_0^2\left(\varphi_T + \varphi_S N\text{Du}\right)\text{Le}^2 + A\left(\varphi_T - \varphi_S\text{Le}N\text{Du}\right)}{\left(2a\psi_0^2\right)^2 + 2a\psi_0^2 B + A^2}, \tag{9a}$$

$$C_S = -\frac{8}{15}\psi_0 \cdot \frac{2a\psi_0^2\left(\varphi_T(\text{LeSr}/N) + \varphi_S\right)\text{Le} + A\text{Le}\left(\varphi_S - \varphi_T(\text{Sr}/N)\right)}{\left(2a\psi_0^2\right)^2 + 2a\psi_0^2 B + A^2}, \tag{9b}$$

where $a = 64/315$, $A = 1-\text{LeSrDu}$, and $B = \text{Le}^2 + 2\text{Le}^2\text{DuSr} + 1$.

Introducing the expressions of C_T and C_S into (8) yields a fourth-order polynomial in terms of ψ_0 for which the following solutions are obtained:

$$\begin{aligned}
&\psi_0 = 0,\\
&\psi_0 = \mp\frac{1}{2}\left[-d_1 \mp \sqrt{d_1^2 - d_2}\right]^{1/2},
\end{aligned} \tag{10}$$

with

$$\begin{aligned}
d_1 &= \frac{1}{2a\text{Le}^2}\left[B - \frac{R_T}{720}F_1\right],\\
d_2 &= \frac{1}{a^2\text{Le}^2}\left[A^2 - \frac{R_T}{720}F_2\right],\\
F_1 &= \text{Le}\left(N + \text{Le}\right),\\
F_2 &= NC_1 + C_2,
\end{aligned} \tag{11}$$

where $C_1 = \text{Du}[\text{Le}(\text{Sr}-1) - 1] + \text{Le}$ and $C_2 = \text{SrLe}[\text{Le}(\text{Du}-1) - 1] + 1$.

From a mathematical point of view, (10) may exhibit, in addition to the rest state solution, two types of bifurcations depending on the sign within the square root. Although it should be mentioned that several numerical tests were performed, only the solutions corresponding to positive sign were obtained numerically. Hence, the convective solution with the negative sign within the square root is termed "unstable solution" and the convective solution corresponding to positive sign within the square root is similarly qualified "stable solution."

The new expressions of Nusselt and Sherwood numbers are obtained as follows:

$$\mathrm{Nu} = \frac{1}{1-(8/15)\psi_0 C_T}, \qquad \mathrm{Sh} = \frac{1}{1-(8/15)\psi_0 \mathrm{Le} C_S}. \tag{12}$$

Foremost, it may be remarked from (10) that the parallel flow solutions exist only when the following two conditions are satisfied:

$$-d_1 \mp \sqrt{d_1^2 - d_2} > 0, \qquad d_1^2 - d_2 > 0. \tag{13}$$

The resolution of the inequalities of (13) is performed in the N-Du plane with Sr, Le, and R_T as parameters. Depending of the sign of B (parameter in the expression of d_1 given before), two main cases are possible.

3.2.1. The Parameter $B > O$. This condition is satisfied if $\mathrm{SrDu} > -(1+\mathrm{Le}^2)/2\mathrm{Le}^2$. Depending on the signs of F_1 and F_2, four cases are to be distinguished.

Case 1 ($F_1 \le 0$ and $F_2 \le 0$). For this case, $F_1 \le 0 \Rightarrow d_1 \ge 0$ and $F_2 \le 0 \Rightarrow d_2 \ge 0$, which means that the parallel flow solution is not existing regardless the values of Sr, Le and R_T. In Figure 1, the domain in the N-Du plane where the convective parallel flow is not possible is denoted as region 1. This region is defined by $F_1 \le 0$ and $F_2 \le 0$.

Case 2 ($F_1 \le 0$ and $F_2 > 0$). For this case, $F_1 \le 0 \Rightarrow d_1 \ge 0$ and $F_2 > 0 \Rightarrow d_2 < 0$ for $R_T \ge R_{TC} = R_0(A^2/F_2)$. Thus, the unstable solution is not possible and the stable solution is possible only when R_T exceeds the critical value R_{TC}. The convective flow bifurcates from the rest state through a zero amplitude convection which indicates that the solution corresponds to a supercritical bifurcation. The critical value of R_{TC} represents the supercritical Rayleigh number, R_{TC}^{sup}, marking the onset of the supercritical convection. The expression of this threshold value is given by

$$R_{TC}^{\mathrm{sup}} = R_0 \frac{A^2}{F_2}. \tag{14}$$

In Figure 1, the domain in the N-Du plane for which the conditions of Case 2 are satisfied is denoted as region 2 and it is defined by $F_1 \le 0$ and $F_2 > 0$.

Case 3 ($F_1 > 0$ and $F_2 > 0$). For these conditions, the signs of d_1 and d_2 depend on the Rayleigh number as follows:

$$d_1 \ge 0 \quad \text{for } R_T \le R_{TC_1} = R_0 \frac{B}{F_1}, \qquad d_2 \ge 0 \quad \text{for } R_T \le R_{TC_2} = R_0 \frac{A^2}{F_2}. \tag{15}$$

As a result, two subcases are to be considered.

Case 3.1 ($R_{TC_2} \le R_{TC_1}$). The development of the above inequality leads to

$$F_3 = N\left(A^2\mathrm{Le} - BC_1\right) + A^2\mathrm{Le}^2 - BC_2 \le 0. \tag{16}$$

Note that, for $R_T \le R_{TC_2}$, $d_1 \ge 0$ and $d_2 \ge 0$ which means that there is no parallel flow solution. On the other hand, for $R_T > R_{TC_2}$, $d_1 \ge 0$ and $d_2 < 0$ indicating the existence of a supercritical bifurcation and the corresponding supercritical Rayleigh number is given by (14). The domain corresponding to these conditions in the N-Du plane is denoted region 3 in Figure 1. This region is defined by $F_1 > 0$, $F_2 > 0$, and $F_3 = N(A^2\mathrm{Le} - BC_1) + A^2\mathrm{Le}^2 - BC_2 \le 0$.

Even if regions 2 and 3 seem to have similar characteristics, they are presented separately because they differ in terms of some asymptotical behaviors at large values of R_T.

Case 3.2 ($R_{TC_2} > R_{TC_1}$). For this case $F_3 = N(A^2\mathrm{Le} - BC_1) + A^2\mathrm{Le}^2 - BC_2 > 0$). Hence, for $R_T < R_{TC_1}$, $d_1 \ge 0$ and $d_2 \ge 0$. It follows that there is no parallel flow solution. For $R_{TC_1} < R_T \le R_{TC_2}$, $d_1 < 0$ and $d_2 \ge 0$ which means that both stable and unstable solutions are existing within this range of R_T. The stable solution corresponds to a subcritical bifurcation, which occurs through finite amplitude convection at a saddle node point. The subcritical threshold, R_{TC}^{sub}, is given by

$$R_{TC}^{\mathrm{sub}} = R_0 \frac{\left(BF_1 - 2\mathrm{Le}^2F_2\right) + 2\mathrm{Le}\sqrt{A^2F_1^2 + \mathrm{Le}^2F_2^2 - BF_1F_2}}{F_1^2}. \tag{17}$$

In addition, for $R_T > R_{TC_2}$, $d_1 < 0$ and $d_2 < 0$ which implies that the unstable solution disappears for this range of R_T. The domain corresponding to these conditions in the N-Du plane is termed region 4 in Figure 1. This region is defined by $F_1 > 0$, $F_2 > 0$, and $F_3 = N(A^2\mathrm{Le}-BC_1)+A^2\mathrm{Le}^2-BC_2 > 0$.

Case 4 ($F_1 > 0$ and $F_2 < 0$). For these conditions, $d_1 \ge 0$ for $R_T \le R_{TC_1} = R_0(B/F_1)$ and $d_2 \ge 0$ independently of the values of the Rayleigh number. It follows that two subcases are possible.

Case 4.1 ($R_T \le R_{TC_1}$). Note that, for this range of R_T, $d_1 \ge 0$ and $d_2 \ge 0$ which means that there is no parallel flow solution.

Case 4.2 ($R_T > R_{TC_1}$). This condition is satisfied for $d_1 < 0$ and $d_2 \ge 0$. This means that both stable and unstable

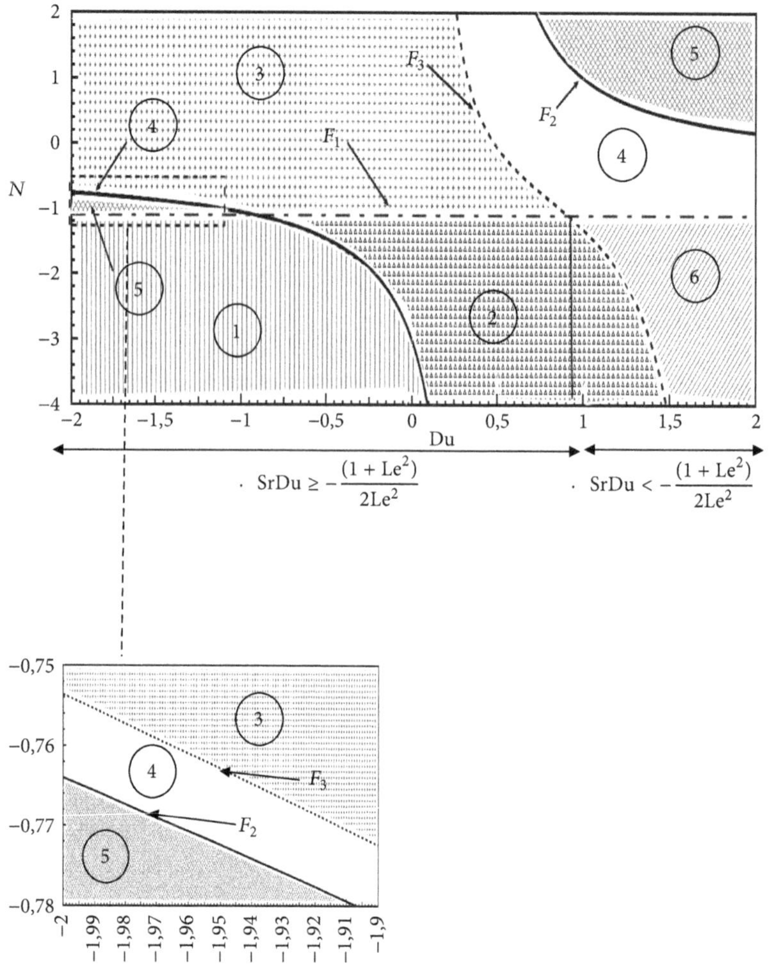

FIGURE 1: Regions corresponding to different bbehaviours for Sr = 0.1 and Le = 1.1.

solutions are existing for any value of R_T greater than the subcritical threshold, R_{TC}^{sub}, given by the same expression as that for region 4. The corresponding domain in the N-Du plane is termed region 5 in Figure 1 and it is defined by $F_1 > 0$ and $F_2 < 0$.

3.2.2. The Parameter B < O. This inequality is satisfied for $\text{SrDu} < -(1 + \text{Le}^2)/2\text{Le}^2$. For this case also, four cases are possible depending of the signs of F_1 and F_2.

Case 1 ($F_1 > 0$ and $F_2 < 0$). These conditions are the same as those defining region 5. The corresponding domain in the N-Du plane is termed region 5 in Figure 1 and it is defined by $F_1 > 0$ and $F_2 < 0$.

Case 2 ($F_1 > 0$ and $F_2 \geq 0$). For these conditions, $d_1 < 0$ independently of the values of the Rayleigh number and $d_2 \geq 0$ for $R_T \leq R_{TC_2} = R_0(A^2/F_2)$. This means that both stable and unstable solutions are existing in this range for $R_T \geq R_{TC}^{sub}$. The supercritical threshold Rayleigh number is given by (17). In addition, for $R_T > R_{TC_2}$, $d_1 < 0$, and $d_2 < 0$ which implies that the unstable solution disappears for this range of R_T. The corresponding domain in the N-Du plane is termed region 4 in Figure 1, defined by $F_1 > 0$ and $F_2 \geq 0$.

Case 3 ($F_1 \leq 0$ and $F_2 \geq 0$). For these conditions, the signs of d_1 and d_2 depend on the Rayleigh number as follows:

$$\begin{aligned} d_1 \geq 0 \quad &\text{for } R_T \geq R_{TC_1} = R_0 \frac{B}{F_1}, \\ d_2 \geq 0 \quad &\text{for } R_T \leq R_{TC_2} = R_0 \frac{A^2}{F_2}. \end{aligned} \tag{18}$$

Two subcases emerge from these conditions.

Case 3.1 ($R_{TC_2} \leq R_{TC_1}$). This inequality is verified for $F_3 = N(A^2\text{Le}-BC_1)+A^2\text{Le}^2-BC_2 \geq 0$. Thus, for $R_T \leq R_{TC_2}$, $d_1 \leq 0$ and $d_2 \geq 0$, which means that both stable and unstable solutions are existing. The bifurcation is of a subcritical nature and the subcritical Rayleigh number is given by (17). For $R_T > R_{TC_2}$, the unstable solution disappears within this range of R_T. The domain in the N-Du plane satisfying these conditions is indicated by region 6 in Figure 1. This region is defined by $F_1 \leq 0$, $F_2 \geq 0$, and $F_3 \geq 0$. It is to underline that, even if regions 5 and 6 seem to have the same characteristics, they are presented separately because they differ in terms of some asymptotical behaviors at large values of R_T.

Case 3.2 ($R_{TC_2} > R_{TC_1}$). For this case, $F_3 < 0$. Then, for $R_T < R_{TC_1}$, $d_1 \geq 0$ and $d_2 \geq 0$ and the parallel flow solution is not existing. On the other side, for $R_T > R_{TC_2}$, $d_1 \geq 0$ and $d_2 < 0$ indicating the existence of a supercritical bifurcation and the corresponding supercritical Rayleigh number is given by (14). The domain corresponding to these conditions in the N-Du plane is termed region 2 in Figure 1 and it is defined by $F_1 \leq 0$, $F_2 \geq 0$, and $F_3 < 0$.

Case 4 ($F_1 \leq 0$ and $F_2 \leq 0$). For this case, the inequality $\mathrm{SrDu} < -(1+\mathrm{Le}^2)/2\mathrm{Le}^2$ is verified and there is no corresponding domain in the N-Du plane.

In summary, the N-Du plane can be divided into five regions for $\mathrm{SrDu} > -(1+\mathrm{Le}^2)/2\mathrm{Le}^2$ and into four regions for $\mathrm{SrDu} < -(1+\mathrm{Le}^2)/2\mathrm{Le}^2$. In region 1, the parallel flow is not possible while in regions 2 and 3, the only possible flow is the supercritical stationary one. For the remaining regions (regions 4, 5, and 6) both subcritical and supercritical convections are possible.

4. Results and Discussion

4.1. Effect of Du and N on Convection Thresholds. This section is devoted to analyze the combined effects of the Dufour parameter and the buoyancy ratio on thresholds of convection. In addition, some interesting behaviors of the fluid flow are illustrated and discussed.

The parameter Du represents the relative importance of the cross-diffusion due to solute concentration component with respect to that due to the thermal gradient. It could be varied by changing the mass flux intensity or by considering various working binary fluids. For a given N, different regions could be crossed in Figure 1 by incrementing Du, depending on whether $N \geq N_L^{\mathrm{Sup}}$ or $N < N_L^{\mathrm{Sup}}$. The parameter N_L^{Sup} is calculated analytically and its expression is given by $N_L^{\mathrm{Sup}} = -\mathrm{SrLe}^2/(\mathrm{Le}(\mathrm{Sr}-1)-1)$. The effect of Du on thresholds of stationary convection is exemplified in Figures 2(a) and 2(b) illustrating, respectively, the cases $N \geq N_L^{\mathrm{Sup}}$ (case illustrated with $N = 1.5$, $\mathrm{Le} = 1.1$, and different Sr) and $N < N_L^{\mathrm{Sup}}$ (case illustrated with $N = -1.5$, $\mathrm{Le} = 1.1$, and different Sr). The value $\mathrm{Le} = 1.1$ was chosen within the experimental range of the Lewis number where both cross-phenomena of Soret and Dufour effects are significant.

For $N \geq N_L^{\mathrm{Sup}}$, the threshold of stationary convection is characterized by a decrease toward a minimum (well visible in Figure 2(a) for the negative values of Sr but not visible for the positive values of this parameter due to the restricted range of Du). It is seen in Figure 2(a) that the threshold of stationary convection decreases first by increasing Du to reach a minimum value at $\mathrm{Du} = \mathrm{Du}_{\min}$ and increases afterwards quickly toward a vertical asymptote (infinite value) for $\mathrm{Du} = \mathrm{Du}_C^{\mathrm{sup}}$. However, for the positive values of Sr, the variations of R_{TC}^{Sup} versus Du are moderate for negative values of the latter parameter. The increase of R_{TC}^{Sup} accompanying the increase of Du in its negative range is characterized by an augmentation which becomes more and more slow by decreasing Sr within its positive range. In the positive range of Du, the fast increase toward the vertical asymptote is also observed for the positive values of Sr. The critical value of Du leading to this asymptote increases by decreasing Sr. For $N < N_L^{\mathrm{Sup}}$, an opposite behavior to that described in Figure 2(a) is observed. In fact, an increase of Du from the limiting value $\mathrm{Du}_C^{\mathrm{sup}}$ induces first a sharp decrease of R_{TC}^{sup} from infinite toward a minimum value obtained at $\mathrm{Du} = \mathrm{Du}_{\min}$. Afterward, R_{TC}^{sup} increases slowly (with different rates depending on Sr). These results show that, for $N \geq N_L^{\mathrm{Sup}}/(N < N_L^{\mathrm{Sup}})$, the Dufour parameter could play a destabilizing effect if $\mathrm{Du} < \mathrm{Du}_C^{\mathrm{sup}}/(\mathrm{Du}_C^{\mathrm{sup}} < \mathrm{Du} < \mathrm{Du}_{\min})$ or a stabilizing one if $\mathrm{Du}_{\min} < \mathrm{Du} < \mathrm{Du}_C^{\mathrm{sup}}/(\mathrm{Du} < \mathrm{Du}_{\min})$. The analytical expressions of $\mathrm{Du}_{\min}$ and $\mathrm{Du}_C^{\mathrm{sup}}$ are, respectively, given by

$$\mathrm{Du}_C^{\mathrm{sup}} = \frac{\mathrm{SrLe}(\mathrm{Le}+1) - N\mathrm{Le} - 1}{\mathrm{SrLe}(N+\mathrm{Le}) - N(\mathrm{Le}+1)},$$

$$\mathrm{Du}_{\min} = -\frac{\mathrm{LeSr}(-2\mathrm{SrLe}(\mathrm{Le}+1) + N(2\mathrm{Le}+1) + \mathrm{Le} + 2) - N(\mathrm{Le}+1)}{\mathrm{LeSr}(\mathrm{LeSr}(N+\mathrm{Le}) - N(\mathrm{Le}+1))}. \quad (19)$$

The Dufour effect on R_{TC}^{Sub} is illustrated for different values of Soret parameter, Sr, for the three possible cases (determined analytically) that lead to different behaviors depending on the crossing region. Thus, Figure 3(a) illustrates the case $N_L^{\mathrm{Sub}} \geq N > -\mathrm{Le}$, Figure 3(b) illustrates the case $N > N_L^{\mathrm{Sub}}$, and the third case illustrated in Figure 3(c) corresponds to $N \leq -\mathrm{Le}$ and $\mathrm{Sr} < \mathrm{Sr}_C$. The analytical expression of N_L^{Sub} is given by

$$N_L^{\mathrm{Sub}} = \frac{\mathrm{Sr}\cdot\mathrm{Le}^2}{\mathrm{Le}\cdot(2-\mathrm{Sr})+2}. \quad (20)$$

For the first case, Figure 3(a) shows that the subcritical Rayleigh number, R_{TC}^{Sub}, decreases linearly by increasing Du and vanishes when Du approaches a critical value D_C^{Sub}. This critical value of Du increases by the increase of Sr. For the second case, Figure 3(b) indicates that the subcritical convection starts from D_C^{Sub}. More increase of Du from this threshold leads to a linear increase of R_{TC}^{Sub} and stabilizes more the system. Note that for the case of Figure 3(b), the effect of Sr on R_{TC}^{Sub} is seen to be limited, but its effect of the limit $\mathrm{Du}_C^{\mathrm{sub}}$ is important. The analytical expression of $\mathrm{Du}_C^{\mathrm{sub}}$ is given by $\mathrm{Du}_C^{\mathrm{sub}} = (2+\beta_1 \mp \sqrt{\beta_1^2 - 4\beta_2})/2\mathrm{SrLe}$, where β_1 and β_2 are functions of the governing parameters. The opposite tendencies observed by increasing the Dufour parameter in Figures 3(a) and 3(b) are explained by the fact that the increase of Du leads to the successive cross of regions 5 and region 4 in the case of Figure 3(a), while in the case of Figure 3(b) it leads to the successive cross of regions 4 and 5. For $N \leq -\mathrm{Le}$, the subcritical convection corresponding to region 6 appears only for $\mathrm{Sr} < \mathrm{Sr}_C$ as shown in Figure 1.

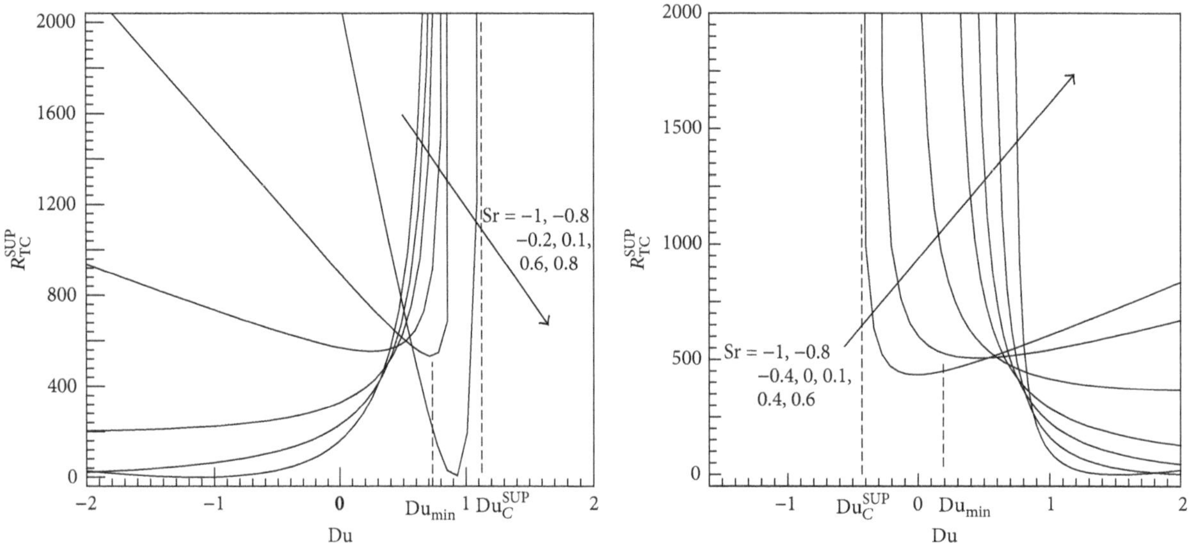

(a) The evolution of R_{TC}^{sup} with Du for $N = 1.5$ and Le = 1.1 (case of $N \geq N_L^{Sup}$)

(b) The evolution of R_{TC}^{sup} with Du for $N = -1.5$ and Le = 1.1 (case of $N < N_L^{Sup}$)

FIGURE 2

For this region, the subcritical convection starts from D_{ud} and increases linearly with Du inducing a stabilizing effect. The Soret parameter has an important effect on the subcritical Rayleigh number in this region as indicated in Figure 3(c).

The buoyancy ratio N measures the relative importance of solutal buoyancy force (due to combined contributions of imposed flux of mass and Soret effect) and thermal buoyancy force (induced by the imposed flux of heat and Dufour effect). It could be varied by changing the imposed flux of heat and/or mass or by considering various fluids with different properties.

The influence of the buoyancy ratio N on the thresholds of stationary convection R_{TC}^{Sup} is illustrated in Figures 4(a) and 4(b) for Le = 1.1 and various Sr. In Figure 4(a), Du = −1.5 to illustrate the case $Du < Du_L^{sup}$, while in Figure 4(b) Du = 1.5 to illustrate the case $Du \geq Du_L^{sup}$. As shown from Figure 4(a), the evolution of R_{TC}^{sup} with N changes drastically depending on whether $Du < Du_L^{sup}$ or $Du \geq Du_L^{sup}$. For $Du \geq Du_L^{sup}$, an increase of N from a supercritical value N_C^{sup} induces a destabilizing effect characterized first by a sharp decrease of R_{TC}^{sup} followed by a monotonous and moderate decrease. The trend is inverted for $Du \geq Du_L^{sup}$ as it can be seen in Figure 4(b). In fact, R_{TC}^{sup} increases first monotonously and in a moderate way. A change in the tendency of the increase occurs when the value of N becomes close to a critical value N_C^{sup}, leading to a quick increase toward the infinite value of the asymptote. For this case, it is clear that the increase of N has a stabilizing role. This behavior can be explained by the fact that, for Du = −1.5, case corresponding to $Du < Du_L^{sup}$, the increase of N in Figure 1 leads to cross region 1 first which requires an infinite Rayleigh number to trigger the flow. However, for Du = 1.5, case corresponding to $Du \geq Du_L^{sup}$, region 3 is crossed first, then region 1. The latter corresponds to the case where the parallel flow is impossible, which explains why an infinite Rayleigh number is required to start the flow. The critical value N_C^{sup} can be calculated from the following analytical expression:

$$N_C^{sup} = -\frac{\mathrm{SrLe}\left(\mathrm{LeDu} - \mathrm{Le}^2 - \mathrm{Le}\right) + 1}{\mathrm{Du}\left(\mathrm{LeSr} - \mathrm{Le}^2 - \mathrm{Le}\right) + \mathrm{Le}}. \tag{21}$$

The expression of Du_L^{Sup} is also calculated analytically to obtain

$$\mathrm{Du}_L^{Sup} = -\frac{\mathrm{Le}}{1 - \mathrm{Le}\left(\mathrm{Sr} - 1\right)}. \tag{22}$$

The evolution of the subcritical Rayleigh number, R_{TC}^{Sub}, versus N is depicted in Figures 5(a) and 5(b). The case of $Du < Du_L^{sub}$ is illustrated in Figure 5(a) for Du = −1.5, Le = 1.1, and different values of Sr. It is to note that, for this case, the subcritical flow exists only for negative values of N. By increasing N from a subcritical value N_C^{sub}, the subcritical Rayleigh number R_{TC}^{Sub} undergoes a sharp decline from its infinite value, indicating that the increase of N has a destabilizing effect. For this case, by increasing N, we cross first region 1 (rest state region) in Figure 1 which explains why an infinite Rayleigh number is required to start the flow for $N = N_C^{Sub}$. The case $Du \geq Du_L^{Sub}$ is exemplified in Figure 5(b) for Du = 2, Le = 1.1, and different values of Sr. The examination of this figure shows a big change in the behavior compared with the first case illustrated in Figure 5(a). In fact, for Sr = −0.8, an increase of N induces first an increase of R_{TC}^{Sub} toward a maximum value and undergoes after that a decrease in the remaining range of N. Note that, for Sr = 0.1, the fluid remains at rest as long as $N < N_C^{Sub}$.

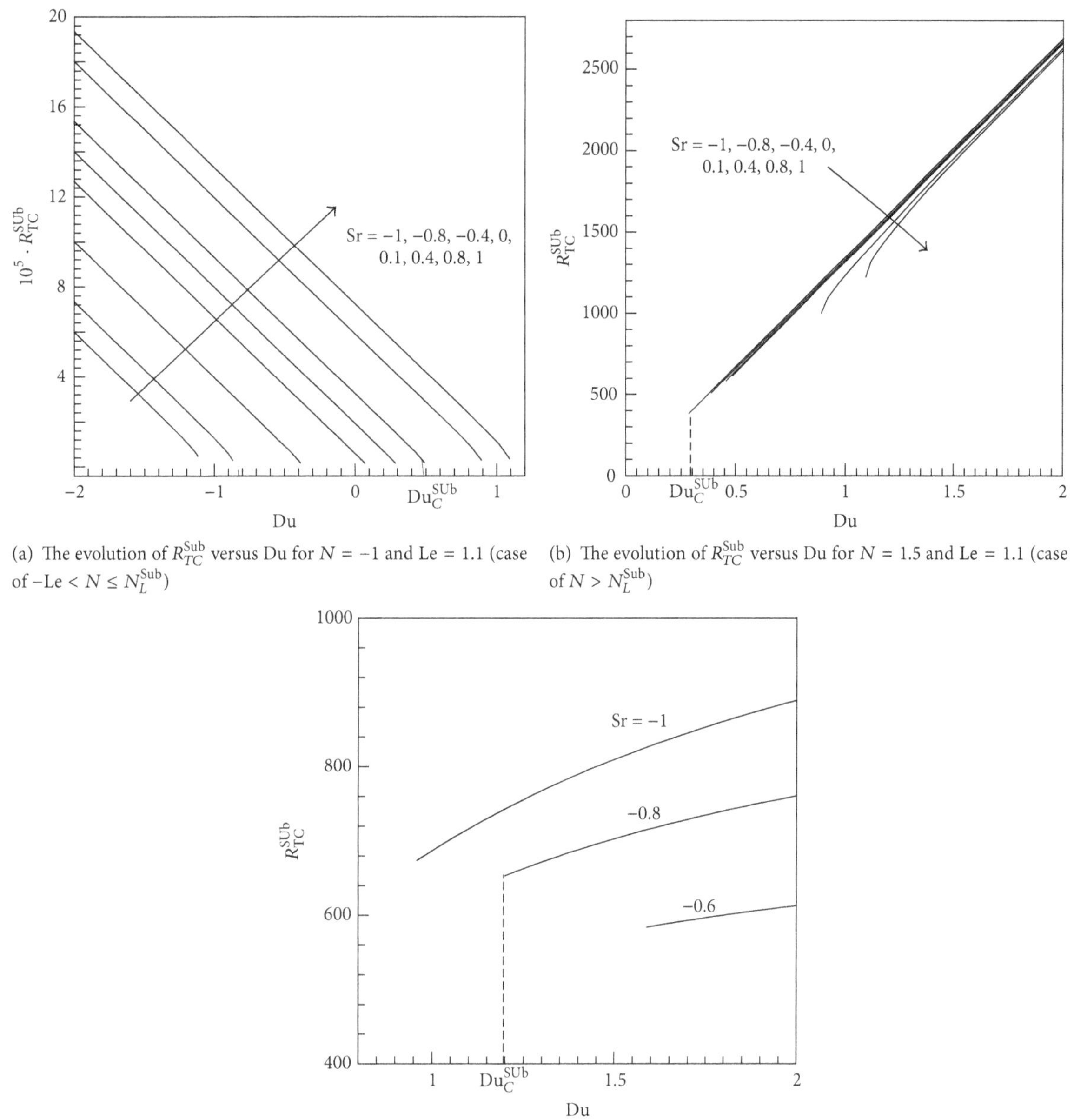

(a) The evolution of R_{TC}^{Sub} versus Du for $N = -1$ and $Le = 1.1$ (case of $-Le < N \leq N_L^{Sub}$)

(b) The evolution of R_{TC}^{Sub} versus Du for $N = 1.5$ and $Le = 1.1$ (case of $N > N_L^{Sub}$)

(c) The evolution of R_{TC}^{Sub} with Du for $N = -1.2$ and $Le = 1.1$ (case of $N \leq -Le$)

FIGURE 3

The analytical expressions of N_C^{Sub} and Du_L^{Sub} are, respectively, given by

$$N_C^{SuB} = \frac{B\mathrm{SrLe}\left(\mathrm{LeDu} - \mathrm{Le}^2 - \mathrm{Le}\right) + B - A^2\mathrm{Le}^2}{A^2\mathrm{Le} - B\mathrm{Du}\left(\mathrm{SrLe} - \mathrm{Le} - 1\right) - B\mathrm{Le}},$$
$$\mathrm{Du}_C^{sub} = \frac{2 + \alpha_1 \mp \sqrt{\alpha_1^2 - 4\alpha_2}}{2\mathrm{SrLe}}, \quad (23)$$

where $\alpha_1 = ((\mathrm{LeSr} - \mathrm{Le} - 1)(\mathrm{Le}^2 + 4\mathrm{Le} + 1) + 2\mathrm{SrLe}^3)/\mathrm{Le}(2\mathrm{Le} + 2 - \mathrm{SrLe})$ and $\alpha_2 = -(\mathrm{Le} + 1)^3(\mathrm{SrLe} - 1)/\mathrm{Le}(2\mathrm{Le} + 2 - \mathrm{SrLe})$.

4.2. Effect of Du on the Flow Intensity and Heat and Mass Transfer. In this subsection, the influence of the Dufour effect on the flow intensity and heat and mass transfer is examined. The effect of Du on ψ_0, Nu, and Sh is depicted in Figures 6(a)–6(c) for $Le = 1.1$, $A_r = 10$, different values of Sr, and two cases: $R_T = 2730$ and $N = 1.2$ (case I) and $R_T = 889.3$, $N = -1.2$ (case II). The choice of these two cases was dictated by the fact that, by varying Du, two types of traversed regions are possible depending on whether $N \geq -Le$ (case I) or $N < -Le$ (case II). In these figures, it can be seen that the numerical results are in excellent agreement with the analytical ones corresponding to the stable branches.

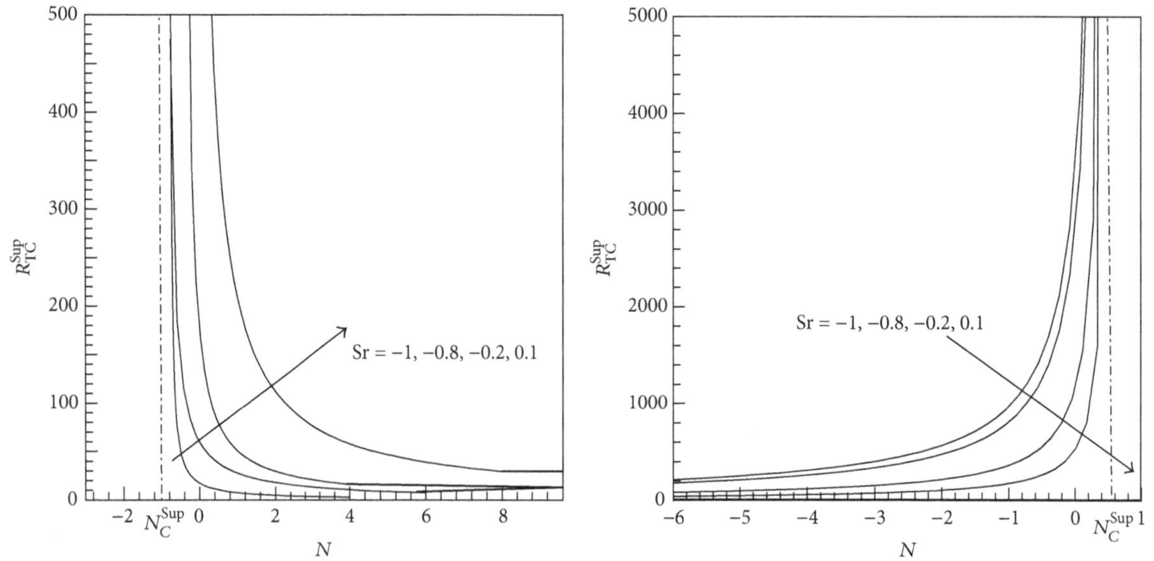

(a) The evolution of R_{TC}^{sup} with N for Du = −1.5 and Le = 1.1 (case of $Du < Du_L^{sup}$)

(b) The evolution of R_{TC}^{sup} with N for Du = 1.5 and Le = 1.1 (case of $Du \geq Du_L^{sup}$)

FIGURE 4

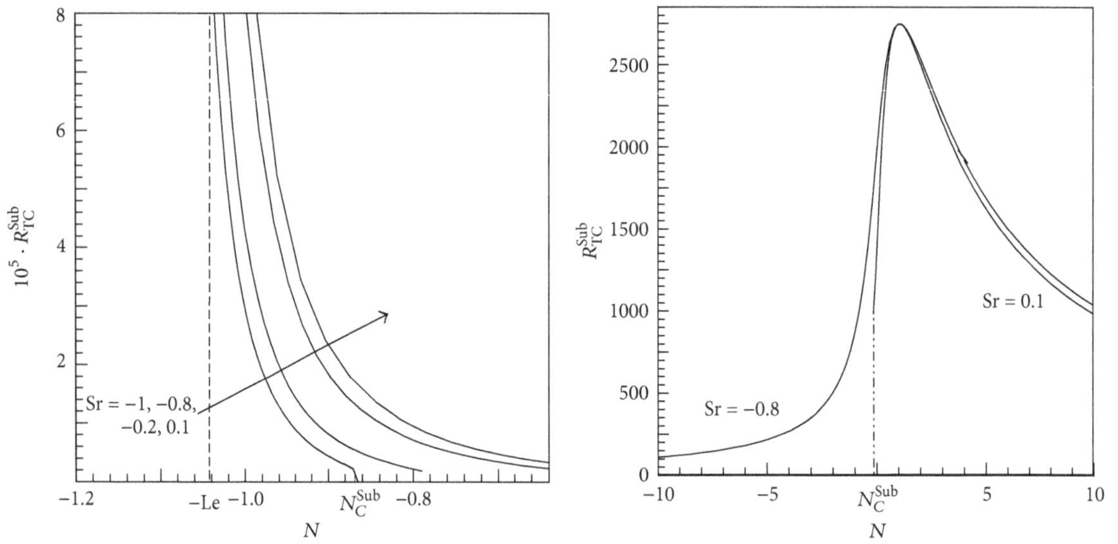

(a) The evolution of R_{TC}^{sub} with N for Du = −1.5 and Le = 1.1 (case of $Du < Du_L^{sub}$)

(b) The evolution of R_{TC}^{sub} versus N for Du = 2 and Le = 1.1 (case of $Du \geq Du_L^{sub}$)

FIGURE 5

For case I, both stable and unstable branches are existing regardless of the Soret parameter. For the stable branch, the evolutions of ψ_0 and Nu versus Du are characterized by a monotonous decrease (the decrease occurs with a higher rate for the negative value of Sr) and vanishes when Du exceeds a critical value Du_c which depends on Sr. The unstable branch exists only between two critical values of Du, that is, for $Du_u < Du \leq Du_c$. Within this range of Du, the evolution of ψ_0 is characterized by an important increase toward a maximum reached for $Du = Du_c$. This critical value Du_c marks also the limit of the existence of the unstable branch. This means that, before their disappearance, the unstable solutions become identical to the ones corresponding to the stable branches.

For case II, the evolutions of ψ_0 and Nu versus Du depend on the Soret parameter. In fact, for $Sr = -1$ (for this value of Sr the corresponding regions in N-Du plane are not presented here), both stable and unstable solutions exist and the quantities ψ_0 and Nu increase with Du to reach a maximum and decrease afterward before vanishing. The unstable branch exists for $Du_u < Du \leq Du_c$. For $Sr = 0.1$, only the stable solution exists and the quantities ψ_0 and Nu

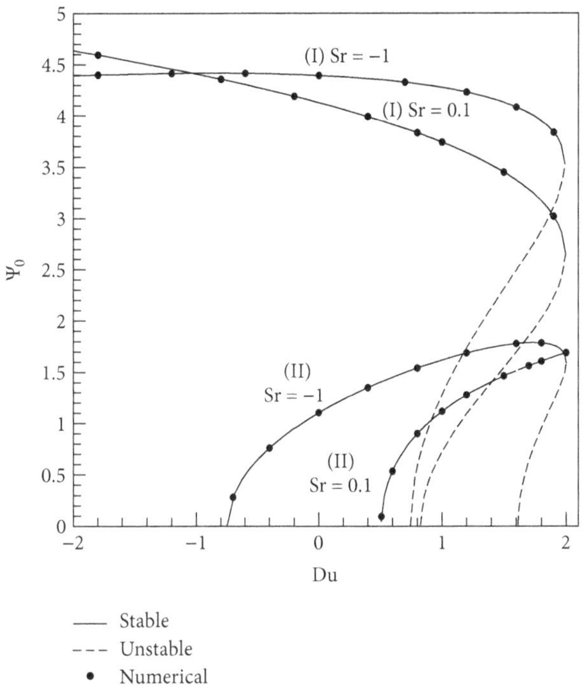

(a) Influence of the Dufour number on ψ_0 for Le = 1.1 and Ar = 10: (I) $R_T = 2730$ and $N = 1.2$ (case of $N \geq -\text{Le}$); (II) $R_T = 889$ and $N = -1.2$ (case of $N < -\text{Le}$)

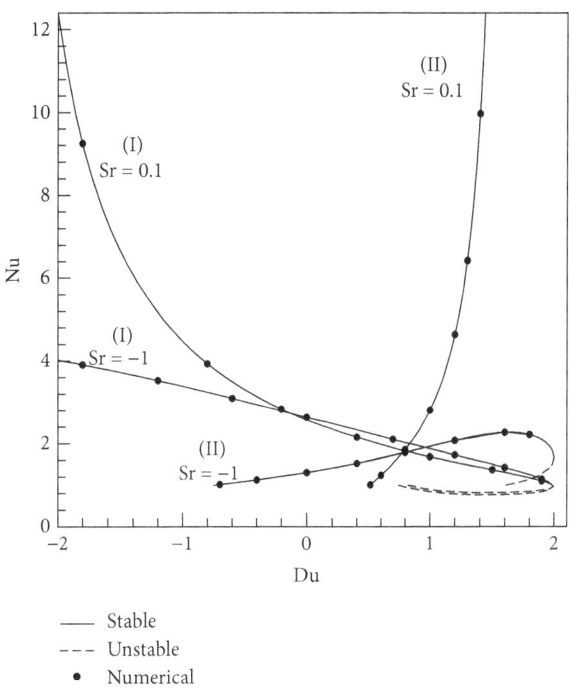

(b) Influence of the Dufour number on Nu for Le = 1.1 and Ar = 10: (I) $R_T = 2730$ and $N = 1.2$ (case of $N \geq -\text{Le}$); (II) $R_T = 889$ and $N = -1.2$ (case of $N < -\text{Le}$)

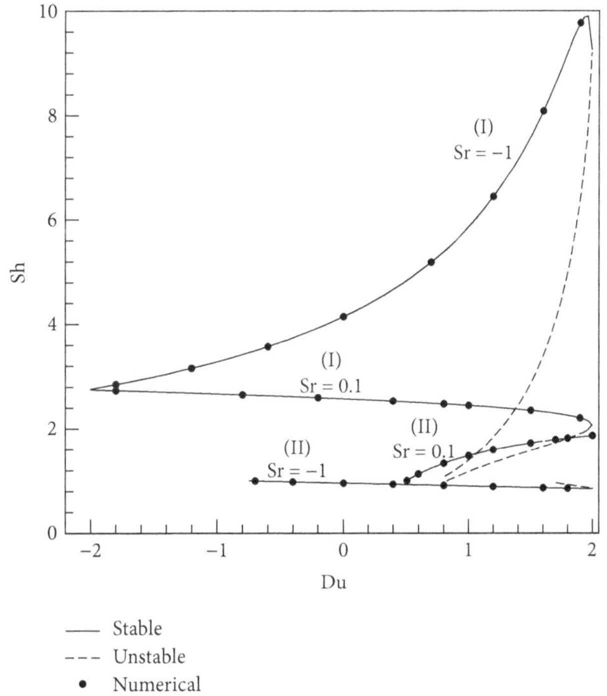

(c) Influence of the Dufour number on Sh for Le = 1.1 and Ar = 10: (I) $R_T = 2730$ (case of $N \geq -\text{Le}$); (II) $R_T = 890$ (case of $N < -\text{Le}$)

Figure 6

exhibit monotonous increases with Du. The dependence of the flow intensity vis-à-vis Sr can be explained by the fact that, by varying Du, the crossed regions are region 2 and then region 6 for Sr = −1 while for Sr = 0.1 the crossed regions are region 1 and then region 2. From these observations, we can conclude that the Dufour effect enhances/reduces the flow intensity and heat transfer for $N < -\text{Le}/N \geq -\text{Le}$.

The evolution of Sh versus Du exhibits a different tendency in comparison with ψ_0 and Nu. In fact, the Sh variations with Du show similar tendencies independently of the crossed regions for both cases I and II and Sr = −1. For this case, Sh varies slightly with Du. The behavior changes for Sr = 0.1 since the evolution of Sh is characterized by a monotonous increase with Du to reach a maximum marking the limit of the stable branch. For case I, by increasing Du the crossed regions are successively regions 3, 4, and 5 while for case II the crossed regions are 2 (crossed first) and 6.

5. Conclusion

Thermosolutal natural convection within a shallow fluid layer subjected to vertical gradients of temperature and solute is investigated analytically and numerically in the presence of Soret and Dufour effects. The analytical solution based on the parallel flow assumption is found to be in good agreement with the numerical solution based on a finite-difference method used to solve the full governing equations in their transient forms. In view of the results discussed, the main conclusions are as follows:

(i) The thresholds of stationary and subcritical convections are derived analytically versus the governing parameters.

(ii) We have demonstrated analytically that the plane N-Du can be divided into different regions (up to six regions) with specific flow regimes.

(iii) The number of these regions, their extension, and their locations depend on whether the product SrDu = $f(\mathrm{Le})$ is lower or higher than $-(1+\mathrm{Le}^2)/2\mathrm{Le}^2$.

(iv) The influence of the Dufour effect on the thresholds of stationary and subcritical convections may be stabilizing or destabilizing, depending on the buoyancy ratio and the Soret parameter.

(v) The important effect of the thermodiffusion and diffusion-thermo on fluid flow and heat and mass transfer characteristics is well demonstrated.

Nomenclature

A_r: Aspect ratio of the enclosure $(= L'/H')$
Du: Dufour parameter $(= \beta_T \kappa_{12}/\beta_S \kappa_{11})$
g: Gravitational acceleration
H': Height of the enclosure
J': Constant mass flux per unit area
L': Length of the enclosure
Le: Lewis number $(= \kappa_{11}/\kappa_{22})$
N: Buoyancy ratio $(= \beta_S \Delta S'/\beta_T \Delta T')$
Nu: Thermal Nusselt number
q': Constant heat flux per unit area
Pr: Prandtl number $(= \nu/\kappa_{11})$
R_T: Thermal Rayleigh number $(= g\beta_T \Delta T' H'^3/\kappa_{11}\nu)$
S: Dimensionless solute concentration $(= (S' - S'_0)/\Delta S')$
$\Delta S'$: Concentration solute difference $(= j' H'/\kappa_{22}\rho C_p)$
Sr: Soret parameter $(= \beta_S \kappa_{21}/\beta_T \kappa_{11})$
Sh: Sherwood number
T: Dimensionless temperature $(= (T' - T'_0)/\Delta T')$
$\Delta T'$: Temperature difference $(= q' H'/\kappa_{11}\rho C_p)$
t: Dimensionless time $(= t'/H'^2/\kappa_{11})$
u: Dimensionless horizontal velocity $(= u' H'/\kappa_{11})$
v: Dimensionless vertical velocity $(= v' H'/\kappa_{11})$
x: Dimensionless distance along the x-axis $(= x'/H')$
y: Dimensionless distance along the y-axis $(= y'/H')$

Greek Symbols

β_S: Solute expansion coefficient
β_T: Thermal expansion coefficient
κ_{11}: Thermal diffusivity
κ_{12}: Cross-diffusion due to S-component
κ_{21}: Cross-diffusion due to T-component
κ_{22}: Solute diffusivity
ρCp: Heat capacity
ν: Kinematic viscosity of the fluid
Ψ: Dimensionless stream function, Ψ'/κ_{11}

Subscripts

c: Critical value
0: Reference value.

Superscripts

$'$: Dimensional variables
Sub: Subcritical
Sup: Supercritical.

Conflicts of Interest

The authors declare that there are no conflicts of interest regarding the publication of this paper.

References

[1] O. E. Tewfikf and Y. Ji-Wu, "The thermodynamic coupling between heat and mass transfer in free convection with helium injection," *International Journal of Heat and Mass Transfer*, vol. 6, no. 10, pp. 915–923, 1963.
[2] D. A. Nield and A. Bejan, *Convection in Porous Media*, vol. 3rd, Spring, New York, NY, USA, 2006.
[3] S. R. De Groot and P. Mazur, *Non-Equilibrium Thermodynamics*, Dover Publications, New York, NY, USA, 1984.
[4] M. Ouriemi, P. Vasseur, A. Bahloul, and L. Robillard, "Natural convection in a horizontal layer of a binary mixture," *International Journal of Thermal Sciences*, vol. 45, no. 8, pp. 752–759, 2006.
[5] M. Er-Raki, M. Hasnaoui, A. Amahmid, and M. Bourich, "Subcritical convection in the presence of Soret effect within a horizontal porous enclosure heated and salted from the short sides," *International Journal of Numerical Methods for Heat & Fluid Flow*, vol. 21, no. 2, pp. 150–167, 2011.

[6] A. Amahmid, M. Hasnaoui, M. Mamou, and P. Vasseur, "Double-diffusive parallel flow induced in a horizontal Brinkman porous layer subjected to constant heat and mass fluxes: Analytical and numerical studies," *Wärme- und Stoffübertragung*, vol. 35, no. 5, pp. 409–421, 1999.

[7] M. A. Rahman and M. Z. Saghir, "Thermodiffusion or Soret effect: historical review," *International Journal of Heat and Mass Transfer*, vol. 73, pp. 693–705, 2014.

[8] M. Bourich, M. Hasnaoui, M. Mamou, and A. Amahmid, "Soret effect inducing subcritical and Hopf bifurcations in a shallow enclosure filled with a clear binary fluid or a saturated porous medium: a comparative study," *Physics of Fluids*, vol. 16, no. 3, pp. 551–568, 2004.

[9] M. S. Malashetty, "Anisotropic thermoconvective effects on the onset of double diffusive convection in a porous medium," *International Journal of Heat and Mass Transfer*, vol. 36, no. 9, pp. 2397–2401, 1993.

[10] R. G. Mortimer and H. Eyring, "Elementary transition state theory of the Soret and Dufour effects," *Proceedings of the National Acadamy of Sciences of the United States of America*, vol. 77, no. 4, pp. 1728–1731, 1980.

[11] S. N. Gaikwad, M. S. Malashetty, and K. Rama Prasad, "Linear and non-linear double diffusive convection in a fluid-saturated anisotropic porous layer with cross-diffusion effects," *Transport in Porous Media*, vol. 80, no. 3, pp. 537–560, 2009.

[12] N. Nithyadevi and R. J. Yang, "Double diffusive natural convection in a partially heated enclosure with Soret and Dufour effects," *International Journal of Heat and Fluid Flow*, vol. 30, no. 5, pp. 902–910, 2009.

[13] O. D. Makinde, K. Zimba, and O. A. Bég, "Numerical study of chemically-reacting hydromagnetic boundary layer flow with soret/dufour effects and a convective surface boundary condition," *International Journal of Thermal and Environmental Engineering*, vol. 4, no. 1, pp. 89–98, 2012.

[14] D. Pal and H. Mondal, "Effects of Soret Dufour, chemical reaction and thermal radiation on MHD non-Darcy unsteady mixed convective heat and mass transfer over a stretching sheet," *Communications in Nonlinear Science and Numerical Simulation*, vol. 16, no. 4, pp. 1942–1958, 2011.

[15] C.-Y. Cheng, "Soret and Dufour effects on natural convection boundary layer flow over a vertical cone in a porous medium with constant wall heat and mass fluxes," *International Communications in Heat and Mass Transfer*, vol. 38, no. 1, pp. 44–48, 2011.

[16] R. Tsai and J. S. Huang, "Heat and mass transfer for Soret and Dufour's effects on Hiemenz flow through porous medium onto a stretching surface," *International Journal of Heat and Mass Transfer*, vol. 52, no. 9-10, pp. 2399–2406, 2009.

[17] T. Hayat, M. Mustafa, and I. Pop, "Heat and mass transfer for Soret and Dufour's effect on mixed convection boundary layer flow over a stretching vertical surface in a porous medium filled with a viscoelastic fluid," *Communications in Nonlinear Science and Numerical Simulation*, vol. 15, no. 5, pp. 1183–1196, 2010.

[18] J. Wang, M. Yang, and Y. Zhang, "Onset of double-diffusive convection in horizontal cavity with Soret and Dufour effects," *International Journal of Heat and Mass Transfer*, vol. 78, pp. 1023–1031, 2014.

[19] A. Lagra, M. Hasnaoui, A. Amahmid, and M. Bourich, "Double diffusive convection in a shallow horizontal binary fluid in the presence of Soret and Dufour effects," in *Proceedings of the 12ème Congrès de Mécanique*, Casablanca, Morocco, April 2015.

19

The Influence of Pits on the Tribological Behavior of Grey Cast Iron under Dry Sliding

Risheng Long,[1,2,3] Peter Kelly,[4] Shaoni Sun,[5] Jiling Feng,[4] Xuewen Wang,[2,3] and Wenyue Li[2,3]

[1] *College of Mechanical Engineering, Shenyang University of Chemical Technology, Shenyang, Liaoning 110142, China*
[2] *College of Mechanical Engineering, Taiyuan University of Technology, Taiyuan, Shanxi 030024, China*
[3] *Shanxi Key Laboratory of Fully Mechanized Coal Mining Equipment, Taiyuan, Shanxi 030024, China*
[4] *Surface Engineering Group, Manchester Metropolitan University, Manchester M1 5GD, Lancs, UK*
[5] *School of Mechanical Engineering and Automation, Northeastern University, Shenyang, Liaoning 110819, China*

Correspondence should be addressed to Peter Kelly; peter.kelly@mmu.ac.uk

Academic Editor: Matteo Aureli

Inspired by the nonsmooth surface of the head of the dung beetle, grey cast iron (GCI) samples with pit textured surfaces were designed and fabricated, based on pin-on-disc friction tester. Using a tribology wear testing rig and APDL programming, the tribological behavior of smooth and textured samples was investigated and reported, both experimentally and numerically. The results show that pits can significantly change the thermal stress and temperature distribution on the surface, which will result in either positive or negative effects on the wear resistance of GCI samples, depending on the parameters. When diameter of the pit (DOP) equals 0.8 mm and distance between pits (DBP) is 1.0 mm, the pit textured surface provided the best wear resistance among all samples tested.

1. Introduction

Grey cast iron (GCI) is one of the most commonly used materials in industrial applications, including, for example, the base and guideways of machine tools, cylinder block of automobile engine, bodies of valves and pumps, reduction gear shells, and trains brake discs, owing to its high friction coefficient, low cost, good damping property, and fine castability and machinability, as well as other properties [1]. For any mechanical system, the relative movement of different parts in contact is inevitable, such as the guideway of a lathe and the bearing mounting position of a gearbox, wherein friction and wear can only be reduced, instead of eliminated or avoided, especially under dry sliding conditions. Therefore, to meet the higher demand for accuracy, speed, and reliability of future applications, like the heavy-duty braking systems of high-speed trains, it is of great significance and urgency to further improve and enhance the wear resistance of GCI parts.

Surface texturing (ST) and biomimetic nonsmooth surfaces (BNSS) provide us two choices from different viewpoints. ST, firstly presented by Hamilton et al. in 1966, is a means for enhancing the tribological performance of mechanical components, through material-removal methods, like electrical discharge machining (EDM), industrial etching/chemical milling, laser surface texturing (LST), or other techniques [2–4]. BNSS was proposed and developed by Ren et al. in Jilin University since the 1980s and is inspired by the unique functional surfaces of natural animals (see Figure 1) and plants, which have gradually developed to best adapt to different environments, through billions of years of natural selection and evolution [5]. Those surface structures (pits, streaks, and rectangles), called 'nonsmooth construction units', are mainly obtained by material remelting/cladding techniques and can provide excellent wear resistance, fatigue resistance, and crack resistance [6, 7].

In the past years, great efforts have been made to study the influence of different nonsmooth surfaces on the thermal

FIGURE 1: Dung beetle and the pit nonsmooth surface of its head.

fatigue resistance of GCI parts [1, 8–14]. In contrast, studies of the antiwear performance or tribological behavior of GCI parts with textured surfaces are quite few, whether in experiment or simulation, especially under dry wear conditions. Based on the previous research on the thermal behavior of pin-on-disc friction systems [15, 16], a finite element model with linearly distributed pits in the radial direction (LDRD) was first established through APDL (ANSYS parametric design language) programming. The influence of different pit parameters on the tribological behavior was researched. The friction coefficient curves and mass losses of samples were measured and reported. The worn surfaces were characterized by stereo-optical microscope. All the obtained data could provide an important reference for the design of heavy-duty braking system, rotating thrust bearing, heavy-load sliding bearing, and so on.

2. Modelling and Testing Procedure

2.1. Thermodynamics Theory and Conditions. For certain conditions, the heat conduction during friction processes can be described by the classical heat conduction equation. In the Cartesian coordinate system, the classical heat conduction equation can be expressed as follows [16]:

$$k_e\left(\frac{\partial^2 T}{\partial x^2}+\frac{\partial^2 T}{\partial y^2}+\frac{\partial^2 T}{\partial z^2}\right)+q+h\left(T-T_h\right)+\sigma\varepsilon\left(T^4-T_h^4\right)=\rho\cdot c\frac{\partial T}{\partial t} \quad (1)$$

where k_e is the thermal conductivity of material, c is the specific heat capacity of material, q is the heat flux through the contact area, ρ is the density of material, h is the convective heat transfer coefficient, T_h is the ambient temperature, σ is the radiation constant of a blackbody, and ε is the radiation rate of the actual object.

To solve (1), the initial condition and boundary conditions are necessary. In this work, the initial condition is the original temperature of the pin-on-disc system at the beginning of the simulation, equal to 20°C. The boundary condition is the heat exchange condition between the outer surface of the pin-on-disc system and its surrounding environment [15].

The heat flux, q_w, at the boundary of the objects, is a known constant or also a function of position and time. For pin-on-disc friction systems, it can be expressed as

$$-k(T)\left.\frac{\partial T}{\partial \vec{n}}\right|_s=q_w$$
$$\text{or } -k(T)\left.\frac{\partial T}{\partial \vec{n}}\right|_s=q_w(x,y,z,t) \quad (2)$$

$$q_w(x,y,z,t)=\eta\cdot\mu\cdot P\cdot v(x,y,z,t) \quad (3)$$

where q_w is the heat flux through the contact surface of the pin-on-disc system, a known constant; $q_w(x,y,z,t)$ is the expression of input heat flux, a function of position and time; η is the conversion efficiency, 0.85; μ is the friction coefficient of the system, 0.38; P is the normal pressure of the pin; $v(x,y,z,t)$ is the relative linear velocity between disc and pin. Meanwhile, the lower surface of the disc can be regarded as being in the adiabatic state, whose boundary condition can be written as

$$-k_e\frac{\partial T}{\partial z}=0 \quad (4)$$

For pin-on-disc friction systems, the temperature distribution in the disc or the pin is very uneven during dry wear processes. As a result, a huge temperature gradient exists in the disc, which further induces high thermal stresses. Thus, the thermal strain is composed of two parts: one is caused by temperature variation, and the other is caused by stress. According to Hooke's law, the thermal strain can be written as

$$\varepsilon_{xx}=\frac{\partial u_x}{\partial x}=\frac{1}{E}\left[\sigma_{xx}-v\left(\sigma_{yy}+\sigma_{zz}\right)\right]+\alpha\tau$$
$$=\frac{1}{2G}\left(\sigma_{xx}-\frac{v}{1+v}\Theta_s\right)+\alpha\tau$$
$$\varepsilon_{yy}=\frac{\partial u_y}{\partial y}=\frac{1}{E}\left[\sigma_{yy}-v\left(\sigma_{xx}+\sigma_{zz}\right)\right]+\alpha\tau$$

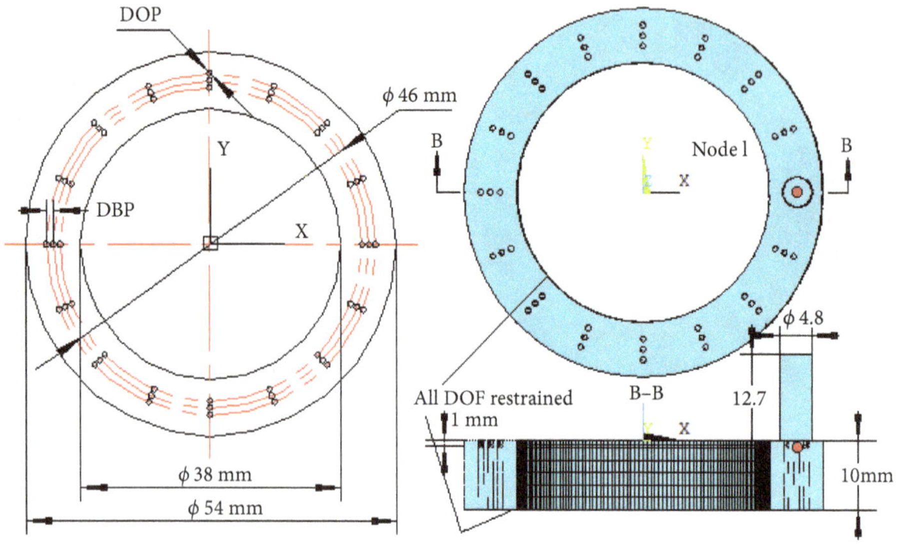

FIGURE 2: The finite element model with pit textured surface and the location of node 1.

$$
\begin{aligned}
&= \frac{1}{2G}\left(\sigma_{yy} - \frac{\nu}{1+\nu}\Theta_s\right) + \alpha\tau \\
\varepsilon_{zz} &= \frac{\partial u_z}{\partial z} = \frac{1}{E}\left[\sigma_{zz} - \nu\left(\sigma_{xx} + \sigma_{yy}\right)\right] + \alpha\tau \\
&= \frac{1}{2G}\left(\sigma_{zz} - \frac{\nu}{1+\nu}\Theta_s\right) + \alpha\tau \\
\Theta_s &= \sigma_{xx} + \sigma_{yy} + \sigma_{zz}, \\
\varepsilon_{xy} &= \frac{\sigma_{xy}}{2G}, \\
\varepsilon_{yz} &= \frac{\sigma_{yz}}{2G}, \\
\varepsilon_{zx} &= \frac{\sigma_{zx}}{2G}
\end{aligned}
\tag{5}
$$

where $\tau = T\text{-}T_0$ is the temperature variation of the object and α is the thermal expansion coefficient. Equation (5) can also be expressed as follows:

$$
\begin{aligned}
\sigma_{xx} &= 2G\varepsilon_{xx} + \lambda\theta - \frac{\alpha ET}{1-2\nu}, \\
\sigma_{xy} &= 2G\varepsilon_{xy} \\
\sigma_{yy} &= 2G\varepsilon_{yy} + \lambda\theta - \frac{\alpha ET}{1-2\nu}, \\
\sigma_{yz} &= 2G\varepsilon_{yz} \\
\sigma_{zz} &= 2G\varepsilon_{zz} + \lambda\theta - \frac{\alpha ET}{1-2\nu}, \\
\sigma_{zx} &= 2G\varepsilon_{zx}
\end{aligned}
\tag{6}
$$

where E is the elastic modulus, ν is Poisson ratio, $\theta = \varepsilon_{xx} + \varepsilon_{yy} + \varepsilon_{zz}$ is the volumetric strain, T is the temperature variation range, $G = E/(2\times(1+\nu))$ is the modulus of elasticity in shear, and $\lambda = \nu E/((1+\nu)\times(1-2\nu))$ is the Lame constant.

TABLE 1: The chemical composition of HT250 gray cast iron.

Elements	C	Si	Mn	P	S	Cu	Cr	Fe
Composition (wt.%)	3.25	1.57	0.92	0.06	0.059	0.50	0.27	Bal.

TABLE 2: Thermal/physical properties of GCI and EN1A tool steel.

Thermal/physical parameter	GCI	EN1A Steel
Thermal conductivity, $K_e/[\mathrm{W}/(\mathrm{m}\cdot\mathrm{K})]$	48	32.2
Density, $\rho/(\mathrm{kg/m^3})$	7200	7800
Specific heat capacity, $c/[\mathrm{J}/(\mathrm{kg}\cdot\mathrm{K})]$	480	460
Thermal expansion coefficient, $\alpha/(\times10^{-5}\mathrm{K}^{-1})$	1.2	0.91
Elastic modulus, E/Gpa	120	210
Poisson ratio, ν	0.25	0.30

2.2. Modelling and Testing Procedure. The dimensions of the finite element (FE) model, established according to the actual pin-on-disc friction system, are as follows: outer diameter of Φ 54mm and inner diameter of Φ 38mm, with the thickness of 10 mm (see Figure 2). The material of the disc is GCI, codenamed HT250 (ISO 250), whose chemical composition is listed in Table 1, and its microstructure is displayed in Figure 3(a). The pin diameter is Φ 4.8 mm × 12.7 mm, manufactured from EN1A steel. The thermal physical properties of GCI and EN1A steel are both listed in Table 2. The main parameters of the pits include diameter of the pit, DOP, and distance between pits, DBP. In order to reveal the thermal behavior of different samples intuitively and visually, node 1 (see Figure 2) was introduced and used to reflect the thermal stress and temperature variation in simulation. The distance between node 1 and the upper surface of the sample is 1.0 mm, which is equal to the depth of the pits.

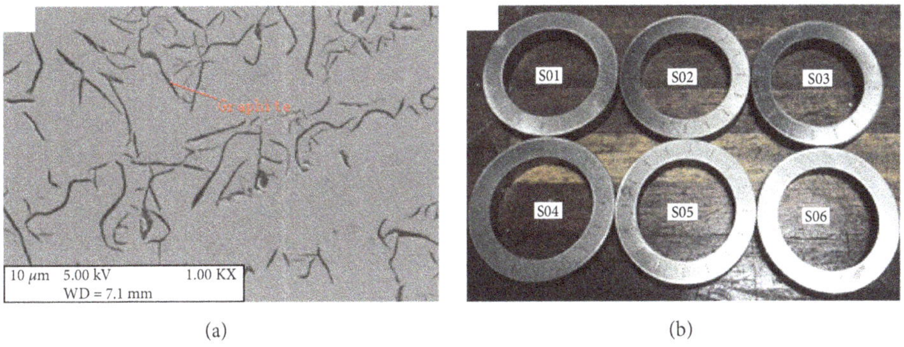

FIGURE 3: SEM photo of GCI and the photo of samples. (a) SEM photo of GCI; (b) photo of S01-S06.

TABLE 3: The parameters of different samples S01-S06.

Sample No.	Diameter of pit, DOP (mm)	Distance between pits, DBP (mm)
S01	0.8	1.00
S02	0.8	1.25
S03	0.8	1.50
S04	0.4	1.00
S05	0.6	1.00
S06	-	-

A laser marking system (CL-FLS-30W, Shenyang Qianshi Laser CO., LTD.) was used to process the pits on the surface of the GCI discs. The laser processing parameters were as follows: 1064 nm wavelength, laser power, 30 W, diameter of laser beam, 0.05 mm, and scanning speed, 5 mm/s. As shown in Table 3, there were six samples fabricated in all. To imitate the head surface of a dung beetle, the samples with 16 sets of pits along the radial direction were called 'nonsmooth' samples and named from S01 to S05 according to the pit parameters. The other sample with a polished surface was introduced for comparison and called the "smooth" sample and marked as S06. Figure 3(b) shows the photo of all those samples.

Wear tests were conducted using a tribology wear testing rig (courtesy of the Surface Engineering Group, Manchester Metropolitan University, MMU) under dry sliding condition, with a normal load of 70 N, equivalent to a pressure of approximately 0.98 MPa, at room temperature, 20°C. The relative rotation speed was 200 rpm, equivalent to 21 rad s^{-1} (about 0.966 m/s at the centre of the pin), and the duration of the tests was designated as 500 seconds. To ensure the accuracy of the result, friction and wear tests were repeated 3 times for each specimen. Prior to the wear tests, the specimens and self-designed counterface (EN1A steel, with a hardness of approximately 800 HV) were mechanically polished with SiC papers (from No. 200 to No. 2000), followed by ultrasonic cleaning in acetone. Before and after testing, samples were cleaned and weighed by an electronic balance (Ohaus T2914) with a precision of 0.1 mg. The wear resistance of the specimens was measured and evaluated by its mass loss, which was the average of three measurements. The worn surfaces were observed by optical microscopy (S1000 Industrial Stereo Microscope, Spectrographic Limited, with Vimage software of Suzhou Vezu Opto Technology, CO., LTD.). The average surface hardness and roughness of GCI samples were 0.111 μm and 122.4 $HV_{0.5}$, respectively, obtained by THV-5 vickers hardness tester and TR300 roughness profile rig (TIME3230 module).

3. Tribological Behavior of Pit Textured Surfaces

3.1. Simulation Results. Based on the mechanical and thermodynamic models in the documents [17, 18], the FE model was established by APDL programming through an ANSYS/LS-DYNA module. The surface between the pit textured disc and the pin was set as a friction pair, wherein the upper surface of the disc was the target surface and the lower surface of the column pin was the contact surface. During the simulation process, the friction pair are always kept in contact with each other under a pressure of 0.98 MPa. To intuitively display the relative positions of disc and pin, based on the equivalent principle, the disc was constricted in all DOFs instead of the pin, as shown in Figure 2, and kept motionless in the simulation. An angular velocity, 21 rad/s (in the anticlockwise direction, about 200 RPM), was applied to the pin, which could rotate counterclockwise. The duration of the simulation is 0.5 s (100 substeps, i.e., 0.005 second per substep); thus, only heat conduction of the friction interface was considered in the simulation. The element type used in this work includes SOLID226, TARGE170, CONTA174, and MPC184. The number of elements is 32951 (S01).

Figure 4 shows the temperature-time curves of node 1 of S01-S06. It is evident that (1) there were three temperature peaks for all curves, which was consistent with the rotated times of the pin; (2) for S01-S03, their temperature curves were almost the same, with the lowest average temperature; (3) the average temperature of S04 was the highest among S01-S06, followed by S05; (4) the average temperature of S06 was higher than S01-S03, but lower than S04-S05.

Figure 5 exhibits the temperature contours of S01-S06 at the 0.5 s. The term "SMX" in the figure refers to the maximum value of temperature. The following results can be obtained from the figure: (1) compared with the smooth surface, the pits had a significant influence on the temperature

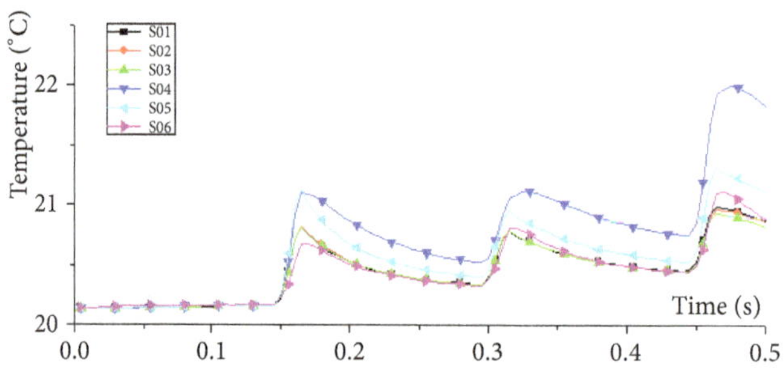

FIGURE 4: Temperature-time curves of node 1 of S01-S06.

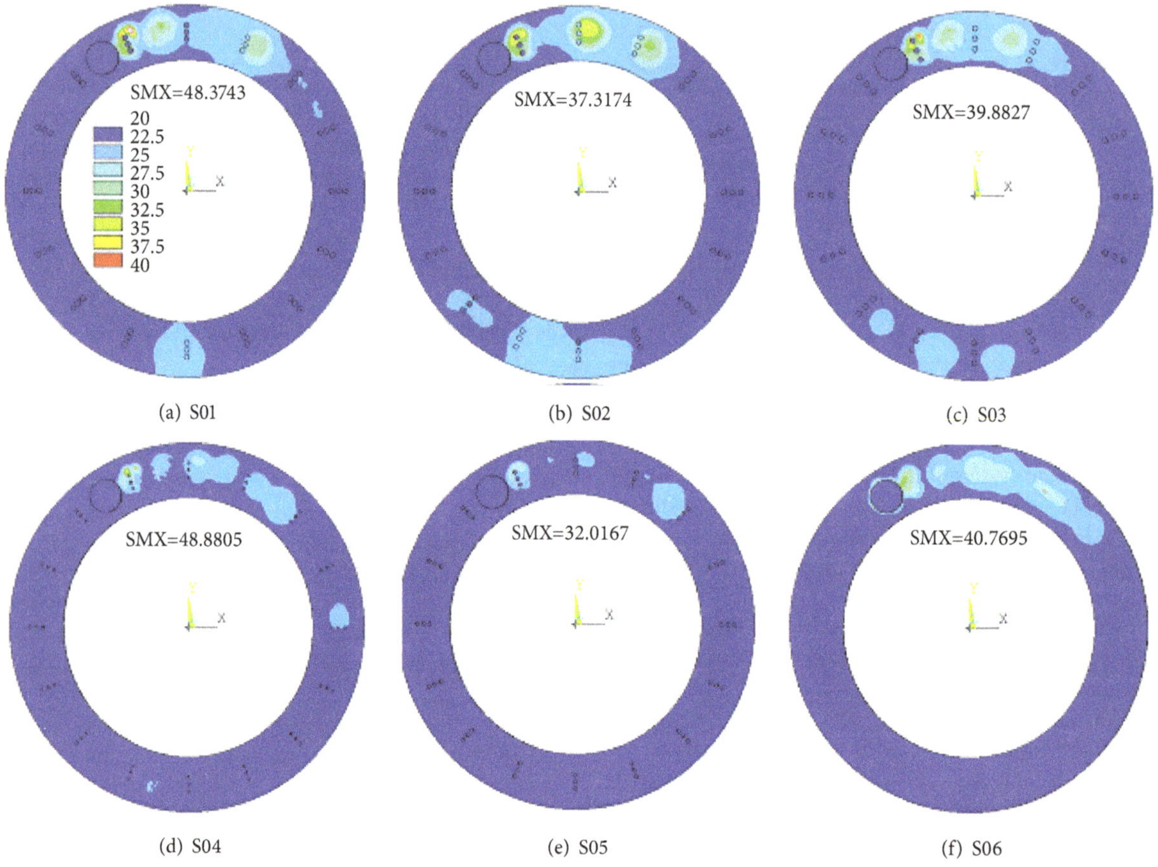

FIGURE 5: Temperature contours of S01-S06 at the 0.5 s.

distribution of the samples; (2) the heat affected area of S01 was the largest, while the high temperature area of S05 was the smallest; (3) for S01-S03, there were obvious high temperature regions near the pits, as well as the "temperature crawling" phenomena; (4) for S04 and S05, there was no evident temperature concentration or crawling phenomena on the surfaces; (5) the temperature distribution of S06 displayed a "long tail" along the rotating direction, without any "temperature crawling" phenomenon.

Figure 6 displays the *von Mises* stress contours of the samples at the 0.5 s. The term "SMX" in the figure refers to the maximum value of *von Mises* stress. It is obvious that (1) the pits can influence the *von Mises* stress distribution of the samples, especially at locations near the pits. There were obvious low *von Mises* stress areas near the pits along the rotating direction for all pit textured specimens; (2) among S01-S03, the *von Mises* stress affected area of S01 was the smallest; (3) the high *von Mises* stress areas of S04 and 05 were both larger than that of S01, with the high *von Mises* stress area of S05 being the largest; (4) evidently, the high *von Mises* stress distribution of S06 was smoother and more continuous.

3.2. Tribological Behavior. Figure 7 shows the friction coefficient curves and mass loss curve of samples S01-S06. It is clear to see that (1) among all the specimens, the friction coefficient and wear rate of S01 were the lowest, while the wear rate of S03 was the largest. The friction coefficient and wear loss rate of S06 were also quite low; (2) for S01-S03, the

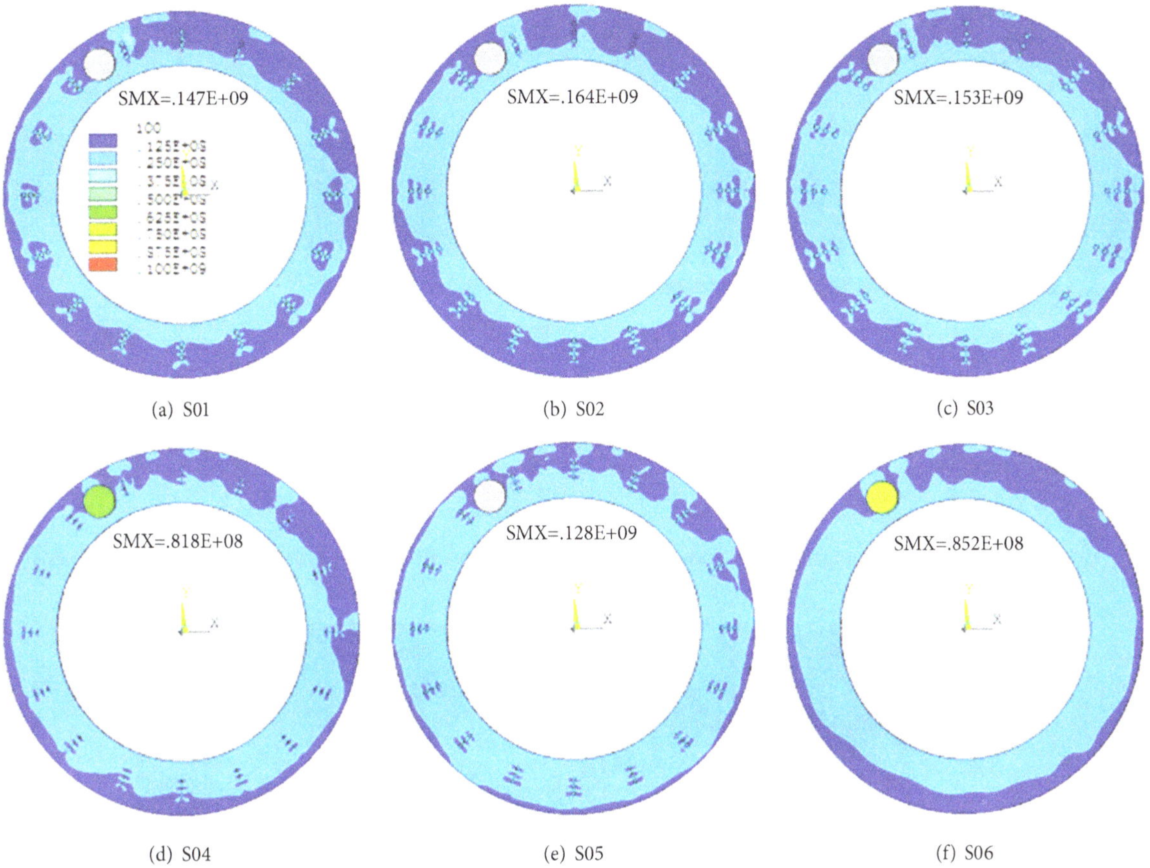

FIGURE 6: The *von Mises* equivalent stress contours of S01-S06 at the 0.5 s.

wear rate increased with the variation of DBP from 1.0 mm to 1.50 mm, while the friction coefficient increased first and then decreased; (3) for S04, S05, and S01, with the increase of DOP, the wear rate increased first and then decreased, but the friction coefficient kept increasing.

The worn surfaces of S01-S06 are shown in Figure 8. The surface conditions of the nonsmooth and smooth specimens are consistent with the results of Figure 7. It is also obvious that (1) there was only slight wear tracks on the surface of S01; (2) the wear of S03 was very serious, and there were very deep and continuous furrows (see the red circle of S03), indicating the poor antiwear performance; (3) for S02, S04-S06, there were obvious grinding phenomena on the surface (see the red circle regions of them), especially S02.

4. Discussion

For the pit textured surfaces, DOP and DBP are the key parameters which affect their final antiwear properties. When DOP was constant, as shown in Figure 8, the wear resistance of the pit textured samples gradually deteriorated with the increase of DBP from 1.0 mm to 1.50 mm. By comparing Figures 5 and 6, it is evident that increasing of DBP could significantly enlarge the heat affected zone and high thermal stress region of the textured surface, which was detrimental to obtaining good antiwear performance. In addition, increasing of DBP was also not conducive to the collection of debris during the dry wear process, which would cause a transition from "adhesive wear" to "three-body abrasion" and to, therefore, deteriorate the wear resistance of the pit textured surfaces. Conversely, when DBP remained constant, with an increase of DOP, the heat affected area gradually enlarged, but the high thermal stress region gradually reduced. Correspondingly, the friction coefficient and wear volume also decreased with the change of DOP from 0.4 mm to 0.8 mm. In general, the parameters of S01 are the best among all samples, i.e., DOP of 0.8 mm, and DBP of 1.0 mm, which are consistent with the results published previously [15, 16].

As shown in Figures 5 and 6, compared with the smooth sample, nonsmooth pit structures can clearly alter the thermal stress and temperature distribution of the textured surfaces, inducing extremely nonuniform thermal stresses and temperature fields, even causing a 'temperature crawling' phenomenon, which are disadvantageous factors for pit textured surfaces to achieve good antiwear properties. In the meantime, the pits in the nonsmooth samples can trap and hold the debris (see Figure 9) produced in the friction process under dry wear condition, which is an advantageous factor for pit textured surfaces to obtaining excellent wear resistance. The final influence of the pit textured surfaces on the wear resistance of nonsmooth specimens depends on the combined result of the two aspects mentioned above.

Furthermore, for the actual pin-on-disc friction system, the pin is fixed during the wear testing, and the disc rotates

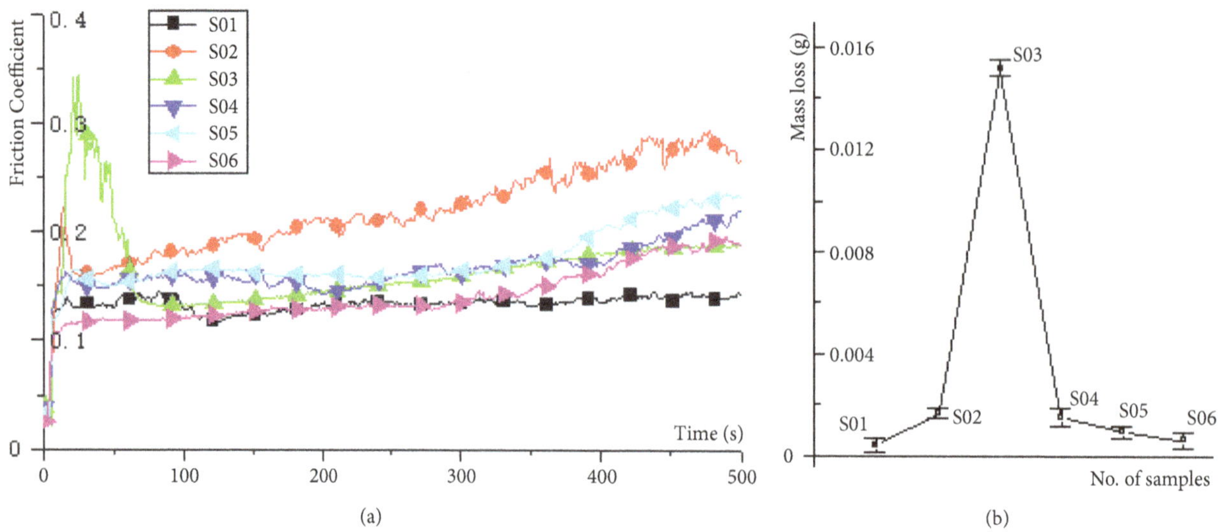

FIGURE 7: Friction coefficient curves and mass loss curve of S01-06. (a) Friction coefficient curves of S01-06; (b) mass loss curve of S01-06.

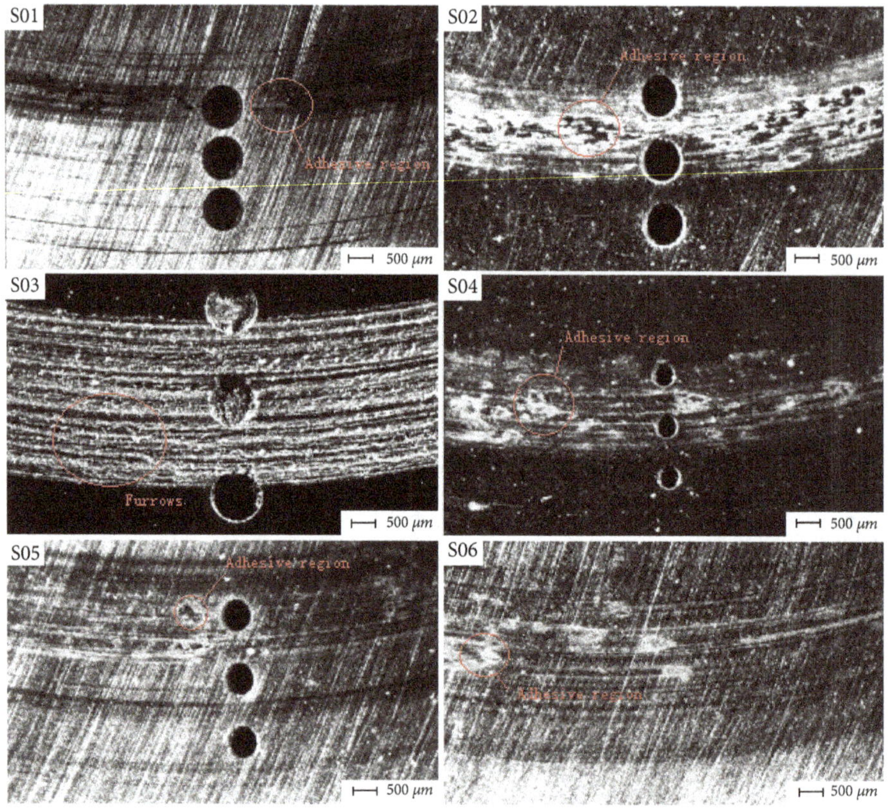

FIGURE 8: Worn surfaces of S01-S06 after wear testing.

anticlockwise or clockwise. The debris left on the textured surface will be subjected to the action of two kinds of forces; one is the pressure of the pin, and the other is the centrifugal force due to the rotation of the disc (see Figure 9). When the rotating speed is high enough, the centrifugal force may be greater than the pushing force of the pin; some debris may move inward firstly, and then move outward possibly. The change of the motion direction of the debris will cause a "secondary grinding phenomenon" on textured or smooth surfaces. This is the reason for the serious adhesive wear (see Figure 8) of the pit textured samples.

5. Conclusions

Through numerical simulation and experimental analysis, the influence of pit textured surfaces on the tribological

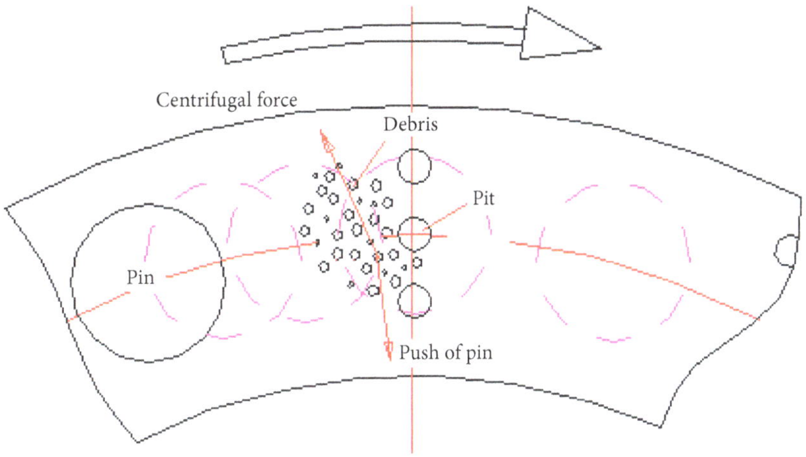

FIGURE 9: Influence mechanism of pit textured surface.

behavior of a pin-on-disc system was investigated and reported in this work. An underlying mechanism was also proposed and discussed, and the following conclusions could be drawn.

(1) DOP and DBP are two important parameters that affect the wear resistance of pit textured surfaces. The simulated data shows that pits can significantly change the thermal stress and temperature distribution of the textured surfaces during friction and wear processes and result in low thermal stress zones near the dimples. In addition, when the DOP is 0.8 mm, there are obvious 'temperature crawling' phenomena on the pit textured surfaces.

(2) Compared with the smooth sample, the pit textured structure is able to either improve or worsen the surface abrasion resistance of the specimen, depending on the combined effect of the capacity of the pits to contain debris and the ability of the pits to change the thermal stress and temperature distribution of the textured surface. According to the above results, when DOP is 0.8 mm and DBP is 1 mm, the pit textured surface has the best wear resistance under dry friction conditions.

The influence of other texture patterns (streaks, rectangles, diamonds, and their combinations) on the tribological behavior will also be reported and shared in the future.

Conflicts of Interest

The authors declare that there are no conflicts of interest regarding the publication of this paper.

Acknowledgments

The research is supported by China Scholarship Council, as well as the 2014 Shanxi coal-based key scientific & technological projects (MJ2014-06). Besides the Shanxi Key Laboratory of Fully Mechanized Coal Mining Equipment, China, the authors would like to thank the coworkers of MMU.

References

[1] N. Sun, H. Shan, H. Zhou et al., "Friction and wear behaviors of compacted graphite iron with different biomimetic units fabricated by laser cladding," *Applied Surface Science*, vol. 258, no. 19, pp. 7699–7706, 2012.

[2] G. Ryk, Y. Kligerman, and I. Etsion, "Experimental investigation of laser surface texturing for reciprocating automotive components," *Tribology Transactions*, vol. 45, no. 4, pp. 444–449, 2002.

[3] I. Etsion, "State of the art in laser surface texturing," *Journal of Tribology*, vol. 127, no. 1, pp. 248–253, 2005.

[4] Y. Wan and D.-S. Xiong, "Study of laser surface texturing for improving tribological properties," *Mocaxue Xuebao/Tribology*, vol. 26, no. 6, pp. 603–607, 2006.

[5] H. Zhou, X. Tong, Z. Zhang, X. Li, and L. Ren, "The thermal fatigue resistance of cast iron with biomimetic non-smooth surface processed by laser with different parameters," *Materials Science and Engineering: A Structural Materials: Properties, Microstructure and Processing*, vol. 428, no. 1-2, pp. 141–147, 2006.

[6] X. Tong, H. Zhou, Z.-H. Zhang, N. Sun, H.-Y. Shan, and L.-Q. Ren, "Effects of surface shape on thermal fatigue resistance of biomimetic non-smooth cast iron," *Materials Science and Engineering: A Structural Materials: Properties, Microstructure and Processing*, vol. 467, no. 1-2, pp. 97–103, 2007.

[7] X. Tong, H. Zhou, M. Liu, and M.-J. Dai, "Effects of striated laser tracks on thermal fatigue resistance of cast iron samples with biomimetic non-smooth surface," *Materials and Corrosion*, vol. 32, no. 2, pp. 796–802, 2011.

[8] Z. Jing, H. Zhou, P. Zhang, C. Wang, C. Meng, and D. Cong, "Effect of thermal fatigue on the wear resistance of graphite cast iron with bionic units processed by laser cladding WC," *Applied Surface Science*, vol. 271, pp. 329–336, 2013.

[9] B. Kim, Y. H. Chae, and H. S. Choi, "Effects of surface texturing on the frictional behavior of cast iron surfaces," *Tribology International*, vol. 70, pp. 128–135, 2014.

[10] D. Braun, C. Greiner, J. Schneider, and P. Gumbsch, "Efficiency of laser surface texturing in the reduction of friction under mixed lubrication," *Tribology International*, vol. 77, pp. 142–147, 2014.

[11] R. Bathe, V. Sai Krishna, S. K. Nikumb, and G. Padmanabham, "Laser surface texturing of gray cast iron for improving tribological behavior," *Applied Physics A: Materials Science & Processing*, vol. 117, no. 1, pp. 117–123, 2014.

[12] X. Yao, H. Huo, M. Bao, and L. Tian, "Impact fatigue behavior of ZrN/Zr-N/Zr coatings on H13 steel by CFUBMSIP," *Journal of Wuhan University of Technology-Mater. Sci. Ed.*, vol. 29, no. 1, pp. 137–142, 2014.

[13] Z.-K. Chen, S.-C. Lu, X.-B. Song, H. Zhang, W.-S. Yang, and H. Zhou, "Effects of bionic units on the fatigue wear of gray cast iron surface with different shapes and distributions," *Optics & Laser Technology*, vol. 66, pp. 166–174, 2014.

[14] L. Zheng, J. Wu, S. Zhang et al., "Bionic Coupling of Hardness Gradient to Surface Texture for Improved Anti-wear Properties," *Journal of Bionic Engineering*, vol. 13, no. 3, pp. 406–415, 2016.

[15] S.-N. Sun, L.-Y. Xie, and Y.-C. Zhang, "Finite element analysis on friction and wear properties of non-smooth surface brake disc," *Dongbei Daxue Xuebao*, vol. 35, no. 11, pp. 1597–1601, 2014.

[16] S. Sun, L. Xie, P. Kelly, R. Long, M. Li, and J. Feng, "The Influence of Rotating Direction on the Tribological Behavior of Grey Cast Iron with Curve Distributed Pit Textured Surface," *Mathematical Problems in Engineering*, vol. 2017, Article ID 8095916, 10 pages, 2017.

[17] L. Ri-Sheng, S. Shao-Ni, and L. Zi-Sheng, "The influence of scanning methods on the cracking failure of thin-wall metal parts fabricated by laser direct deposition shaping," *Engineering Failure Analysis*, vol. 59, pp. 269–278, 2016.

[18] R. Long, S. Sun, and Z. Lian, "Research on the Hard-Rock Breaking Mechanism of Hydraulic Drilling Impact Tunneling," *Mathematical Problems in Engineering*, vol. 2015, Article ID 153648, 34 pages, 2015.

Permissions

All chapters in this book were first published in MPE, by Hindawi Publishing Corporation; hereby published with permission under the Creative Commons Attribution License or equivalent. Every chapter published in this book has been scrutinized by our experts. Their significance has been extensively debated. The topics covered herein carry significant findings which will fuel the growth of the discipline. They may even be implemented as practical applications or may be referred to as a beginning point for another development.

The contributors of this book come from diverse backgrounds, making this book a truly international effort. This book will bring forth new frontiers with its revolutionizing research information and detailed analysis of the nascent developments around the world.

We would like to thank all the contributing authors for lending their expertise to make the book truly unique. They have played a crucial role in the development of this book. Without their invaluable contributions this book wouldn't have been possible. They have made vital efforts to compile up to date information on the varied aspects of this subject to make this book a valuable addition to the collection of many professionals and students.

This book was conceptualized with the vision of imparting up-to-date information and advanced data in this field. To ensure the same, a matchless editorial board was set up. Every individual on the board went through rigorous rounds of assessment to prove their worth. After which they invested a large part of their time researching and compiling the most relevant data for our readers.

The editorial board has been involved in producing this book since its inception. They have spent rigorous hours researching and exploring the diverse topics which have resulted in the successful publishing of this book. They have passed on their knowledge of decades through this book. To expedite this challenging task, the publisher supported the team at every step. A small team of assistant editors was also appointed to further simplify the editing procedure and attain best results for the readers.

Apart from the editorial board, the designing team has also invested a significant amount of their time in understanding the subject and creating the most relevant covers. They scrutinized every image to scout for the most suitable representation of the subject and create an appropriate cover for the book.

The publishing team has been an ardent support to the editorial, designing and production team. Their endless efforts to recruit the best for this project, has resulted in the accomplishment of this book. They are a veteran in the field of academics and their pool of knowledge is as vast as their experience in printing. Their expertise and guidance has proved useful at every step. Their uncompromising quality standards have made this book an exceptional effort. Their encouragement from time to time has been an inspiration for everyone.

The publisher and the editorial board hope that this book will prove to be a valuable piece of knowledge for researchers, students, practitioners and scholars across the globe.

List of Contributors

Shurong Li
Automation School, Beijing University of Posts and Telecommunications, Beijing 100876, China

Yulei Ge
College of Information and Control Engineering, China University of Petroleum (East China), Qingdao 266580, China

Ricardo Aguilar-López
Department of Biotechnology and Bioengineering, Centro de Investigación y de Estudios Avanzados del I.P.N. (CINVESTAV), Av. Instituto Politćnico Nacional, No. 2508, Colonia San Pedro Zacatenco, C.P. 07360, Ciudad de México, Mexico

Fengping Li
School of Aerospace Engineering, Xiamen University, Xiamen 361005, China
Zhejiang Provincial Engineering Lab of Laser and Optoelectronic Intelligent Manufacturing, Wenzhou University, Wenzhou 325035, China

Zhengya Zhang, Yao Xue, Sijia Zhou and Yuqing Zhou
Zhejiang Provincial Engineering Lab of Laser and Optoelectronic Intelligent Manufacturing, Wenzhou University, Wenzhou 325035, China

Xuebin Qin, Pai Wang, Lang Liu, Mei Wang and Jie Xin
School of Electrical and Control Engineering, Xi'an University of Science and Technology, 58 Yanta Rd, Xi'an, Shaanxi, China

Mauro Henrique Alves de Lima Jr., Carl Horst Albrecht, Beatriz Souza Leite Pires de Lima and Breno Pinheiro Jacob
Laboratory of Computer Methods and Offshore Systems (LAMCSO), PEC/COPPE/UFRJ, Civil Engineering Department, Post-Graduate Institute of the Federal University of Rio de Janeiro, Avenida Pedro Calmon, S/N, Cidade Universitária, IlhadoFundão, 21941-596 Rio de Janeiro, RJ, Brazil

Juliana Souza Baioco
Laboratory of Computer Methods and Offshore Systems (LAMCSO), PEC/COPPE/UFRJ, Civil Engineering Department, Post-Graduate Institute of the Federal University of Rio de Janeiro, Avenida Pedro Calmon, S/N, Cidade Universitária, Ilhado Fundão, 21941-596 Rio de Janeiro, RJ, Brazil
Chemical and Petroleum Engineering Department (TEQ), UFF, Rua Passo da Pátria No. 156, São Domingos, 24210-240 Niterói, RJ, Brazil

Djalene Maria Rocha
Petróleo Brasileiro S.A. (Petrobras), Research & Development Center (CENPES), Avenida Horacio Macedo 950, Cidade Universitária, Ilhado Fundão, 21941-915 Rio de Janeiro, RJ, Brazil

Kang Liu, Daiyong Cao and Yingchun Wei
College of Geoscience and Surveying Engineering, China University of Mining & Technology, Beijing 100083, China

Zhongyue Lin
Beijing Dadi Gaoke Coalbed Methane Engineering Technology Research Institute, Beijing 100040, China
National Administration of Coal Geology in China, Beijing 100038, China

Wenyin Zhang and Kongwei Zhu
School of Information Science and Engineering, Linyi University, Linyi 276005, China
Key Laboratory of Complex Systems and Intelligent Computing in Universities of Shandong (Linyi University), Linyi 276005, China

Jianbao Zhang and Chengdong Yang
School of Information Science and Engineering, Linyi University, Linyi 276005, China
Key Laboratory of Complex Systems and Intelligent Computing in Universities of Shandong (Linyi University), Linyi 276005, China
Department of Mathematics, Southeast University, Nanjing 210096, China

Jianlong Qiu
Key Laboratory of Complex Systems and Intelligent Computing in Universities of Shandong (Linyi University), Linyi 276005, China
Department of Electrical and Computer Engineering, Faculty of Engineering, King Abdulaziz University, Jeddah 21589, Saudi Arabia

Denghua Zhang
College of Science and Technology, North China Electric Power University, Baoding 071000, China

Chunyu Zhao and Xiyue Tang
School of Business, Sichuan Normal University, Chengdu 610101, China

Lijuan Yuan
School of Mathematical Sciences, Sichuan Normal University, Chengdu 610101, China

Binglin Li and Wen You
School of Electrical and Electronic Engineering, Changchun University of Technology, Changchun 130012, China

Wanxiu Xu and Guanyu Zhu
Jiangnan University, Wuxi, Jiangsu, China

Chunfang Song, Shaogang Hu and Zhenfeng Li
Jiangnan University, Wuxi, Jiangsu, China
Jiangsu Key Laboratory of Advanced Food Manufacturing Equipment and Technology, Wuxi, Jiangsu, China

Jovan Trifunovic
Faculty of Electrical Engineering, University of Belgrade, Bulevar Kralja Aleksandra 73, 11000 Belgrade, Serbia

Yong Xiao
China Zhenhua Oil Co., Ltd, Beijing, China
State Key Laboratory on Oil and Gas Reservoir Geology and Exploitation, Southwest Petroleum University, Chengdu, China

Hehua Wang and Lize Lu
China Zhenhua Oil Co., Ltd, Beijing, China
Chengdu Northern Petroleum Exploration and Development Technology Co., Ltd., Chengdu, Sichuan, China

Jianchun Guo
State Key Laboratory on Oil and Gas Reservoir Geology and Exploitation, Southwest Petroleum University, Chengdu, China

John McLennan
Energy & Geoscience Institute, University of Utah, UT, USA

Mengting Chen
Borehole Operation Branch Office of Sinopec Southwest Petroleum Engineering Co., Ltd., Sichuan, China

Xiaoqi Song, Dezhi Xu, Weilin Yang and Yan Xia
School of Internet of Things Engineering, Jiangnan University, Wuxi 214122, China

Bin Jiang
College of Automation Engineering, Nanjing University of Aeronautics and Astronautics, Nanjing 211106, China

Shuanghong Li, Chengjun Zhan and Yupu Yang
Department of Automation, Key Laboratory of System Control and Information Processing, Ministry of Education, Shanghai Jiao Tong University, Shanghai, China

N. Kanagaraj
Electrical Engineering Department, College of Engineering at Wadi Addawasir, Prince Sattam Bin Abdulaziz University, Wadi Addawasir, Saudi Arabia

M. Al-Dhaifallah
Electrical Engineering Department, College of Engineering at Wadi Addawasir, Prince Sattam Bin Abdulaziz University, Wadi Addawasir, Saudi Arabia
Systems Engineering Department, King Fahd University of Petroleum and Minerals, Dhahran, Saudi Arabia

K. S. Nisar
Department of Mathematics, College of Arts & Science at Wadi Addawasir, Prince Sattam Bin Abdulaziz University, AlKharj, Saudi Arabia

Fayiz Abu Khadra
Mechanical Engineering Department, King Abdulaziz University, Rabigh, Saudi Arabia

Shaozhen Jin, Zhizhong Mao and Hongru Li
School of Information Science & Engineering, Northeastern University, Shenyang 110819, China

Wenhai Qi
School of Engineering, Qufu Normal University, Rizhao 276800, China

A. Lagra, M. Hasnaoui and A. Amahmid
Faculty of Sciences Semlalia, Department of Physics, LMFE, BP 2390, Marrakesh, Morocco

M. Bourich
National School of Applied Sciences, Physics Department, LMFE, BP 575, Marrakesh, Morocco

M. Er-Raki
High School of Technology, Cadi Ayyad University, LMFE, Essaouira, Morocco

Risheng Long
College of Mechanical Engineering, Shenyang University of Chemical Technology, Shenyang, Liaoning 110142, China
College of Mechanical Engineering, Taiyuan University of Technology, Taiyuan, Shanxi 030024, China
Shanxi Key Laboratory of Fully Mechanized Coal Mining Equipment, Taiyuan, Shanxi 030024, China

Wenyue Li and Xuewen Wang
College of Mechanical Engineering, Taiyuan University of Technology, Taiyuan, Shanxi 030024, China
Shanxi Key Laboratory of Fully Mechanized Coal Mining Equipment, Taiyuan, Shanxi 030024, China

Peter Kelly and Jiling Feng
Surface Engineering Group, Manchester Metropolitan University, Manchester M1 5GD, Lancs, UK

Shaoni Sun
School of Mechanical Engineering and Automation, Northeastern University, Shenyang, Liaoning 110819, China

Index

www.ingramcontent.com/pod-product-compliance
Lightning Source LLC
Chambersburg PA
CBHW082043190326
41458CB00010B/3443
9781647261320